METHODS IN MOLECULAR BIOLOGY

For further volumes:
http://www.springer.com/series/7651

For over 35 years, biological scientists have come to rely on the research protocols and methodologies in the critically acclaimed *Methods in Molecular Biology* series. The series was the first to introduce the step-by-step protocols approach that has become the standard in all biomedical protocol publishing. Each protocol is provided in readily-reproducible step-by-step fashion, opening with an introductory overview, a list of the materials and reagents needed to complete the experiment, and followed by a detailed procedure that is supported with a helpful notes section offering tips and tricks of the trade as well as troubleshooting advice. These hallmark features were introduced by series editor Dr. John Walker and constitute the key ingredient in each and every volume of the *Methods in Molecular Biology* series. Tested and trusted, comprehensive and reliable, all protocols from the series are indexed in PubMed.

Human Respiratory Syncytial Virus

Methods and Protocols

Edited by

Bo Liang

Department of Biochemistry, Emory University School of Medicine, Atlanta, GA, USA

Rachel Fearns

Department of Virology, Immunology & Microbiology, Boston University Chobanian & Avedisian School of Medicine, Boston, MA, USA

Humana Press

Editors
Bo Liang
Department of Biochemistry
Emory University School of Medicine
Atlanta, GA, USA

Rachel Fearns
Department of Virology, Immunology & Microbiology
Boston University Chobanian & Avedisian
School of Medicine
Boston, MA, USA

ISSN 1064-3745 ISSN 1940-6029 (electronic)
Methods in Molecular Biology
ISBN 978-1-0716-4665-6 ISBN 978-1-0716-4666-3 (eBook)
https://doi.org/10.1007/978-1-0716-4666-3

This Humana imprint is published by the registered company Springer Science+Business Media, LLC, part of Springer Nature.
The registered company address is: 1 New York Plaza, New York, NY 10004, U.S.A.

Preface

Respiratory syncytial virus (RSV) is a leading cause of respiratory infections, particularly affecting infants, young children, older adults, and immunocompromised individuals. Despite its global prevalence and significant health burden, effective vaccines and targeted antiviral therapies remained elusive for decades. However, recent advancements in molecular virology, immunology, and structural biology have provided critical insights into RSV replication, pathogenesis, and immune evasion. This work has led to effective vaccines and an antibody that can be administered prophylactically to infants. In addition, this research has helped to progress the development of small molecule antivirals currently in clinical development.

This volume, *Human Respiratory Syncytial Virus: Methods and Protocols*, compiles cutting-edge methodologies that have propelled RSV research forward. It provides a comprehensive overview of the latest techniques used to study RSV at different scales, from the atomic details of RSV molecular biology to the analysis of RSV infection in various animal models. This book is designed as a comprehensive resource for new and experienced researchers engaged in RSV studies. It will be particularly valuable to virologists investigating viral genome replication, protein interactions, and host-pathogen dynamics; immunologists studying immune responses and vaccine strategies; and structural biologists employing biophysical and imaging techniques to characterize viral components. Additionally, translational researchers, pharmaceutical scientists developing antivirals and vaccines, clinicians, and epidemiologists examining RSV pathogenesis and disease outcomes will find this volume highly relevant.

This book aims to accelerate progress toward effective RSV prevention and treatment strategies by compiling state-of-the-art methodologies and insights from leading experts. We hope it serves as a vital resource for the scientific community, inspiring future discoveries and facilitating the development of novel therapeutic approaches against this significant respiratory pathogen.

Atlanta, GA, USA *Bo Liang*
Boston, MA, USA *Rachel Fearns*

Contents

Contributors

MARK ACKERMANN • *National Animal Disease Center, Ames, IA, USA*

XINGBIN AI • *Department of Pediatrics, Division of Newborn Medicine, Massachusetts General Hospital, Boston, MA, USA*

SARHAD ALNAJJAR • *Department of Comparative Biomedical Sciences, School of Veterinary Medicine, Faculty of Health and Medical Sciences, University of Surrey, Guildford, UK; Lambcure LLC, Corvallis, OR, USA*

GAURANG M. AMONKAR • *Department of Pediatrics, Division of Newborn Medicine, Massachusetts General Hospital, Boston, MA, USA*

LARRY J. ANDERSON • *Department of Pediatrics, School of Medicine, Emory University, Atlanta, GA, USA*

HARRISON C. BERGERON • *Department of Infectious Diseases, University of Georgia, Athens, GA, USA*

NAMITA BHATTARAI • *Center for Microbe and Immunity Research, Abigail Wexner Research Institute at Nationwide Children's Hospital, Columbus, OH, USA; Infectious Diseases Institute, The Ohio State University, Columbus, OH, USA*

JUSTIN W. BRENNAN • *Department of Microbiology and Immunology, University of Rochester Medical Center, Rochester, NY, USA*

DONGDONG CAO • *Department of Biochemistry, Emory University School of Medicine, Atlanta, GA, USA*

MICHELLE CHAMBLEE • *Infectious Diseases Institute, The Ohio State University, Columbus, OH, USA; Department of Veterinary Biosciences, College of Veterinary Medicine, The Ohio State University, Columbus, OH, USA*

PHYLIP CHEN • *Center for Microbe and Immunity Research, Abigail Wexner Research Institute at Nationwide Children's Hospital, Columbus, OH, USA; Infectious Diseases Institute, The Ohio State University, Columbus, OH, USA*

ZHENHANG CHEN • *Department of Biochemistry, Emory University School of Medicine, Atlanta, GA, USA*

GONZALO DE PRAT-GAY • *Instituto Leloir, IIBBA-CONICET, Buenos Aires, Argentina*

REBECCA DUTCH • *Department of Molecular and Cellular Biochemistry, University of Kentucky, Lexington, KY, USA*

KEARSTIN EDMONDS • *Department of Molecular and Cellular Biochemistry, University of Kentucky, Lexington, KY, USA*

JEAN-FRANÇOIS ELÉOUËT • *Unité de Virologie et Immunologie Moléculaires (VIM) UMR0892, INRAE-UVSQ-Université Paris-Saclay, Jouy-en-Josas, France*

SEBASTIAN ESPERANTE • *Center for Protein Redesign and Engineering (CRIP), Instituto de Investigaciones Biotecnologicas, Universidad Nacional de San Martin, Buenos Aires, Argentina*

RACHEL FEARNS • *Department of Virology, Immunology & Microbiology, Boston University Chobanian & Avedisian School of Medicine, Boston, MA, USA*

MARIE GALLOUX • *Unité de Virologie et Immunologie Moléculaires (VIM) UMR0892, INRAE-UVSQ-Université Paris-Saclay, Jouy-en-Josas, France*

JACK GALLUP • *Vaxxinova, Worthington, MN, USA*
YUNRONG GAO • *Department of Biochemistry, Emory University School of Medicine, Atlanta, GA, USA*
BINH HA • *Department of Pediatrics, School of Medicine, Emory University, Atlanta, GA, USA*
SARA HÄGGLUND • *Department of Clinical Science, Host Pathogen Interaction Group, Section of Ruminant Medicine, Swedish University of Agricultural Sciences (SLU), Uppsala, Sweden*
CHUAN HE • *Department of Chemistry, Biochemistry and Molecular Biology, Institute for Biophysical Dynamics, University of Chicago, Chicago, IL, USA; Howard Hughes Medical Institute, University of Chicago, Chicago, IL, USA*
NATALIE HORGAN • *Department of Biochemistry, Emory University School of Medicine, Atlanta, GA, USA*
CHENG CHIH HSU • *Infectious Diseases Institute, The Ohio State University, Columbus, OH, USA; Department of Veterinary Biosciences, College of Veterinary Medicine, The Ohio State University, Columbus, OH, USA*
SAMADHAN J. JADHAO • *Department of Pediatrics, School of Medicine, Emory University, Atlanta, GA, USA*
NICHOLAS KEUNG • *Department of Biochemistry, Emory University School of Medicine, Atlanta, GA, USA*
HEESU KIM • *Department of Virology, Immunology & Microbiology, Boston University Chobanian & Avedisian School of Medicine, Boston, MA, USA*
PAUL H. LEROU • *Department of Pediatrics, Division of Newborn Medicine, Massachusetts General Hospital, Boston, MA, USA*
DAISY W. LEUNG • *Department of Medicine, Washington University School of Medicine, St. Louis, MO, USA*
JIANRONG LI • *Department of Veterinary Biosciences, College of Veterinary Medicine, The Ohio State University, Columbus, OH, USA; Infectious Disease Institute, The Ohio State University, Columbus, OH, USA; Center for RNA Biology, The Ohio State University, Columbus, OH, USA*
BO LIANG • *Department of Biochemistry, Emory University School of Medicine, Atlanta, GA, USA*
BARBARA LUDEKE • *Department of Virology, Immunology & Microbiology, Boston University Chobanian & Avedisian School of Medicine, Boston, MA, USA*
RONAN MACLOUGHLIN • *Research and Development, Science and Emerging Technologies, Aerogen Ltd., Galway Business Park, Galway, Ireland; School of Pharmacy and Biomolecular Science, Royal College of Surgeons in Ireland, Dublin, Ireland; School of Pharmacy and Pharmaceutical Sciences, Trinity College, Dublin, Ireland*
MAHESH K. C. • *Center for Microbe and Immunity Research, Abigail Wexner Research Institute at Nationwide Children's Hospital, Columbus, OH, USA; Infectious Diseases Institute, The Ohio State University, Columbus, OH, USA*
JONATHAN MILLER • *Department of Veterinary Biosciences, College of Veterinary Medicine, The Ohio State University, Columbus, OH, USA*
IVANA MOLINA • *Instituto Leloir, IIBBA-CONICET, Buenos Aires, Argentina*
STEFAN NIEWIESK • *Department of Veterinary Biosciences, College of Veterinary Medicine, The Ohio State University, Columbus, OH, USA*

SARAH L. NOTON • *Department of Virology, Immunology & Microbiology, Boston University Chobanian & Avedisian School of Medicine, Boston, MA, USA*
MARK E. PEEPLES • *Center for Microbe and Immunity Research, Abigail Wexner Research Institute at Nationwide Children's Hospital, Columbus, OH, USA; Infectious Diseases Institute, The Ohio State University, Columbus, OH, USA; Department of Pediatrics, The Ohio State University, Columbus, OH, USA*
RAMON PERALTA-MARTINEZ • *Instituto Leloir, IIBBA-CONICET, Buenos Aires, Argentina*
RICHARD K. PLEMPER • *Center for Translational Antiviral Research, Georgia State University Institute for Biomedical Sciences, Atlanta, GA, USA*
DAVID A. PRICE • *Department of Medicine, Washington University School of Medicine, St. Louis, MO, USA*
MARIE-ANNE RAMEIX-WELTI • *Unit M3P Institut Pasteur, Université Paris-Saclay, Université de Versailles St. Quentin, Université Paris Cité, UMR 1173 (2I), INSERM; Assistance Publique des Hôpitaux de Paris, Centre National de Référence Virus des Infections Respiratoire (CNR VIR), Paris, France*
RINCHEVAL V. • *Université Paris-Saclay, UVSQ, LGBC, Versailles, France*
RISSO-BALLESTER J. • *Institute of Virology, School of Medicine, Technical University of Munich, Munich, Germany*
ELISEO SALAS • *Department of Biochemistry, Emory University School of Medicine, Atlanta, GA, USA*
RANJINI SANKARANARAYANAN • *Department of Pediatrics, School of Medicine, Emory University, Atlanta, GA, USA*
AMIT SHARMA • *Department of Population Health and Pathobiology, College of Veterinary Medicine, North Carolina State University, Raleigh, NC, USA*
MICHAEL R. SIRRINE • *Center for Translational Antiviral Research, Georgia State University Institute for Biomedical Sciences, Atlanta, GA, USA*
YAN SUN • *Department of Microbiology and Immunology, University of Rochester Medical Center, Rochester, NY, USA*
GERALDINE TAYLOR • *The Pirbright Institute, Pirbright, Woking, UK*
MICHAEL N. TENG • *Division of Allergy and Immunology, Department of Internal Medicine, University of South Florida Morsani College of Medicine, Tampa, FL, USA*
ILADA THONGPAN • *Center for Microbe and Immunity Research, Abigail Wexner Research Institute at Nationwide Children's Hospital, Columbus, OH, USA; Infectious Diseases Institute, The Ohio State University, Columbus, OH, USA*
KIM C. TRAN • *Division of Allergy and Immunology, Department of Internal Medicine, University of South Florida Morsani College of Medicine, Tampa, FL, USA*
RALPH A. TRIPP • *Department of Infectious Diseases, College of Veterinary Medicine, University of Georgia, Athens, GA, USA*
JEAN FRANÇOIS VALARCHER • *Department of Clinical Science, Host Pathogen Interaction Group, Section of Ruminant Medicine, Swedish University of Agricultural Sciences (SLU), Uppsala, Sweden*
NICOLE D. WAGNER • *Department of Medicine, Washington University School of Medicine, St. Louis, MO, USA; Department of Chemistry, Washington University in St. Louis, St. Louis, MO, USA*
M. A. RAMEIX WELTI • *Institut Pasteur, Université Paris-Saclay, Université de Versailles St. Quentin, Université Paris Cité, UMR 1173 (2I), INSERM; Assistance Publique des Hôpitaux de Paris, Hôpital Ambroise Paré, Paris, France*

LI-SHENG ZHANG • *Department of Chemistry, Biochemistry and Molecular Biology, Institute for Biophysical Dynamics, University of Chicago, Chicago, IL, USA*
YUEXIU ZHANG • *Department of Veterinary Biosciences, College of Veterinary Medicine, The Ohio State University, Columbus, OH, USA*
CAIQI ZHAO • *Department of Pediatrics, Division of Newborn Medicine, Massachusetts General Hospital, Boston, MA, USA*

Chapter 1

Introduction to Respiratory Syncytial Virus

Rachel Fearns and Bo Liang

Abstract

Human respiratory syncytial virus (RSV) is a significant human pathogen in infants, young children, older adults, and the immunocompromised. Over the last decade, many important mechanisms contributing to RSV infection, replication, and disease pathogenesis have been revealed, and a body of information has accumulated regarding how RSV proteins modulate molecular signaling and immune responses to infection. In this chapter, we briefly introduce RSV replication and pathogenesis and summarize the state of the RSV vaccine and antiviral compounds in the clinical stages of development. This chapter frames the features of this book and the molecular methods used for understanding RSV interaction with the host.

Key words Respiratory syncytial virus, Introduction, Overview, Replication, Pathogenesis, Vaccine, Antiviral compounds

1 Introduction

For decades, respiratory syncytial virus (RSV) has been recognized as a major cause of respiratory disease, particularly in infants and young children, older adults, and those with comorbidities [1]. It has a global distribution and is widely prevalent. RSV disease manifestations range from self-limiting cold-like symptoms to severe bronchiolitis and pneumonia [1]. The toll of RSV has been significant. Based on a meta-analysis, it has been estimated that in 2019, RSV was responsible for 33 million acute lower respiratory tract infections, 3.6 million RSV-associated acute lower respiratory infection hospital admissions, and 101,400 RSV-attributable overall deaths in children aged 0–60 months [2]. One in every 28 deaths in children aged 28 days to 6 months is attributable to RSV [2], with the largest toll exerted in low- and middle-income countries. Although RSV is typically associated with disease in very young children, the virus also has a significant impact on older adults [3, 4]. For example, it has been estimated that RSV could be responsible for 787,000 annual hospitalizations of individuals ≥65 in industrialized countries [5].

Bo Liang and Rachel Fearns (eds.), *Human Respiratory Syncytial Virus: Methods and Protocols*, Methods in Molecular Biology, vol. 2948, https://doi.org/10.1007/978-1-0716-4666-3_1,

2 RSV Epidemiology

RSV does not have a known animal reservoir and is sustained in the population by human–human transmission [6]. It is spread through direct contact, large droplets, and contaminated fomites, with nosocomial infections presenting a significant concern [7]. There are two antigenic subgroups of RSV that are distinguished based on the surface glycoprotein (G). The two groups co-circulate globally, with subgroup A tending to be predominant. However, the emergence of some novel subgroup B genotypes has increased the dominance of subgroup B infections in recent years [8]. RSV has a seasonal distribution. In temperate regions, the RSV epidemic emerges and peaks during the autumn and winter, whereas in tropical areas, it peaks during the rainy seasons [9].

3 RSV Disease and Pathogenesis

RSV infection can result in a range of outcomes, from mild cold-like symptoms to severe bronchitis and pneumonia [6]. The prevalence of RSV means that almost everyone is infected within the first 2 years of life, and individuals can be reinfected throughout life [6]. The first infection is typically the most severe RSV childhood infection, with severity diminishing in subsequent infections through childhood and adulthood [10]. RSV infections become more severe in older adults due to diminished CD4 and CD8 T cells and increased comorbidities [11]. RSV infects the upper respiratory tract through mucosal surfaces in the nasopharynx or conjunctiva and then spreads to the lower respiratory tract [12]. Analysis of pediatric postmortem samples showed that RSV infects the bronchiolar epithelium and type 1 and 2 alveolar pneumocytes [13]. Airway obstruction can occur with the accumulation of epithelial and inflammatory cell debris and mucus [13]. RSV replicates predominantly in ciliated epithelial cells [14] but has also been reported to infect other cell types, including neuronal and immune cells, albeit at a much lower efficiency than epithelial cells [6, 12]. Thus, it is ciliated epithelial cells that drive the infection. RSV is only released from the apical surface of epithelial cells. For this reason, it does not typically infect basal cells, although it can do so if the epithelium is damaged [15]. For the same reason, RSV is not normally associated with syncytia formation in human infections, although syncytia have been observed occasionally [16, 17].

RSV infection is sensed by toll-like receptors TLR 2, TLR 4, TLR 6, RIG-I receptors, and NOD 2, which trigger the expression of type I interferon [18, 19]. Type I interferons serve a protective role against RSV infection and are key for stimulating B cell responses. Still, RSV has evolved mechanisms to antagonize interferon induction and interferon responses [19]. RSV also stimulates type III interferon expression, which may correlate with protection against severe disease [20, 21]. In addition to eliciting protective innate immune responses, RSV infection elicits extensive inflammation associated with immunopathology. RSV stimulates the expression of IL-8, leading to the recruitment of neutrophils, which are strongly associated with severe disease [11, 12, 22]. Other cytokines and chemokines secreted by RSV-infected and/or bystander cells also play a pathogenic role [11]. Resolution of RSV infection depends on dendritic cells and alveolar macrophages, key antigen-presenting cells, and cytotoxic T cells, which are associated with a reduced viral load [11, 23]. IgG and IgA antibodies elicited against the viral fusion (F), and G proteins play a protective role [11].

Severe RSV infection in infancy is associated with wheezing and asthma development later in life [16]. One possible mechanism for this is that the airway epithelial cell barrier is immature at birth and continues to develop during the first 2 years of life [10]. RSV infection causes significant changes to the transcriptome of infected cells [24–26], which could perturb the development of the airway epithelium [10]. In addition, although RSV is typically associated with acute respiratory infection, there is evidence suggesting that it can be shed for months. These findings, coupled with the long-term sequelae of RSV infection, indicate that RSV might persist, although the molecular basis for viral persistence remains to be fully explored [27].

4 The RSV Replication Cycle

RSV is a non-segmented negative-strand RNA virus. Its formal species name is *Orthopneumovirus hominis*, and it is classified in the family *Pneumoviridae* in the order *Mononegavirales* [28]. RSV encodes ten genes and eleven proteins [29]. The RSV particle consists of the viral RNA genome encapsidated along its length with a chain of nucleoprotein (N) protomers to form a helical nucleocapsid. Associated with the nucleocapsid are the viral polymerase proteins, the large subunit (L) and the phosphoprotein (P), and a transcription elongation factor (M2-1). Surrounding the nucleocapsid is the viral matrix protein (M) and a viral envelope studded with three proteins: the attachment glycoprotein (G), the fusion protein (F), and a small hydrophobic protein (SH) [30, 31]. The virus codes for three intracellular nonstructural proteins, nonstructural proteins 1 and 2 (NS1 and NS2), and M2-2 protein [32–34]. It also codes for a truncated, secreted form of the G protein [35, 36].

The RSV replication cycle is shown in Fig. 1. RSV attaches to a susceptible cell and enters by fusion through a series of

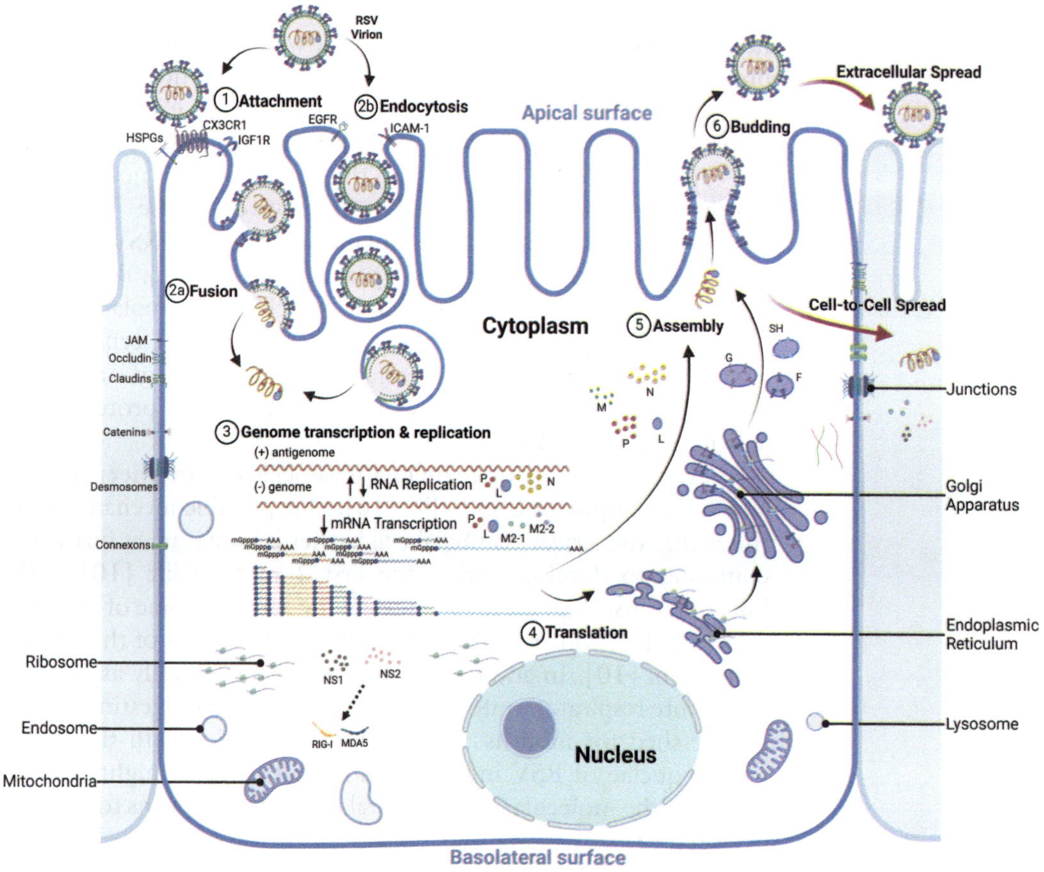

Fig. 1 Schematic illustration of the respiratory syncytial virus (RSV) life cycle. (**1**) RSV initiates infection by the attachment to human epithelial cells via receptors predominantly localized on the apical surface, including HSPGs, CX3CR1, IGF1R, EGFR, and ICAM-1. The G glycoprotein binds primarily to CX3CR1 and HSPGs, tethering the virion to the cell surface, while the F glycoprotein interacts with ICAM-1, EGFR, and IGF1R to mediate entry. RSV enters the host cell through two mechanisms: (**2a**) direct membrane fusion, or (**2b**) macropinocytosis, a form of endocytosis followed by fusion. Fusion, mediated by the trimeric class I F protein, releases the RNP complex into the cytoplasm, initiating viral replication. (**3**) RSV transcription and replication occur within cytoplasmic inclusion bodies. The RSV polymerase complex, comprising the L and P proteins, synthesizes viral mRNAs and genomic RNA using the N-RNA template, with M2-1 enabling transcription. (**4**) Viral mRNAs are processed by host ribosomes in the cytoplasm to produce structural and nonstructural proteins. Most of the translated RSV proteins contribute to the replication, regulation, virion reassembly, and budding, including structural and accessory proteins such as the F, G, SH, N, P, L, M, M2-1, and M2-2 proteins. In contrast, the nonstructural proteins NS1 and NS2 primarily target the host immune system, suppressing interferon responses, disrupting RIG-I and MDA5 signaling pathways, and facilitating immune evasion to support viral replication. (**5**) Surface and structural proteins undergo glycosylation and maturation in the ER and Golgi before transport to the apical plasma membrane. Genomic RNA is encapsidated by the N protein to form RNP complexes, which are delivered to the assembly sites at the apical membrane. The M protein organizes viral components, enabling the assembly of mature filamentous RSV virions. (**6**) After assembly, virions bud from the apical plasma membrane, acquiring a lipid envelope embedded with the F, G, and SH proteins during the process. RSV spreads within the respiratory epithelium through two main mechanisms. Extracellular virions infect neighboring cells via direct fusion, while cell-to-cell spread through intercellular junctions allows RSV to traverse the basolateral surface, evading immune detection. This direct propagation, mediated by the F protein and the host cytoskeleton, ensures efficient infection of adjacent cells and limits exposure to host immune defenses. Abbreviations: The RSV proteins include (**a**) surface

well-coordinated steps [6, 37–45]. The process begins with the virus binding to specific receptors on the surface of the host cell. RSV utilizes its surface glycoproteins, primarily the G and F proteins, to facilitate this attachment [6, 37–40]. The G protein helps the virus bind to receptor molecules on the cell membrane, such as heparan sulfate proteoglycans (HSPGs) and other potential protein receptors, like CX3C-chemokine receptor 1 (CX3CR1) [46–48]. This attachment allows RSV to make initial contact with the host cell. The F protein interacts with intercellular adhesion molecule-1 (ICAM-1), epidermal growth factor receptor (EGFR), or insulin-like growth factor-1 receptor (IGF1R), eliciting a signaling cascade that transposes nucleolin from cell nuclei to the plasma membrane, where it serves as an F coreceptor [49, 50]. Following attachment, the F protein undergoes a conformational change, which is crucial for the fusion of the viral envelope with the cell membrane [51, 52]. This fusion brings the viral and cellular membranes close enough to merge, creating an entry pathway for the viral nucleocapsid into the cytoplasm of the host cell [38]. Fusion typically occurs directly at the plasma membrane. However, it can sometimes involve endocytosis, where the cell engulfs the virus, and fusion occurs within an endosome as it acidifies [39].

Following delivery into the cytoplasm of a host cell, the nucleocapsid and associated polymerase can engage in viral gene expression and genome replication. The 15.2 kb viral genome codes for ten mRNAs (one of these mRNAs, *M2*, codes for both M2-1 and M2-2) [53]. The ten genes are arranged linearly along the genome, and each is flanked with conserved elements called *gene start* and *gene end* signals; between each gene is a short intergenic region [53]. At the 3′ and 5′ ends of the genome are extragenic regions called the *leader* and *trailer*, respectively [54] (Fig. 2). The *leader* region contains a promoter that directs the initiation of transcription and genome replication [55]. To transcribe the genome, the polymerase moves from the 3′ to the 5′ end of the genome, recognizing and responding to the *gene start* and *gene end* signals to produce subgenomic mRNAs [56, 57]. The transcribing polymerase initiates RNA synthesis at position 3 of the *leader* region and first synthesizes a short RNA complementary to the 3′ end of

Fig. 1 (continued) glycoproteins: attachment glycoprotein (G), fusion glycoprotein (F), and small hydrophobic protein (SH); (**b**) core and replication-associated proteins: nucleoprotein (N), phosphoprotein (P), and large polymerase protein (L); (**c**) matrix and regulatory proteins: matrix protein (M), matrix protein 2-1 (M2-1), matrix protein 2-2 (M2-2), nonstructural protein 1 (NS1), and nonstructural protein 2 (NS2). CX3CR1 CX3C-chemokine receptor 1, EGFR epidermal growth factor receptor, ICAM-1 intercellular adhesion molecule-1, IGF1R insulin-like growth factor-1 receptor, HSPGs heparan sulfate proteoglycans, RIG-I retinoic acid-inducible gene I, MDA5 melanoma differentiation-associated protein 5, RNP ribonucleoprotein, ER endoplasmic reticulum. Note that the figure is not drawn to scale. Created in BioRender. Liang, B. (2025)

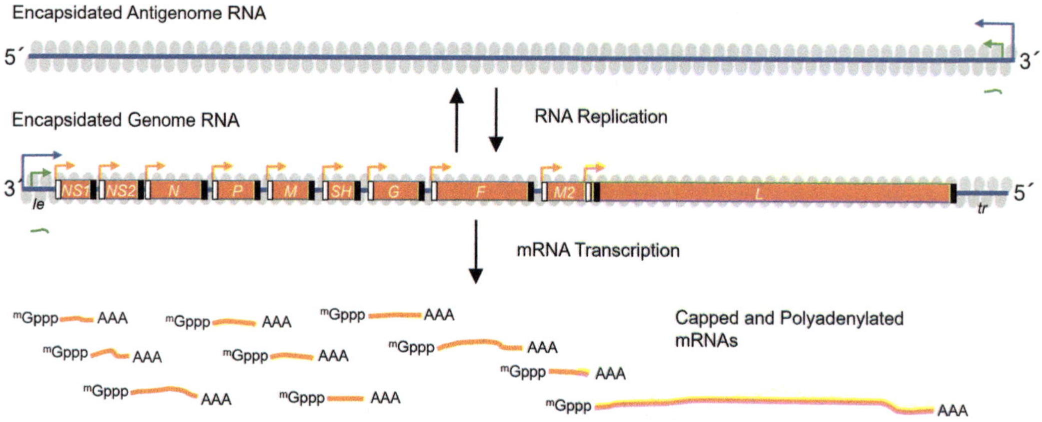

Fig. 2 Schematic diagram showing the organization of the RSV genome and the transcription and replication products that it encodes. The genome and antigenome RNAs are shown as blue lines with the N coating illustrated in gray. Each gene on the RSV genome is shown as an orange rectangle and is flanked with conserved *gene start* and *gene end* signals (white and black boxes, respectively). At the 3′ and 5′ ends of the genome are the *leader* (*le*) and *trailer* (*tr*) regions, respectively. The RNA synthesis start sites at position 3 of the *leader* and *trailer* regions are shown as green arrows, RNA synthesis start sites for each of the mRNAs are shown as orange arrows, and RNA synthesis start sites for the antigenome and genome RNAs are shown as blue arrows. The small RNAs that are generated from position 3 of the *leader* and *trailer* regions are shown as short green lines, and the mRNAs are shown as orange lines. Each mRNA has a methylguanosine cap (^{m}Gppp) and poly A tail (AAA). Note that although there is an RNA synthesis start site at the beginning of each gene, the polymerase begins the process of transcription from the 3′ end of the genome, at position 3 of the *leader* region. Note that the figure is not drawn to scale. Fearns, R. (2025)

the leader, le(+) RNA [58]. The polymerase releases the le(+) RNA after ~25 nt and then scans the nucleocapsid template to locate the *NS1 gene start* signal [58]. Here, it reinitiates RNA synthesis, caps the mRNA, and adds methyl groups to create a cap 1 structure at the 5′ end of the mRNA [58, 59]. Having capped the mRNA, the polymerase can elongate to the end of the gene, with efficient elongation depending on the RSV M2-1 protein [60, 61]. The M2-1 protein seems to associate with the transcribing polymerase and contact the nascent mRNA as it is extruded from the polymerase [62]. When the polymerase reaches a *gene end* signal, it polyadenylates the mRNA, most likely by stuttering on a U tract in the gene end signal, and then releases the mRNA [57, 63]. The polymerase can then scan the genome until it locates the next *gene start* signal, where it reinitiates mRNA synthesis [64, 65]. Continuing in this vein, the polymerase generates monocistronic mRNAs corresponding to each of the ten genes.

To replicate the genome, the polymerase begins at the 3′ *leader* promoter but initiates RNA synthesis at position 1 [58, 66]. In this case, the polymerase does not release the le(+) RNA and continues elongation to the end of the genome to produce a positive-sense antigenome RNA [58, 66]. The complement of the *trailer* region that lies at the 3′ end of the antigenome RNA serves as a promoter for the polymerase to initiate genome RNA synthesis [67]. Like the *leader* promoter, the *trailer* promoter directs the polymerase to initiate at positions 1 and 3. Initiation at position 3 results in a short ~25 nt RNA that might be involved in counteracting the cellular stress granule response [68]. Initiation at position 1 results in the production of genome RNA. In contrast to the mRNAs, the antigenome and genome RNAs are encapsidated as they are synthesized [60, 69]. Encapsidation begins with the binding of soluble N protein to a sequence at the 5′ end of the antigenome and genome RNAs, followed by repeated addition of soluble N protein to the growing N oligomeric chain. Concurrent encapsidation is thought to enable the polymerase to efficiently elongate the RNA to produce the long antigenome and genome RNAs [58, 66]. Occasionally, the polymerase also produces another type of replicative RNA, termed a defective viral genome (DVG), which lacks most of the viral genome but contains the terminal promoter elements, which allows the DVG to be replicated [70]. DVGs can compete with the viral genome and antigenome RNA for polymerase binding and thus can affect the efficiency of viral replication [71]. They are also associated with antiviral responses and appear to play a significant role in viral pathogenesis [72].

RSV transcription and replication are accompanied by viral mRNA translation, packaging, assembly, and release to produce new virions [40, 73]. The viral RNA genome is transcribed into mRNAs, translated by the host's ribosomes to produce the full complement of viral proteins (NS1, NS2, N, P, M, SH, G, F, M2-1, M2-2, and L). F, SH, and G proteins are processed in the endoplasmic reticulum and Golgi before being transported to the cell membrane [73]. Simultaneously, new copies of viral RNA are produced, encapsidated by N proteins, and organized into the ribonucleoprotein complex, stabilizing the RNA for packaging. Viral assembly occurs at the plasma membrane, where structural proteins and N-RNA complexes converge [73, 74]. The M protein is critical in organizing components, linking envelope proteins with RNA, and facilitating their incorporation into the budding virion [75]. During budding, RSV particles acquire a lipid envelope from the host plasma membrane, incorporating F, SH, and G proteins. Finally, the fully formed virions are released from the cell surface, allowing the virus to infect nearby cells and spread throughout the respiratory tract [73].

The nonstructural proteins NS1, NS2, and the SH protein, along with the secreted form of the G protein, play crucial roles in RSV's ability to infect and evade the host's immune response [76]. NS1 and NS2 are primarily involved in counteracting the host's immune defenses [77–80]. They specifically target the RIG-I and MDA5 pathways and inhibit the interferon (IFN) response, a critical part of the antiviral defense mechanism, by interfering with the signaling pathways that activate IFN production and targeting key signaling molecules for degradation [77–80]. This suppression of the IFN response helps RSV evade early detection and enhances viral replication within the host cells [81]. The SH protein, a small hydrophobic protein embedded in the viral envelope, has multiple functions related to immune evasion and viral persistence. It forms ion channels in the host cell membrane that may alter cell signaling and membrane permeability [82, 83]. Additionally, SH helps to modulate apoptosis, or programmed cell death, allowing infected cells to survive longer and produce more viral progeny [82]. The secreted form of the G protein, distinct from the membrane-bound G protein, acts as a decoy to further dampen the host's immune response [84]. By mimicking the viral attachment protein, the secreted G protein binds to antibodies and other immune components, reducing their ability to target RSV virions [85–87]. This diversion helps the virus to persist in the host by preventing effective antibody-mediated neutralization. NS1, NS2, SH, and the secreted G protein contribute to RSV's ability to replicate efficiently, evade immune responses, and establish infection within the respiratory tract [12].

5 Structures of RSV Virions and RSV Proteins

Advanced structural techniques, including X-ray crystallography, nuclear magnetic resonance (NMR) spectroscopy, cryo-electron microscopy (cryo-EM), and cryo-electron tomography (cryo-ET), have provided detailed views of RSV virions and their protein components, shedding light on the virus's mechanisms of entry, replication, immune evasion, and assembly. RSV virions are flexible, enveloped particles that cryo-ET has mapped extensively, revealing the lipid-bilayer envelope that houses the G and F glycoproteins and the M protein beneath the envelope [30, 31, 88, 89]. F protein in both its pre-fusion and post-fusion states has been visualized by X-ray crystallography, capturing structural shifts critical for entry [38, 42, 90–92]. Recently, cryo-EM has been extensively used for engineered F protein with its epitopes that are targeted by neutralizing antibodies, pivotal for vaccine design [93–96]. M protein alone has also been crystalized, revealing its surface features [97, 98]. Due to its small size, the SH protein has been studied through NMR, providing insight into its role in host immune modulation and cell survival [83].

Inside the virion, the viral RNA genome is wrapped by the N protein in a helical ribonucleoprotein (RNP) complex, stabilizing the genome and protecting it from degradation. X-ray crystallography and cryo-EM have revealed the helical structure of the N protein [99–101]. Due to its large size, the structures of L protein, the main enzyme for RSV RNA synthesis, and its cofactor P protein have been successfully determined by cryo-EM and revealed a complex architecture suited for transcription and replication [102–104]. Crystal structures of M2-1 and its complex with RNA shed light on the role of M2-1 in preventing premature termination of transcription [105, 106]. Finally, crystal structures of RSV's non-structural proteins, NS1 and NS2, have shown how these proteins interact with host immune pathways, providing insight into RSV's immune evasion strategy [80, 107]. Together, these structural insights into RSV's virion and protein organization provide a foundation for developing targeted antiviral therapies and vaccines, opening new avenues for effectively combating RSV infections.

6 Countermeasures Against RSV

Significant progress has been made in the development of countermeasures against RSV. In 1998, a recombinant humanized antibody, palivizumab, became available for prophylactic use. Palivizumab targets the F protein of RSV, which is highly conserved between subgroups A and B. Although palivizumab effectively inhibits RSV infection, it is costly and requires monthly injections to be effective, limiting its use [108]. Recently, new countermeasures have come to the fore. Nirsevimab is also a monoclonal antibody against the RSV F protein that has been approved for prophylactic use [108]. In contrast to palivizumab, nirsevimab binds to the pre-fusion form of F, preventing F from fulfilling its fusion activity, and has modifications in its Fc region, which prolong antibody half-life [109]. Nirsevimab can protect against RSV for 5 months following a single intramuscular administration, thus providing protection for an RSV season [110–113]. In addition to a prophylactic antibody to protect infants, there is now a vaccine for administration to pregnant women. This vaccine provides passive infant immunization by transplacental transfer of antibodies and efficacy against RSV-associated lower respiratory tract illness and RSV-associated hospitalization in infants up to 6 months of age [114]. In addition to these advances in protecting infants against RSV, two vaccines are now available to prevent severe RSV disease in older adults, which reduce the severity of disease in adults ≥60 years of age [115, 116]. In tandem with the development of these countermeasures that prevent infection, there are multiple antiviral therapies in development [117, 118]. Thus, the landscape of RSV disease is likely to change significantly over the next few years.

7 Future Directions

Although there are now effective countermeasures against RSV, the virus warrants further study. Questions remain regarding the mechanisms by which RSV replicates and spreads, and its mechanisms of pathogenesis. Addressing these questions could aid the development of antiviral drugs and therapeutics that provide symptom relief for those individuals who succumb to RSV infection. In addition, RSV can serve as a prototype for related viruses, including human metapneumovirus, another respiratory virus in the *pneumovirus* family, other respiratory viruses, and viruses in other closely related families, such as the paramyxoviruses and filoviruses. This book provides detailed methodologies for future studies to further this research. The book is organized into a logical sequence of chapters, which cover the following topics: (i) RSV virology, including RSV growth and quantification by plaque assay, methods for generating RSV stocks with relatively high or low proportions of DVGs, and generation of recombinant RSV; (ii) methods to study different aspects of RSV molecular biology, including expression and purification of different viral proteins, fusion and polymerase assays, fluorescence microscopy of newly synthesized RNAs, evaluation of host factors involved in RSV replication, and analysis of posttranscriptional modifications in the viral RNA; (iii) methods for studying RSV pathogenesis and to assess vaccines or antivirals, including protocols for working with RSV in different model systems, including human airway cells, mice, cotton rats, lambs, and calves, and methods for profiling mutations that confer resistance to antiviral compounds and measuring RSV antibodies in human plasma. Thus, this book spans a wide range of topics on RSV, ranging from analysis of its molecular facets to replication in large animal hosts, and should serve as a valuable guide for anyone who is interested in RSV research.

Conflict of Interest

The Fearns laboratory has a sponsored research agreement with Merck & Co.

References

1. Borchers AT, Chang C, Gershwin ME, Gershwin LJ (2013) Respiratory syncytial virus – a comprehensive review. Clin Rev Allergy Immunol 45(3):331–379. https://doi.org/10.1007/s12016-013-8368-9
2. Li Y, Wang X, Blau DM, Caballero MT, Feikin DR, Gill CJ et al (2022) Global, regional, and national disease burden estimates of acute lower respiratory infections due to respiratory syncytial virus in children younger than 5 years in 2019: a systematic analysis. Lancet 399(10340):2047–2064. https://doi.org/10.1016/S0140-6736(22)00478-0
3. Falsey AR, Hennessey PA, Formica MA, Cox C, Walsh EE (2005) Respiratory syncytial virus infection in elderly and high-risk adults. N Engl J Med 352(17):1749–1759. https://doi.org/10.1056/NEJMoa043951
4. Savic M, Penders Y, Shi T, Branche A, Pircon JY (2023) Respiratory syncytial virus disease burden in adults aged 60 years and older in high-income countries: a systematic literature review and meta-analysis. Influenza Other Respir Viruses 17(1):e13031. https://doi.org/10.1111/irv.13031

5. Li Y, Kulkarni D, Begier E, Wahi-Singh P, Wahi-Singh B, Gessner B et al (2023) Adjusting for case under-ascertainment in estimating RSV hospitalisation burden of older adults in high-income countries: a systematic review and modelling study. Infect Dis Ther 12(4): 1137–1149. https://doi.org/10.1007/s40121-023-00792-3
6. Collins PL, Fearns R, Graham BS (2013) Respiratory syncytial virus: virology, reverse genetics, and pathogenesis of disease. Curr Top Microbiol Immunol 372:3–38. https://doi.org/10.1007/978-3-642-38919-1_1
7. Hall CB (2000) Nosocomial respiratory syncytial virus infections: the "Cold War" has not ended. Clin Infect Dis 31(2):590–596. https://doi.org/10.1086/313960
8. Cantu-Flores K, Rivera-Alfaro G, Munoz-Escalante JC, Noyola DE (2022) Global distribution of respiratory syncytial virus A and B infections: a systematic review. Pathog Glob Health 116(7):398–409. https://doi.org/10.1080/20477724.2022.2038053
9. Chadha M, Hirve S, Bancej C, Barr I, Baumeister E, Caetano B et al (2020) Human respiratory syncytial virus and influenza seasonality patterns-early findings from the WHO global respiratory syncytial virus surveillance. Influenza Other Respir Viruses 14(6):638–646. https://doi.org/10.1111/irv.12726
10. Berdnikovs S, Newcomb DC, Hartert TV (2024) How early life respiratory viral infections impact airway epithelial development and may lead to asthma. Front Pediatr 12: 1441293. https://doi.org/10.3389/fped.2024.1441293
11. Russell CD, Unger SA, Walton M, Schwarze J (2017) The human immune response to respiratory syncytial virus infection. Clin Microbiol Rev 30(2):481–502. https://doi.org/10.1128/CMR.00090-16
12. Bergeron HC, Tripp RA (2021) Immunopathology of RSV: an updated review. Viruses 13(12). https://doi.org/10.3390/v13122478
13. Johnson JE, Gonzales RA, Olson SJ, Wright PF, Graham BS (2007) The histopathology of fatal untreated human respiratory syncytial virus infection. Mod Pathol 20(1):108–119. https://doi.org/10.1038/modpathol.3800725
14. Zhang L, Peeples ME, Boucher RC, Collins PL, Pickles RJ (2002) Respiratory syncytial virus infection of human airway epithelial cells is polarized, specific to ciliated cells, and without obvious cytopathology. J Virol 76(11):5654–5666. https://doi.org/10.1128/jvi.76.11.5654-5666.2002
15. Persson BD, Jaffe AB, Fearns R, Danahay H (2014) Respiratory syncytial virus can infect basal cells and alter human airway epithelial differentiation. PLoS One 9(7):e102368. https://doi.org/10.1371/journal.pone.0102368
16. Griffiths C, Drews SJ, Marchant DJ (2017) Respiratory syncytial virus: infection, detection, and new options for prevention and treatment. Clin Microbiol Rev 30(1): 277–319. https://doi.org/10.1128/CMR.00010-16
17. Neilson KA, Yunis EJ (1990) Demonstration of respiratory syncytial virus in an autopsy series. Pediatr Pathol 10(4):491–502. https://doi.org/10.3109/15513819009067138
18. Sabbah A, Chang TH, Harnack R, Frohlich V, Tominaga K, Dube PH et al (2009) Activation of innate immune antiviral responses by Nod2. Nat Immunol 10(10):1073–1080. https://doi.org/10.1038/ni.1782
19. Stephens LM, Varga SM (2020) Function and modulation of type I interferons during respiratory syncytial virus infection. Vaccines (Basel) 8(2). https://doi.org/10.3390/vaccines8020177
20. Taveras J, Garcia-Maurino C, Moore-Clingenpeel M, Xu Z, Mertz S, Ye F et al (2022) Type III interferons, viral loads, age, and disease severity in young children with respiratory syncytial virus infection. J Infect Dis 227(1):61–70. https://doi.org/10.1093/infdis/jiac404
21. Werder RB, Lynch JP, Simpson JC, Zhang V, Hodge NH, Poh M et al (2018) PGD2/DP2 receptor activation promotes severe viral bronchiolitis by suppressing IFN-lambda production. Sci Transl Med 10(440). https://doi.org/10.1126/scitranslmed.aao0052
22. Sebina I, Phipps S (2020) The contribution of neutrophils to the pathogenesis of RSV bronchiolitis. Viruses 12(8). https://doi.org/10.3390/v12080808
23. Jung HE, Kim TH, Lee HK (2020) Contribution of dendritic cells in protective immunity against respiratory syncytial virus infection. Viruses 12(1). https://doi.org/10.3390/v12010102
24. Dong J, Basse V, Bierre M, Peres de Oliveira A, Vidalain PO, Sibille P et al (2022) Respiratory syncytial virus NS1 protein targets the transactivator binding domain

of MED25. J Mol Biol 434(19):167763. https://doi.org/10.1016/j.jmb.2022.167763
25. Pei J, Beri NR, Zou AJ, Hubel P, Dorando HK, Bergant V et al (2021) Nuclear-localized human respiratory syncytial virus NS1 protein modulates host gene transcription. Cell Rep 37(2):109803. https://doi.org/10.1016/j.celrep.2021.109803
26. Van Royen T, Sedeyn K, Moschonas GD, Toussaint W, Vuylsteke M, Van Haver D et al (2022) An unexpected encounter: respiratory syncytial virus nonstructural protein 1 interacts with mediator subunit MED25. J Virol 96(19):e0129722. https://doi.org/10.1128/jvi.01297-22
27. Piedimonte G (2013) Respiratory syncytial virus and asthma: speed-dating or long-term relationship? Curr Opin Pediatr 25(3):344–349. https://doi.org/10.1097/MOP.0b013e328360bd2e
28. Amarasinghe GK, Ayllon MA, Bao Y, Basler CF, Bavari S, Blasdell KR et al (2019) Taxonomy of the order Mononegavirales: update 2019. Arch Virol 164(7):1967–1980. https://doi.org/10.1007/s00705-019-04247-4
29. Huang YT, Collins PL, Wertz GW (1985) Characterization of the 10 proteins of human respiratory syncytial virus: identification of a fourth envelope-associated protein. Virus Res 2(2):157–173. https://doi.org/10.1016/0168-1702(85)90246-1
30. Conley MJ, Short JM, Burns AM, Streetley J, Hutchings J, Bakker SE et al (2022) Helical ordering of envelope-associated proteins and glycoproteins in respiratory syncytial virus. EMBO J 41(3):e109728. https://doi.org/10.15252/embj.2021109728
31. Sibert BS, Kim JY, Yang JE, Ke Z, Stobart CC, Moore ML et al (2024) Assembly of respiratory syncytial virus matrix protein lattice and its coordination with fusion glycoprotein trimers. Nat Commun 15(1):5923. https://doi.org/10.1038/s41467-024-50162-x
32. Collins PL, Hill MG, Johnson PR (1990) The two open reading frames of the 22K mRNA of human respiratory syncytial virus: sequence comparison of antigenic subgroups A and B and expression in vitro. J Gen Virol 71(Pt 12):3015–3020. https://doi.org/10.1099/0022-1317-71-12-3015
33. Elango N, Satake M, Venkatesan S (1985) mRNA sequence of three respiratory syncytial virus genes encoding two nonstructural proteins and a 22K structural protein. J Virol 55(1):101–110. https://doi.org/10.1128/JVI.55.1.101-110.1985
34. Johnson PR, Collins PL (1989) The 1B (NS2), 1C (NS1) and N proteins of human respiratory syncytial virus (RSV) of antigenic subgroups A and B: sequence conservation and divergence within RSV genomic RNA. J Gen Virol 70(Pt 6):1539–1547. https://doi.org/10.1099/0022-1317-70-6-1539
35. Hendricks DA, McIntosh K, Patterson JL (1988) Further characterization of the soluble form of the G glycoprotein of respiratory syncytial virus. J Virol 62(7):2228–2233. https://doi.org/10.1128/JVI.62.7.2228-2233.1988
36. Roberts SR, Lichtenstein D, Ball LA, Wertz GW (1994) The membrane-associated and secreted forms of the respiratory syncytial virus attachment glycoprotein G are synthesized from alternative initiation codons. J Virol 68(7):4538–4546. https://doi.org/10.1128/JVI.68.7.4538-4546.1994
37. Battles MB, McLellan JS (2019) Respiratory syncytial virus entry and how to block it. Nat Rev Microbiol 17(4):233–245. https://doi.org/10.1038/s41579-019-0149-x
38. McLellan JS, Ray WC, Peeples ME (2013) Structure and function of respiratory syncytial virus surface glycoproteins. Curr Top Microbiol Immunol 372:83–104. https://doi.org/10.1007/978-3-642-38919-1_4
39. Cadena-Cruz C, Villarreal Camacho JL, De Avila-Arias M, Hurtado-Gomez L, Rodriguez A, San-Juan-Vergara H (2023) Respiratory syncytial virus entry mechanism in host cells: a general overview. Mol Microbiol 120(3):341–350. https://doi.org/10.1111/mmi.15133
40. Shang Z, Tan S, Ma D (2021) Respiratory syncytial virus: from pathogenesis to potential therapeutic strategies. Int J Biol Sci 17(14):4073–4091. https://doi.org/10.7150/ijbs.64762
41. Duan Y, Liu Z, Zang N, Cong B, Shi Y, Xu L et al (2024) Landscape of respiratory syncytial virus. Chin Med J 137:2953. https://doi.org/10.1097/CM9.0000000000003354
42. Zhou H, Skolnick J (2024) FRAGSITE2: a structure and fragment-based approach for virtual ligand screening. Protein Sci 33(1):e4869. https://doi.org/10.1002/pro.4869
43. Lay MK, Gonzalez PA, Leon MA, Cespedes PF, Bueno SM, Riedel CA et al (2013) Advances in understanding respiratory syncytial virus infection in airway epithelial cells and consequential effects on the immune response. Microbes Infect 15(3):230–242.

https://doi.org/10.1016/j.micinf.2012.11.012

44. Dawre S, Maru S (2021) Human respiratory viral infections: current status and future prospects of nanotechnology-based approaches for prophylaxis and treatment. Life Sci 278: 119561. https://doi.org/10.1016/j.lfs.2021.119561
45. Kaler J, Hussain A, Patel K, Hernandez T, Ray S (2023) Respiratory syncytial virus: a comprehensive review of transmission, pathophysiology, and manifestation. Cureus 15(3): e36342. https://doi.org/10.7759/cureus.36342
46. Hallak LK, Kwilas SA, Peeples ME (2007) Interaction between respiratory syncytial virus and glycosaminoglycans, including heparan sulfate. Methods Mol Biol 379:15–34. https://doi.org/10.1007/978-1-59745-393-6_2
47. Johnson SM, McNally BA, Ioannidis I, Flano E, Teng MN, Oomens AG et al (2015) Respiratory syncytial virus uses CX3CR1 as a receptor on primary human airway epithelial cultures. PLoS Pathog 11(12):e1005318. https://doi.org/10.1371/journal.ppat.1005318
48. Anderson CS, Chu CY, Wang Q, Mereness JA, Ren Y, Donlon K et al (2020) CX3CR1 as a respiratory syncytial virus receptor in pediatric human lung. Pediatr Res 87(5): 862–867. https://doi.org/10.1038/s41390-019-0677-0
49. Griffiths CD, Bilawchuk LM, McDonough JE, Jamieson KC, Elawar F, Cen Y et al (2020) IGF1R is an entry receptor for respiratory syncytial virus. Nature 583(7817): 615–619. https://doi.org/10.1038/s41586-020-2369-7
50. Tayyari F, Marchant D, Moraes TJ, Duan W, Mastrangelo P, Hegele RG (2011) Identification of nucleolin as a cellular receptor for human respiratory syncytial virus. Nat Med 17(9):1132–1135. https://doi.org/10.1038/nm.2444
51. Martin D, Calder LJ, Garcia-Barreno B, Skehel JJ, Melero JA (2006) Sequence elements of the fusion peptide of human respiratory syncytial virus fusion protein required for activity. J Gen Virol 87(Pt 6):1649–1658. https://doi.org/10.1099/vir.0.81715-0
52. Bermingham IM, Chappell KJ, Watterson D, Young PR (2018) The heptad repeat C domain of the respiratory syncytial virus fusion protein plays a key role in membrane fusion. J Virol 92(4). https://doi.org/10.1128/JVI.01323-17
53. Collins PL, Dickens LE, Buckler-White A, Olmsted RA, Spriggs MK, Camargo E et al (1986) Nucleotide sequences for the gene junctions of human respiratory syncytial virus reveal distinctive features of intergenic structure and gene order. Proc Natl Acad Sci USA 83(13):4594–4598. https://doi.org/10.1073/pnas.83.13.4594
54. Mink MA, Stec DS, Collins PL (1991) Nucleotide sequences of the 3′ leader and 5′ trailer regions of human respiratory syncytial virus genomic RNA. Virology 185(2):615–624. https://doi.org/10.1016/0042-6822(91)90532-g
55. Fearns R, Peeples ME, Collins PL (2002) Mapping the transcription and replication promoters of respiratory syncytial virus. J Virol 76(4):1663–1672. https://doi.org/10.1128/jvi.76.4.1663-1672.2002
56. Dickens LE, Collins PL, Wertz GW (1984) Transcriptional mapping of human respiratory syncytial virus. J Virol 52(2):364–369. https://doi.org/10.1128/JVI.52.2.364-369.1984
57. Kuo L, Grosfeld H, Cristina J, Hill MG, Collins PL (1996) Effects of mutations in the gene-start and gene-end sequence motifs on transcription of monocistronic and dicistronic minigenomes of respiratory syncytial virus. J Virol 70(10):6892–6901. https://doi.org/10.1128/JVI.70.10.6892-6901.1996
58. Kleiner VA, Fearns R (2024) How does the polymerase of non-segmented negative strand RNA viruses commit to transcription or genome replication? J Virol 98(8): e0033224. https://doi.org/10.1128/jvi.00332-24
59. Liuzzi M, Mason SW, Cartier M, Lawetz C, McCollum RS, Dansereau N et al (2005) Inhibitors of respiratory syncytial virus replication target cotranscriptional mRNA guanylylation by viral RNA-dependent RNA polymerase. J Virol 79(20):13105–13115. https://doi.org/10.1128/JVI.79.20.13105-13115.2005
60. Collins PL, Hill MG, Cristina J, Grosfeld H (1996) Transcription elongation factor of respiratory syncytial virus, a nonsegmented negative-strand RNA virus. Proc Natl Acad Sci USA 93(1):81–85. https://doi.org/10.1073/pnas.93.1.81
61. Fearns R, Collins PL (1999) Role of the M2-1 transcription antitermination protein of respiratory syncytial virus in sequential transcription. J Virol 73(7):5852–5864. https://doi.org/10.1128/JVI.73.7.5852-5864.1999

62. Braun MR, Noton SL, Blanchard EL, Shareef A, Santangelo PJ, Johnson WE et al (2021) Respiratory syncytial virus M2-1 protein associates non-specifically with viral messenger RNA and with specific cellular messenger RNA transcripts. PLoS Pathog 17(5):e1009589. https://doi.org/10.1371/journal.ppat.1009589
63. Kuo L, Fearns R, Collins PL (1997) Analysis of the gene start and gene end signals of human respiratory syncytial virus: quasi-templated initiation at position 1 of the encoded mRNA. J Virol 71(7):4944–4953. https://doi.org/10.1128/JVI.71.7.4944-4953.1997
64. Fearns R, Collins PL (1999) Model for polymerase access to the overlapped L gene of respiratory syncytial virus. J Virol 73(1):388–397. https://doi.org/10.1128/JVI.73.1.388-397.1999
65. Kuo L, Fearns R, Collins PL (1996) The structurally diverse intergenic regions of respiratory syncytial virus do not modulate sequential transcription by a dicistronic minigenome. J Virol 70(9):6143–6150. https://doi.org/10.1128/JVI.70.9.6143-6150.1996
66. Noton SL, Tremaglio CZ, Fearns R (2019) Killing two birds with one stone: how the respiratory syncytial virus polymerase initiates transcription and replication. PLoS Pathog 15(2):e1007548. https://doi.org/10.1371/journal.ppat.1007548
67. Peeples ME, Collins PL (2000) Mutations in the 5′ trailer region of a respiratory syncytial virus minigenome which limit RNA replication to one step. J Virol 74(1):146–155. https://doi.org/10.1128/jvi.74.1.146-155.2000
68. Hanley LL, McGivern DR, Teng MN, Djang R, Collins PL, Fearns R (2010) Roles of the respiratory syncytial virus trailer region: effects of mutations on genome production and stress granule formation. Virology 406(2):241–252. https://doi.org/10.1016/j.virol.2010.07.006
69. Yu Q, Hardy RW, Wertz GW (1995) Functional cDNA clones of the human respiratory syncytial (RS) virus N, P, and L proteins support replication of RS virus genomic RNA analogs and define minimal trans-acting requirements for RNA replication. J Virol 69(4):2412–2419. https://doi.org/10.1128/JVI.69.4.2412-2419.1995
70. Sun Y, Kim EJ, Felt SA, Taylor LJ, Agarwal D, Grant GR et al (2019) A specific sequence in the genome of respiratory syncytial virus regulates the generation of copy-back defective viral genomes. PLoS Pathog 15(4):e1007707. https://doi.org/10.1371/journal.ppat.1007707
71. Manzoni TB, Lopez CB (2018) Defective (interfering) viral genomes re-explored: impact on antiviral immunity and virus persistence. Future Virol 13(7):493–503. https://doi.org/10.2217/fvl-2018-0021
72. Felt SA, Sun Y, Jozwik A, Paras A, Habibi MS, Nickle D et al (2021) Detection of respiratory syncytial virus defective genomes in nasal secretions is associated with distinct clinical outcomes. Nat Microbiol 6(5):672–681. https://doi.org/10.1038/s41564-021-00882-3
73. Shaikh FY, Crowe JE Jr (2013) Molecular mechanisms driving respiratory syncytial virus assembly. Future Microbiol 8(1):123–131. https://doi.org/10.2217/fmb.12.132
74. Vanover D, Smith DV, Blanchard EL, Alonas E, Kirschman JL, Lifland AW et al (2017) RSV glycoprotein and genomic RNA dynamics reveal filament assembly prior to the plasma membrane. Nat Commun 8(1):667. https://doi.org/10.1038/s41467-017-00732-z
75. Shahriari S, Wei KJ, Ghildyal R (2018) Respiratory syncytial virus matrix (M) protein interacts with actin in vitro and in cell culture. Viruses 10(10). https://doi.org/10.3390/v10100535
76. Thornhill EM, Verhoeven D (2020) Respiratory syncytial virus's non-structural proteins: masters of interference. Front Cell Infect Microbiol 10:225. https://doi.org/10.3389/fcimb.2020.00225
77. Pretel E, Camporeale G, de Prat-Gay G (2013) The non-structural NS1 protein unique to respiratory syncytial virus: a two-state folding monomer in quasi-equilibrium with a stable spherical oligomer. PLoS One 8(9):e74338. https://doi.org/10.1371/journal.pone.0074338
78. Wu W, Tran KC, Teng MN, Heesom KJ, Matthews DA, Barr JN et al (2012) The interactome of the human respiratory syncytial virus NS1 protein highlights multiple effects on host cell biology. J Virol 86(15):7777–7789. https://doi.org/10.1128/JVI.00460-12
79. Ling Z, Tran KC, Teng MN (2009) Human respiratory syncytial virus nonstructural protein NS2 antagonizes the activation of beta interferon transcription by interacting with RIG-I. J Virol 83(8):3734–3742. https://doi.org/10.1128/JVI.02434-08

80. Pei J, Wagner ND, Zou AJ, Chatterjee S, Borek D, Cole AR et al (2021) Structural basis for IFN antagonism by human respiratory syncytial virus nonstructural protein 2. Proc Natl Acad Sci USA 118(10). https://doi.org/10.1073/pnas.2020587118
81. Merritt TN, Pei J, Leung DW (2023) Pathogenicity and virulence of human respiratory syncytial virus: multifunctional nonstructural proteins NS1 and NS2. Virulence 2283897. https://doi.org/10.1080/21505594.2023.2283897
82. Triantafilou K, Kar S, Vakakis E, Kotecha S, Triantafilou M (2013) Human respiratory syncytial virus viroporin SH: a viral recognition pathway used by the host to signal inflammasome activation. Thorax 68(1):66–75. https://doi.org/10.1136/thoraxjnl-2012-202182
83. Gan SW, Tan E, Lin X, Yu D, Wang J, Tan GM et al (2012) The small hydrophobic protein of the human respiratory syncytial virus forms pentameric ion channels. J Biol Chem 287(29):24671–24689. https://doi.org/10.1074/jbc.M111.332791
84. Liang B, Kabatova B, Kabat J, Dorward DW, Liu X, Surman S et al (2019) Effects of alterations to the CX3C motif and secreted form of human respiratory syncytial virus (RSV) G protein on immune responses to a parainfluenza virus vector expressing the RSV G protein. J Virol 93(7). https://doi.org/10.1128/JVI.02043-18
85. Bergeron HC, Tripp RA (2022) RSV replication, transmission, and disease are influenced by the RSV G protein. Viruses 14(11). https://doi.org/10.3390/v14112396
86. Anderson LJ, Jadhao SJ, Paden CR, Tong S (2021) Functional features of the respiratory syncytial virus G protein. Viruses 13(7). https://doi.org/10.3390/v13071214
87. Tripp RA, Moore D, Winter J, Anderson LJ (2000) Respiratory syncytial virus infection and G and/or SH protein expression contribute to substance P, which mediates inflammation and enhanced pulmonary disease in BALB/c mice. J Virol 74(4):1614–1622. https://doi.org/10.1128/jvi.74.4.1614-1622.2000
88. Ke Z, Dillard RS, Chirkova T, Leon F, Stobart CC, Hampton CM et al (2018) The morphology and assembly of respiratory syncytial virus revealed by cryo-electron tomography. Viruses 10(8). https://doi.org/10.3390/v10080446
89. Liljeroos L, Krzyzaniak MA, Helenius A, Butcher SJ (2013) Architecture of respiratory syncytial virus revealed by electron cryotomography. Proc Natl Acad Sci USA 110(27):11133–11138. https://doi.org/10.1073/pnas.1309070110
90. McLellan JS, Yang Y, Graham BS, Kwong PD (2011) Structure of respiratory syncytial virus fusion glycoprotein in the postfusion conformation reveals preservation of neutralizing epitopes. J Virol 85(15):7788–7796. https://doi.org/10.1128/JVI.00555-11
91. Swanson KA, Settembre EC, Shaw CA, Dey AK, Rappuoli R, Mandl CW et al (2011) Structural basis for immunization with postfusion respiratory syncytial virus fusion F glycoprotein (RSV F) to elicit high neutralizing antibody titers. Proc Natl Acad Sci USA 108(23):9619–9624. https://doi.org/10.1073/pnas.1106536108
92. McLellan JS, Chen M, Chang JS, Yang Y, Kim A, Graham BS et al (2010) Structure of a major antigenic site on the respiratory syncytial virus fusion glycoprotein in complex with neutralizing antibody 101F. J Virol 84(23):12236–12244. https://doi.org/10.1128/JVI.01579-10
93. Joyce MG, Zhang B, Ou L, Chen M, Chuang GY, Druz A et al (2016) Iterative structure-based improvement of a fusion-glycoprotein vaccine against RSV. Nat Struct Mol Biol 23(9):811–820. https://doi.org/10.1038/nsmb.3267
94. Che Y, Gribenko AV, Song X, Handke LD, Efferen KS, Tompkins K et al (2023) Rational design of a highly immunogenic prefusion-stabilized F glycoprotein antigen for a respiratory syncytial virus vaccine. Sci Transl Med 15(693):eade6422. https://doi.org/10.1126/scitranslmed.ade6422
95. Liang Y, Shao S, Li XY, Zhao ZX, Liu N, Liu ZM et al (2024) Mutating a flexible region of the RSV F protein can stabilize the prefusion conformation. Science 385(6716):1484–1491. https://doi.org/10.1126/science.adp2362
96. Gonzalez KJ, Huang J, Criado MF, Banerjee A, Tompkins SM, Mousa JJ et al (2024) A general computational design strategy for stabilizing viral class I fusion proteins. Nat Commun 15(1):1335. https://doi.org/10.1038/s41467-024-45480-z
97. Money VA, McPhee HK, Mosely JA, Sanderson JM, Yeo RP (2009) Surface features of a Mononegavirales matrix protein indicate sites

of membrane interaction. Proc Natl Acad Sci USA 106(11):4441–4446. https://doi.org/10.1073/pnas.0805740106

98. Forster A, Maertens GN, Farrell PJ, Bajorek M (2015) Dimerization of matrix protein is required for budding of respiratory syncytial virus. J Virol 89(8):4624–4635. https://doi.org/10.1128/JVI.03500-14
99. Tawar RG, Duquerroy S, Vonrhein C, Varela PF, Damier-Piolle L, Castagne N et al (2009) Crystal structure of a nucleocapsid-like nucleoprotein-RNA complex of respiratory syncytial virus. Science 326(5957):1279–1283. https://doi.org/10.1126/science.1177634
100. Wang Y, Zhang C, Luo Y, Ling X, Luo B, Jia G et al (2023) Cryo-EM structure of the nucleocapsid-like assembly of respiratory syncytial virus. Signal Transduct Target Ther 8(1):323. https://doi.org/10.1038/s41392-023-01602-5
101. Gonnin L, Desfosses A, Bacia-Verloop M, Chevret D, Galloux M, Eleouet JF et al (2023) Structural landscape of the respiratory syncytial virus nucleocapsids. Nat Commun 14(1):5732. https://doi.org/10.1038/s41467-023-41439-8
102. Cao D, Gao Y, Roesler C, Rice S, D'Cunha P, Zhuang L et al (2020) Cryo-EM structure of the respiratory syncytial virus RNA polymerase. Nat Commun 11(1):368. https://doi.org/10.1038/s41467-019-14246-3
103. Gilman MSA, Liu C, Fung A, Behera I, Jordan P, Rigaux P et al (2019) Structure of the respiratory syncytial virus polymerase complex. Cell 179(1):193–204 e14. https://doi.org/10.1016/j.cell.2019.08.014
104. Cao D, Gao Y, Chen Z, Gooneratne I, Roesler C, Mera C et al (2024) Structures of the promoter-bound respiratory syncytial virus polymerase. Nature 625(7995):611–617. https://doi.org/10.1038/s41586-023-06867-y
105. Gao Y, Cao D, Pawnikar S, John KP, Ahn HM, Hill S et al (2020) Structure of the human respiratory syncytial virus M2-1 protein in complex with a short positive-sense gene-end RNA. Structure 28(9):979–90 e4. https://doi.org/10.1016/j.str.2020.07.001
106. Tanner SJ, Ariza A, Richard CA, Kyle HF, Dods RL, Blondot ML et al (2014) Crystal structure of the essential transcription antiterminator M2-1 protein of human respiratory syncytial virus and implications of its phosphorylation. Proc Natl Acad Sci USA 111(4):1580–1585. https://doi.org/10.1073/pnas.1317262111
107. Chatterjee S, Luthra P, Esaulova E, Agapov E, Yen BC, Borek DM et al (2017) Structural basis for human respiratory syncytial virus NS1-mediated modulation of host responses. Nat Microbiol 2:17101. https://doi.org/10.1038/nmicrobiol.2017.101
108. Loe MWC, Soenong H, Lee E, Li-Kim-Moy J, Williams PC, Yeo KT (2024) Nirsevimab: alleviating the burden of RSV morbidity in young children. J Paediatr Child Health 60:489. https://doi.org/10.1111/jpc.16643
109. Zhu Q, McLellan JS, Kallewaard NL, Ulbrandt ND, Palaszynski S, Zhang J et al (2017) A highly potent extended half-life antibody as a potential RSV vaccine surrogate for all infants. Sci Transl Med 9(388). https://doi.org/10.1126/scitranslmed.aaj1928
110. Drysdale SB, Cathie K, Flamein F, Knuf M, Collins AM, Hill HC et al (2023) Nirsevimab for prevention of hospitalizations due to RSV in infants. N Engl J Med 389(26):2425–2435. https://doi.org/10.1056/NEJMoa2309189
111. Griffin MP, Yuan Y, Takas T, Domachowske JB, Madhi SA, Manzoni P et al (2020) Single-dose nirsevimab for prevention of RSV in preterm infants. N Engl J Med 383(5):415–425. https://doi.org/10.1056/NEJMoa1913556
112. Hammitt LL, Dagan R, Yuan Y, Baca Cots M, Bosheva M, Madhi SA et al (2022) Nirsevimab for prevention of RSV in healthy late-preterm and term infants. N Engl J Med 386(9):837–846. https://doi.org/10.1056/NEJMoa2110275
113. Muller WJ, Madhi SA, Seoane Nunez B, Baca Cots M, Bosheva M, Dagan R et al (2023) Nirsevimab for prevention of RSV in term and late-preterm infants. N Engl J Med 388(16):1533–1534. https://doi.org/10.1056/NEJMc2214773
114. Kampmann B, Madhi SA, Munjal I, Simoes EAF, Pahud BA, Llapur C et al (2023) Bivalent Prefusion F vaccine in pregnancy to prevent RSV illness in infants. N Engl J Med 388(16):1451–1464. https://doi.org/10.1056/NEJMoa2216480
115. Papi A, Ison MG, Langley JM, Lee DG, Leroux-Roels I, Martinon-Torres F et al (2023) Respiratory syncytial virus prefusion F protein vaccine in older adults. N Engl J Med 388(7):595–608. https://doi.org/10.1056/NEJMoa2209604
116. Walsh EE, Perez Marc G, Zareba AM, Falsey AR, Jiang Q, Patton M et al (2023) Efficacy

and safety of a bivalent RSV prefusion F vaccine in older adults. N Engl J Med 388(16): 1465–1477. https://doi.org/10.1056/NEJMoa2213836

117. Bonneux B, Jacoby E, Ceconi M, Stobbelaar K, Delputte P, Herschke F (2024) Direct-acting antivirals for RSV treatment, a review. Antivir Res 229:105948. https://doi.org/10.1016/j.antiviral.2024.105948

118. Cockerill GS, Good JAD, Mathews N (2019) State of the art in respiratory syncytial virus drug discovery and development. J Med Chem 62(7):3206–3227. https://doi.org/10.1021/acs.jmedchem.8b01361

Chapter 2

Quantification of Respiratory Syncytial Virus by Immunostaining Plaque Assay

Harrison C. Bergeron and Ralph A. Tripp

Abstract

A useful and reproducible method to determine infectious respiratory syncytial virus (RSV) is to quantify virus plaques using immunostaining of infected Vero E6 cells. Immunostaining, in lieu of counterstaining, allows for accurate detection and quantification of RSV-specific plaques, avoiding spurious and nonspecific results. This method is important because RSV plaque morphology can be atypical in susceptible cell lines and often varies because it is a function of virus replication, cytopathic effect, and viral fitness. This protocol describes determining RSV titers using an immunostaining plaque assay on Vero E6 cells.

Key words Respiratory syncytial virus, RSV, Vero cell plaque assay, Immunostaining, Viral titer determination

1 Introduction

Viruses have intrinsic properties that affect their growth in cells. Cells can be permissive and support virus replication, while others are nonpermissive where replication is blocked. The histological appearance of cellular damage produced following infection by cytocidal viruses defines the cytopathic effects or CPE. A noticeable feature of the infection of cell monolayers by respiratory syncytial virus (RSV) is the production of syncytia associated with a high multiplicity of infections. Syncytia affects virus plaque morphology. Several cell lines have been evaluated as a substrate for RSV replication and propagation, and a few of these have been used to optimize endpoint assays that allow quantitation of virus replication, such as the plaque assay. Specifically, HEp-2 cells (ATCC CL-23), derived from human epithelial cells of a larynx carcinoma, have been used to propagate RSV and in RSV plaque assays [1–9]. Many of these studies have reported substantial diversity in plaque morphology, with RSV strain A plaques being appreciably

Bo Liang and Rachel Fearns (eds.), *Human Respiratory Syncytial Virus: Methods and Protocols*, Methods in Molecular Biology, vol. 2948, https://doi.org/10.1007/978-1-0716-4666-3_2,

larger in the RSV-infected HEp-2 cell lawn [6]. Since RSV infection forms syncytia, determining the RSV titers is difficult and is best performed using anti-RSV antibody immunostaining.

RSV may be grown in several immortalized cell lines including Hep-2 cells (ATCC CCL-23) derived from human epithelial cells, A549 cells (ATCC CCL-185) derived from human lung epithelial cells, and Vero E6 cells (ATCC CRL-1586) isolated from kidney epithelial cells extracted from an African green monkey [1], and end users should consider their downstream experimental needs and that titers may vary depending on the cell line used for propagation [10]. While RSV propagation may be performed in several cell lines, Vero cells typically produce the best plaque phenotype and are therefore recommended in this protocol. Overlaying RSV-infected Vero cells with 2% methylcellulose for 6 days post-infection (pi) facilitates optimum virus plaque formation and minimal cell death. After 6 days, the methylcellulose is removed via aspiration, the infected Vero cells are fixed, and the RSV F and/or G viral surface proteins are detected. For the greatest sensitivity and specificity, monoclonal antibodies (mAbs) against RSV F and/or G protein are used for detection, followed by treatment with alkaline phosphatase (AP)-conjugated secondary anti-species antibody. Once the enzyme substrate is added, a colorimetric reaction occurs, and plaques can be readily counted. We provide an updated protocol for detecting infectious RSV within lung samples using an immunostaining plaque assay [3].

2 Materials

All reagents are sterile.

1. Biosafety level II cabinet.
2. 37 °C incubator with 5% CO_2.
3. Flat-bottom 24-well tissue culture plates.
4. Sample containing RSV, e.g., cell culture supernatant, cell lysate, homogenized lung tissue.
5. Vero E6 cells.
6. Heat-inactivated (56 °C for 30 min) fetal bovine serum (FBS).
7. Dulbecco's modified Eagle's medium (DMEM).
8. Growth media: 5% FBS/DMEM.
9. Overlay medium (2% methylcellulose): 10 g methylcellulose, 100 mL Hank's balanced salt solution (HBSS), antibiotic/antimycotic solution, FBS, and DMEM.
10. Deionized water (diH_2O).
11. Phosphate buffer saline (PBS), pH 7.4.

12. 15 mL and 50 mL conical tubes and 1.5 mL Eppendorf tubes or equivalent.
13. Fixative solution: 60% (v/v) acetone and 40% methanol.
14. Wash buffer: 1× KPL (SeraCare) wash buffer or similar buffer containing imidazole-buffered saline and Tween®-20.
15. Blocking solution: 5% nonfat dry milk diluted in 1× KPL wash buffer.
16. Primary antibodies: mouse anti-RSV F protein monoclonal antibody (mAb) (e.g., clone 131-2A and/or anti-RSV G protein, e.g., clone 131-2G).
17. Secondary antibody: goat anti-mouse-alkaline phosphatase (AP).
18. Colorimetric alkaline phosphatase substrate: 1-step Nitroblue tetrazolium (NBT) 5-bromo-4-chloro-3-indolyl phosphate (BCIP).
19. Serological and micropipette tips and pipettors.

3 Methods

1. Subculture Vero E6 cells in growth media 2 days before the RSV plaque assay, so that cells are healthy and in log-phase (*see* **Note 1**).
2. One day before RSV plaque assay, plate Vero E6 cells to 90% confluency in 24-well tissue culture plates in 1 mL of growth media per well (Fig. 1) (*see* **Note 2**).
3. On the day of the RSV plaque assay, perform all tenfold dilutions of the RSV-containing sample using ice-cold 0% FBS/DMEM (*see* **Notes 3, 12**, and **15**).
4. Starting with the mock wells and working to the most concentrated samples, add 200 μL of sample in quadruplicate to Vero E6 cells in the 24-well plate on ice (Fig. 1). Mock wells receive media only and are the negative control (*see* **Note 4**).

10^{-1}	10^{-2}	10^{-3}	10^{-4}	10^{-5}	Mock
10^{-1}	10^{-2}	10^{-3}	10^{-4}	10^{-5}	Mock
10^{-1}	10^{-2}	10^{-3}	10^{-4}	10^{-5}	Mock
10^{-1}	10^{-2}	10^{-3}	10^{-4}	10^{-5}	Mock

Fig. 1 Plate layout. Each plate contains mock wells (negative control), and each dilution is plated in quadruplicate

5. Using 200 μL of the diluted virus or media, add to Vero E6 cells and incubate plates at 37 °C/5% CO_2 for 2 h (*see* **Note 5**).
6. During incubation, place overlay medium (i.e., 2% methylcellulose) in a 37 °C water bath (*see* **Notes 13** and **14**).
7. After 2 h incubation, without removing the virus overlay, add 1 mL of overlay medium containing 2% methylcellulose to Vero E6 cells and incubate at 37 °C/5% CO_2 for 6 days.
8. Six days after incubation, remove plates from the incubator.
9. Using a vacuum with a Pasteur pipette attached, carefully remove the methylcellulose overlay medium—do not disturb the cell monolayer (*see* **Note 6**).
10. Gently wash wells with 1 mL PBS and carefully decant as above using a vacuum with a Pasteur pipette attached (*see* **Note 7**).
11. With no methylcellulose remaining on the plate, carefully add 500 μL of 60:40 acetone: methanol to each well and incubate for 20 min at room temperature.
12. After 20 min, remove acetone and methanol from the plates and pour them into a chemical waste container. Pat the plates dry with a paper towel and place them on the benchtop to air-dry (overnight).
13. When plates are dry the next day, rinse once with KPL wash buffer.
14. Block cells by adding 500 μL Blotto (5% nonfat dry milk in KPL wash buffer) overnight at 4 °C, 2 h at room temperature, or 1 h at 37 °C.
15. Decant Blotto, dab plates gently on a paper towel, and add 200 μL anti-RSV mAb primary (1°) antibody to each well (*see* **Note 8**).
16. Incubate 1° antibody overnight at 4 °C, 2 h at room temperature, or 1 h at 37 °C.
17. Following the 1° antibody incubation, the decant was plated and rinsed with diH_2O, and then the wells were washed 3× with 1 mL KPL wash buffer (*see* **Note 9**).
18. After decanting the final wash buffer, gently dab plates on a paper towel and add 200 μL of diluted goat anti-mouse-AP in Blotto (1:500–1:2000) based on prior optimization. Incubate 2° antibody overnight at 4 °C, 2 h at room temperature, or 1 h at 37 °C (*see* **Note 10**).
19. After 2° antibody incubation, decant, rinse with diH_2O, and wash wells 3× with 1 mL KPL wash buffer. Each wash should incubate for 5 min for a total of 15 min.

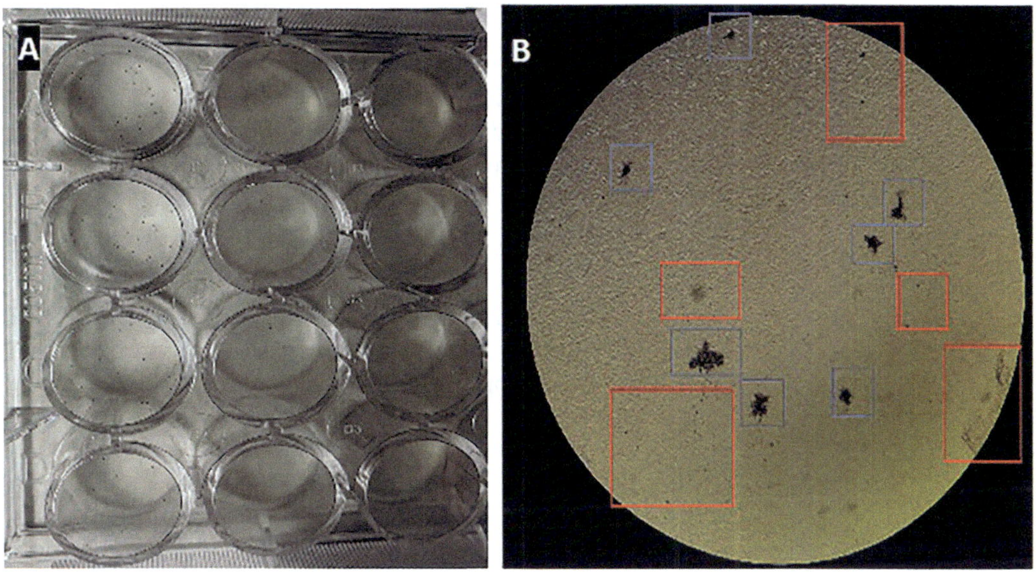

Fig. 2 (**A**) Plaques are visible to the unaided eye and appear as small dots in the wells. RSV was diluted tenfold (left to right) in quadruplicate. (**B**) Dissection microscope view of RSV plaques. True RSV plaques are boxed in blue. Notice plaque sizes vary. In red boxes are backgrounds that should not be enumerated. Images captured with an iPhone

20. After decanting the final wash, gently dab plates on a paper towel and add 200 μL NBT/BCIP substrate/well. As the color development begins, it can be stopped by rinsing wells with diH_2O. Dab plates on a paper towel and air-dry (*see* **Note 11**).
21. The number of virus plaques can be enumerated using a dissection microscope (Fig. 2A, B) (*see* **Notes 16** and **17**).
22. Determine PFU/mL (titer) using the following formula:

$$[(\text{average\#of plaques from 4 wells of last dilution containing plaques}) / (\text{dilution factor} \times \text{volume of virus added to well})].$$

For example, if there were countable plaques in the 10^{-4} dilution but none in 10^{-5}, count and determine the average of the four wells in the 10^{-4} dilution. Divide this by $10^{-4} \times 200$ μL (0.2 mL). If there were a total of 40 plaques counted in four quadruplicates, this would be an average of 10 PFU. Divide this value by $10^{-4} \times 0.2$ mL. This = 5×10^5 PFU/mL.

4 Notes

1. Vero E6 cell morphology changes with cell crowding, so be aware of cell seeding. Plan to split cell cultures in 5% FBS/DMEM and maintain them in a 37 °C/5% CO_2 incubator to monitor cell densities and growth kinetics.
2. Typically, Vero cells are plated at 2–3 × 10^5 cells/well.
3. RSV is heat-sensitive and will inactivate as it warms to room temperature or higher. Remove 24-well plates from the 37 °C/5% CO_2 incubator and check by microscopy to confirm that Vero E6 cells are approximately 90% confluent. Decant media and gently wash cells with 2 mL sterile PBS. Tip: To ensure mixing, perform dilutions using 1.5 mL snap-cap tubes and vortex between dilutions. Add 100 μL of RSV sample to 900 μL of 10% FBS/DMEM, mix, and repeat until appropriate dilutions have been made. This is sufficient volume for quadruplicate wells with overage. Keep dilutions and media on the ice at all times.
4. This assay allows five serial dilutions to be analyzed per plate. It is recommended to test 10^{-5}–10^{-9} for high-titer viral stocks and 10^{-1}–10^{-5} for homogenized tissue (e.g., mouse lungs).
5. Gently rock plates every 30 min.
6. Tilt the plate and avoid contact with the Vero E6 cell lawn.
7. Gently dab the plates on a paper towel to remove excess overlay medium.
8. It is recommended to use an anti-F protein and anti-G protein mAb cocktail (i.e., 131-2A and 131-2G each at 1–2 μg/mL) diluted in Blotto. *Tip:* The exact concentration of antibody used should be determined by checkerboard assay with each new mAb lot.
9. Each wash step should incubate for 5 min for a total of 15 min.
10. The exact concentration of the antibody used should be determined by checkerboard assay with each new antibody lot.
11. It is recommended to stop color development once color changes are noted—this will reduce background staining.
12. RSV is highly sensitive to temperature. All steps during the plaque assay before fixation must be performed on ice.
13. To prepare 500 mL of 2% methylcellulose, bring 100 mL of HBSS to a boil and slowly add 10 g of methylcellulose while stirring on a heated plate warmer. When adding

methylcellulose to HBSS, fill an autoclave tray with diH_2O and store it in the warm autoclave. Once 10 g of methylcellulose has been added, transfer the bottle to the autoclave and place it in a water tray for autoclaving. Assure the water level is above the HBSS level. Autoclave to sterilize according to your institution's liquid cycle. During autoclaving, mix the remaining reagents. Add 10 mL FBS and 5 mL antibiotic/antimycotic solution (100×) to 400 mL DMEM. Heat reagents in a 37 °C water bath. Remove the tray containing methylcellulose from the autoclave and transfer it to near the biosafety cabinet. Transfer the methylcellulose container to a heated plate warmer within the biosafety cabinet. When the bottle is cool enough to touch (i.e., 40–42 °C), add the media to the methylcellulose and stir for approximately 4 h until the solution is clear. After the solution is clear, stir the methylcellulose solution overnight at 4 °C. Methylcellulose may be stored for up to 6 months at 4 °C.

14. Antibiotics/antimycotics are only required for samples derived in vivo. These should be excluded if methylcellulose is only used for in vitro evaluations.
15. As samples are serially diluted, the plaque assay plates should contain a plate with a dilution that would result in too many plaques to count, dilutions with plaques that could be counted, and a dilution with too few plaques to count. Wells with too many plaques to count will have largely disrupted or destroyed Vero E6 cell lawns where no plaques can be counted. Wells that have too few plaques to count are indistinguishable from negative control wells. Wells with sufficient plaques to enumerate will have distinct plaques evenly distributed among the monolayer.
16. As plaque staining fades and the background increases over time, it is recommended to enumerate and collect data as soon as possible. A fine-point Sharpie may be used to mark plaques once they are counted, so laboratorians do not count plaques multiple times.
17. While counterstaining may be used in lieu of immunostaining, Fig. 3 demonstrates the lower sensitivity of counterstaining for enumerating RSV plaques, as distinguishing plaques from background is difficult.

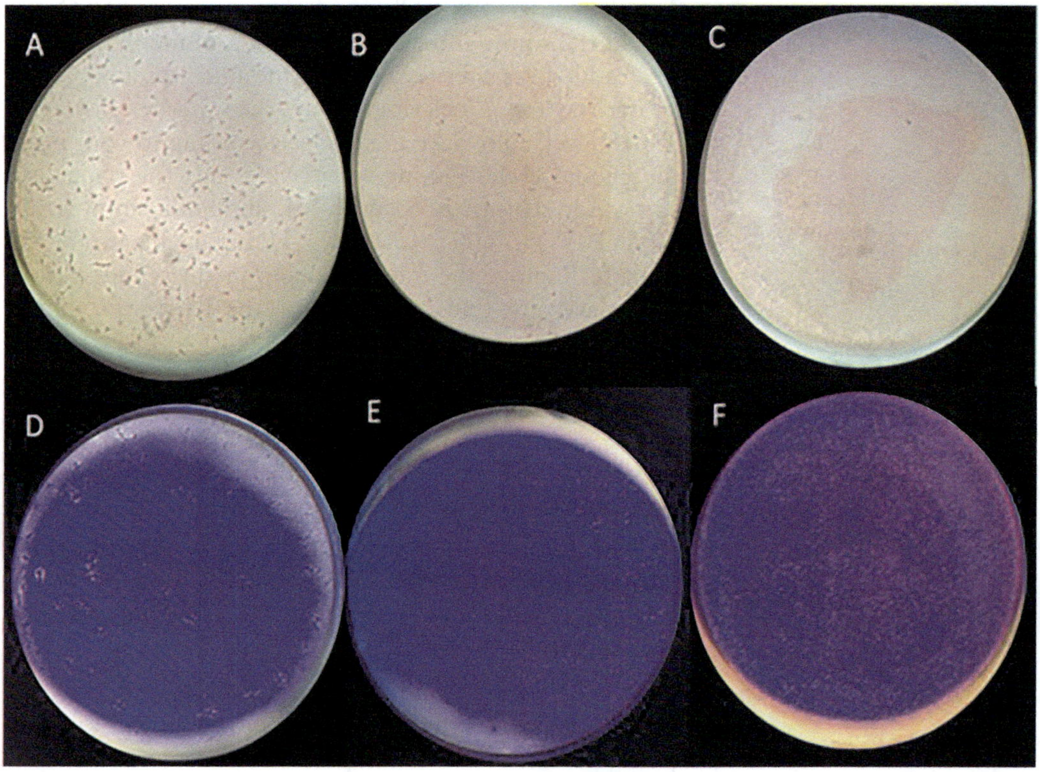

Fig. 3 Counterstained plaque assay. After removing 2% methylcellulose, cells were fixed with acetone: methanol and air-dried overnight. The following day, wells were rinsed once with KPL wash buffer and incubated with 0.05% Neutral Red (**A**–**C**) for 1 h or 1% Crystal Violet (**D**–**F**) for 10 min. After incubation, wells were rinsed with diH_2O, dried, and visualized with a dissection microscope. While plaques can be distinguished at higher concentrations of inoculum (**A** and **D**), counterstains do not offer sensitive enumeration at lower concentrations (**B** and **E**), as distinguishing plaques from background becomes more difficult. Part labels **C** and **F** show uninfected monolayers, and compared to immunostained wells that remain transparent (Fig. 2), these monolayers have various shades of color, which may prevent accurate enumeration. Images captured with an iPhone on a dissection microscope

Acknowledgments

An endowment from the Georgia Research Alliance provided financial support for Ralph Tripp at the University of Georgia.

References

1. Bergeron HC, Tripp RA (2022) RSV replication, transmission, and disease are influenced by the RSV G protein. Viruses 14(11). https://doi.org/10.3390/v14112396
2. Derscheid RJ et al (2013) Human respiratory syncytial virus Memphis 37 grown in HEp-2 cells causes more severe disease in lambs than virus grown in Vero cells. Viruses 5(11):

2881–2897. https://doi.org/10.3390/v5112881

3. Jorquera PA, Tripp RA (2016) Quantification of RSV infectious particles by plaque assay and immunostaining assay. Methods Mol Biol 1442:33–40. https://doi.org/10.1007/978-1-4939-3687-8_3
4. Kim YI et al (2015) Relating plaque morphology to respiratory syncytial virus subgroup, viral load, and disease severity in children. Pediatr Res 78(4):380–388. https://doi.org/10.1038/pr.2015.122
5. Lee SH, Chu KB, Kim MJ, Quan FS (2023) Virus-like particles assembled using respiratory syncytial virus matrix protein elicit protective immunity in mice. Infect Drug Resist 16:6099–6110. https://doi.org/10.2147/IDR.S426039
6. Martinez I et al (2009) Cultures of HEp-2 cells persistently infected by human respiratory syncytial virus differ in chemokine expression and resistance to apoptosis as compared to lytic infections of the same cell type. Virology 388(1):31–41. https://doi.org/10.1016/j.virol.2009.03.008
7. McKimm-Breschkin JL (2004) A simplified plaque assay for respiratory syncytial virus – direct visualization of plaques without immunostaining. J Virol Methods 120(1):113–117. https://doi.org/10.1016/j.jviromet.2004.02.020
8. Rajan A et al (2022) Multiple respiratory syncytial virus (RSV) strains infecting HEp-2 and A549 cells reveal cell line-dependent differences in resistance to RSV infection. J Virol 96(7):e0190421. https://doi.org/10.1128/jvi.01904-21
9. Tristram DA, Hicks W Jr, Hard R (1998) Respiratory syncytial virus and human bronchial epithelium. Arch Otolaryngol Head Neck Surg 124(7):777–783. https://doi.org/10.1001/archotol.124.7.777
10. Widjojoatmodjo MN et al (2010) A highly attenuated recombinant human respiratory syncytial virus lacking the G protein induces long-lasting protection in cotton rats. Virol J 7:114. https://doi.org/10.1186/1743-422X-7-114

Chapter 3

Methods for the Detection, Purification, and Production of Defective Viral Genomes During Respiratory Syncytial Virus Infection

Justin W. Brennan and Yan Sun

Abstract

Respiratory syncytial virus (RSV) is prone to generating the copy-back type of defective viral genomes (cbDVGs). cbDVGs play crucial roles in RSV pathogenesis by modulating innate immunity and directly interfering with infectious virus replication. Clinically, the timing of cbDVG emergence impacts the severity of RSV infection. DVGs are also considered a novel therapeutic approach for combating viral infections. Therefore, detecting DVGs and comprehensively understanding their functions during RSV infection are paramount. To facilitate these studies, in this chapter, we provide a summary of the commonly used methods for the detection, purification, and production of RSV cbDVGs.

Key words RSV, cbDVGs, RT-PCR, RNA-FISH, Bioinformatics, DVG enrichment, DVG purification, Virus-like particles

1 Introduction

Defective viral genomes (DVGs) are truncated forms of viral genomes that are naturally generated during the replication of most RNA viruses [1–3], including respiratory syncytial virus (RSV). DVGs are reported to reduce the pathogenesis of several respiratory viruses by rapidly stimulating interferon (IFN) responses and directly interfering with standard virus replication [4–6]. Thus, DVGs are considered novel antiviral therapy candidates with broad antiviral potential. Two major types of DVGs, deletion and copy-back (cb), have been observed in numerous different RNA viruses. Specifically, RSV predominately generates cbDVGs. cbDVGs constitute rearranged genomes characterized by reverse complementary 5′ and 3′ ends and form a theoretical panhandle structure [7–9]. cbDVG generation is thought to occur when the viral polymerase dissociates from the template strand, at a genomic position referred to as the break point (W), and reinitiates with the

Bo Liang and Rachel Fearns (eds.), *Human Respiratory Syncytial Virus: Methods and Protocols*, Methods in Molecular Biology, vol. 2948, https://doi.org/10.1007/978-1-0716-4666-3_3,

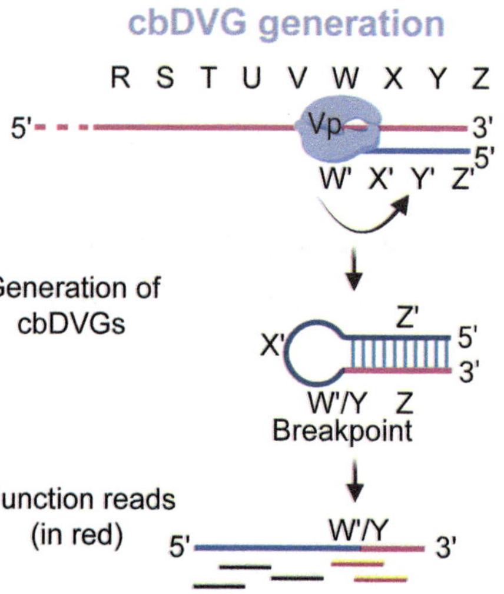

Fig. 1 Copy-back DVG generation during RSV infection. During genomic replication, using the +sense RSV antigenome as the template, the viral polymerase complex (Vp) dissociates from the template strand (pink, +sense) at the position W and then reinitiates at a different position, Y′, of the progeny strand (blue, −sense), resulting in the formation of a cbDVG with the breakpoint W′/Y. For RNA deep sequencing, certain reads will be amplified from this junction region and thus contain breakpoint W′/Y (or W/Y′). Those reads, referred to as junction reads (highlighted in red), can be identified by algorithms like ViReMa, VODKA, and DVGfinder, which are specifically designed to identify DVGs

nascently synthesized strand, at a rejoin point (Y′), and continues elongating through the 5′ end. This process creates the breakpoint, W′/Y, that is unique in DVG (Fig. 1) [10, 11]. Previously, we showed that RSV cbDVGs can rapidly induce robust antiviral responses, reduce viral load, and minimize lung pathology in mice [5]. Importantly, nearly half of pediatric patients hospitalized due to RSV infection contained DVGs in their nasal secretions, and the amount of DVGs positively correlated with type I IFN responses [5]. Interestingly, RSV disease severity was associated with the timing of DVG detection in nasal samples [12]. Given the versatile roles of DVGs in RSV pathogenesis, it is important to summarize the commonly used methods for their detection, purification, and production to further study their functions.

Three approaches can be used to detect DVGs during RSV infection: DVG-specific reverse transcription followed by PCR (DVG-RT-PCR), RNA fluorescence in situ hybridization (FISH), and RNA deep sequencing combined with bioinformatic tools. (1) The principle of DVG-RT-PCR utilizes the unique sequential structure of cbDVGs to distinguish them from their homologous full-length viral genomes. The primer used for RT reaction is designed to target the noncomplementary region of cbDVGs,

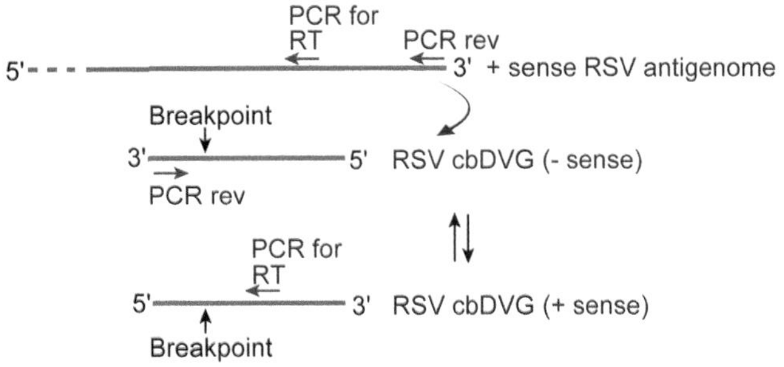

Fig. 2 RT-PCR strategy to specifically detect cbDVGs. To better illustrate this strategy, we denote the DVG strand directly produced from the +sense viral antigenome as −sense, whereas the DVG strand that is further amplified from −sense DVGs is +sense. Pink indicates +sense viral antigenome and blue color indicates −sense viral genome. The RT primer is designed to target the noncomplementary region of +sense cbDVGs, also targeting +sense RSV antigenomic sequences. This primer also serves as the forward primer in the PCR step (PCR for). Conversely, the PCR reverse primer (PCR rev) is designed to target the complementary region of −sense cbDVGs while still targeting +sense RSV antigenomic sequences. Two primers are indicated by red arrows. As depicted, the two primers flank the breakpoint of cbDVGs, resulting in a PCR amplicon produced from cbDVGs. However, since both primers target the +sense viral antigenomic sequences (pink line) in the same orientation, no PCR amplicon will be produced from full-length viral genomes. This approach only detects +sense cbDVGs

whereas the second primer in the PCR targets the complementary region of cbDVGs on the other side of DVG breakpoint (Fig. 2, red arrows). As shown in Fig. 2, due to the "loop-back" structure of cbDVGs, this PCR approach only produces an amplicon from cbDVGs, but not from full-length viral genomes. This strategy allows detection of RSV cbDVGs of various sizes within the detection limit of the primers [5, 13, 14] and has also been commonly used to detect cbDVGs from other viruses in the order of Mononegavirales, such as Sendai [15, 16], VSV [17], measles [18], mumps [19], and Ebola virus [20, 21]. (2) To visualize DVGs in infected cells, dual-color RNA-FISH was recently developed by Carolina Lopez's group to target both RSV cbDVGs and full-length viral genomes [22, 23]. Briefly, two sets of probes, linked with different fluorophores, are designed targeting two different regions of the RSV negative sense genome. One set (set1) of probes targets the 5′ end of negative sense RSV genome, where the majority of cbDVGs are generated. The other set (set2) of probes targets the 3′ end of negative sense RSV genome with a 3–4 kb gap between the two targeted regions (Fig. 3). As a result, full-length viral genomes (−sense) are detected by both sets of probes (thus positive for two colors), whereas cbDVGs can only be detected by set1 (positive for one color). This approach allows us to observe

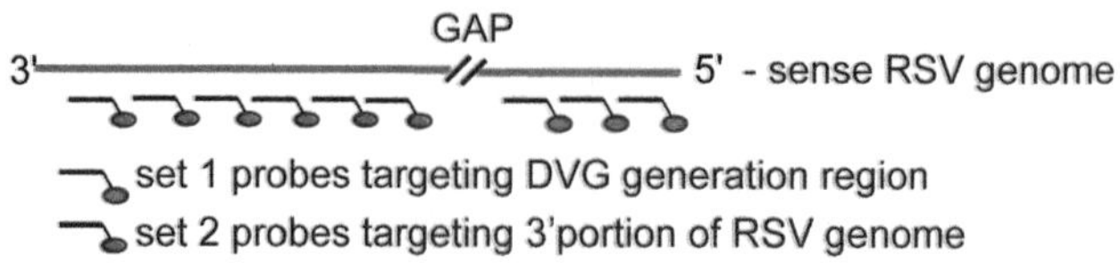

Fig. 3 RNA-FISH probe design targeting RSV cbDVGs and full-length viral genomes. According to published protocols [22, 23], set1 probes are linked with Quasar-670 and target the 5′ end of the RSV genome and DVGs. Set2 probes are linked with Quasar-570 and target the 3′ end of the RSV genome. Polarity of the viral genome is indicated

cbDVGs at the single-cell level. RNA-FISH probes can also be designed to target the positive sense viral antigenome and DVGs. However, positive sense probes also detect RSV mRNAs. (3) To unbiasedly examine the abundance and break/rejoin point distributions of the entire DVG population, several bioinformatic algorithms have been developed to identify DVGs from deep sequencing datasets, including VODKA [24, 25], ViReMa [26], DVGfinder [27], and DI-tector [28]. The principle of these algorithms is to identify sequencing reads spanning the junction region of DVGs, referred to as junction reads, because this region is unique for DVGs and does not exist as continuous sequences in full-length viral genomes (Fig. 1). Here, we list one of the pipelines to identify RSV cbDVG junction reads using ViReMa.

In addition to detection, to study DVG function, they must first be enriched and purified. DVGs were first isolated from an influenza virus stock that was passaged many times at a high multiplicity of infection (MOI) [29, 30]. Using this principle, RSV DVGs can be enriched by passaging the parental virus at MOI of 4–10 $TCID_{50}$/cell for 3–4 passages. Defective viral particles packaging DVGs have smaller particle sizes compared to infectious viral particles, thus, they can be further isolated from standard viral particles by gradient sucrose cushion ultracentrifugation. However, these methods produce a population of DVGs, containing various DVG species, and purified DVG particles are often contaminated with a small amount of infectious virus particles. To perform species-specific assessments of DVG function, we further developed an approach to produce defective viral particles packaging specific DVGs adapted from methods generating RSV virus-like particles [31, 32]. Briefly, plasmids ecoding the essential RSV proteins required for the production of virus-like particles, including the fusion protein (F), matrix protein (M), phosphoprotein (P), and nucleocapsid protein (N), are co-transfected with a plasmid encoding one DVG species of interest. For RSV, packaging of viral genomes depends on the specific interaction between N and M [33]. Consequently, encapsidated DVG RNAs are specifically packaged into RSV particles in cells, producing defective viral particles containing specific DVG species.

In summary, this chapter includes techniques for identifying and producing DVGs at both the population and species levels. We believe that the employment of these methods will facilitate mechanistic investigations into the generation and function of RSV DVGs.

2 Materials

2.1 Detection of RSV DVGs—DI-RT-PCR

1. TRIzol reagent.
2. SuperScript® III First-Strand Synthesis System.
3. Platinum™ Taq DNA Polymerase.
4. Nuclease-free H_2O.
5. dNTP (10 mM each).
6. $MgSO_4$ solution in nuclease-free H_2O (50 mM).
7. Thermo Scientific™ GeneRuler™ 100 bp Plus DNA Ladder 100–3000 bp.
8. Electrophoresis, loading dyes, Thermo Scientific, 6× DNA loading dye.
9. UltraPure agarose.
10. Tris–acetate–EDTA (TAE), 10×.
11. Primers: RSV-small (5′- CGAGAAAAAAAGTGTCAAAAAC TAATATC -3′); RSV DI1 (5′- CTTAGGTAAGGATATGTA GATTCTACC-3′) (*see* **Note 1**).
12. Agarose gel casting tray.
13. Electrophoresis chamber.
14. Power supply.
15. UV light box.
16. Gel DocTM XR+ gel imaging system.
17. NanoDrop spectrophotometer.

2.2 Detection of RSV DVGs—RNA-FISH

1. Formamide (deionized).
2. RNAseZAP.
3. ProLong Diamond Antifade mountant (used as Mounting media).
4. 20× SSC.
5. 1× Phosphate-buffered saline (PBS) (pH 7.4).
6. Nuclease-free H_2O.
7. 37% Formaldehyde.

8. 200 proof ethanol (1 pint).
9. EDTA (0.5 M, pH 8.0).
10. Tris–Cl (1 M, pH 8.0).
11. Glucose.
12. Catalase.
13. Glucose oxidase.
14. RNA-FISH probes (Table 1) [22].
15. 2X SSC: SSC (20×) 5 mL, nuclease-free H_2O 45 mL.
16. Fixation buffer (*see* **Note 2**): 37% formaldehyde 5 mL, PBS (1×) 45 mL.
17. Wash buffer (*see* **Note 2**): SSC (20×) 5 mL, formamide 5 mL, nuclease-free H_2O 40 mL.
18. 70% EtOH: 100% 200 proof ethanol 70 mL, nuclease-free H_2O 30 mL.
19. Hybridization buffer: dextran sulfate 1 g in total volume of 7 mL nuclease-free H_2O (*see* **Note 3**), formamide 1 mL, 20× SSC 1 mL, add nuclease-free H_2O up to total volume of 10 mL (*see* **Note 4**).
20. TE buffer: EDTA (0.5 M, pH 8.0) 0.2 mL, Tris–Cl (1 M, pH 8.0) 1 mL, nuclease-free H_2O to 100 mL. Filter the solution using 0.22 mm filters.
21. Prepare probes as follows: spin down tubes upon receiving and dilute the probes in 12.5 μM using TE buffer via vortex. Aliquot the stock into 20 μL per tube, seal with the parafilm, and store at −20 °C in the dark. To further dilute the stock solution as working solution, first make 1:10 dilution as follows: stock probe (12.5 mM) 20 μL, nuclease-free H_2O 180 μL. Aliquot if necessary. Then make 1:50 working solution: 1:10 dilution (1.25 μM) 10 μL, nuclease-free H_2O 40 μL (*see* **Note 5**).
22. Antifade buffer: nuclease-free H_2O 850 μL, 20× SSC 100 μL, glucose (10% w/v) 40 μL, Tris–HCl (pH 8.0) 10 μL.
23. GLOX antifade solution: antifade buffer 100 μL, glucose oxidase 1 μL, Catalase 1 μL.
24. 18 mm × 18 mm coverslips.
25. Microscope slides.
26. 150 mm × 15 mm culture dish.
27. Six-well plates.
28. Aspirators.
29. Parafilm.

Table 1
Sequences for RSV RNA-FISH probes

Set 1 probes (sequence 5′ to 3′)
tctaattggtttatatgtgt
gttaaacagcttgacaacca
ctacatatccttacctaagt
aaccatttatatatggtaga
gaagttttcagcaataaact
gtgttgttagtggagatata
actgcattgtcaaaactaaa
ataaagagtctattgatgca
atgctaaattgatactatca
ttcccagtatttaatgtagt
cttacaataggtcctgcgaa
gggatcggaggtttacttag
Set 2 probes (sequence 5′ to 3′)
gtgctctatcatcacagatc
ccctagaaattacatgccat
ggactacgtttctatcgtga
acttatccttctttgttgga
ataagtggagctgcagagtt
attgtgtcatgctatggcaa
ttcttcccacaagctgaaac
cagaggatggtactgtgaca
ggcgtaactacacctgtaag
tagtgctctgagaactggtt
agcttcaacaacaccaggag
tattcatagcctcggcaaac
ttgagttaccaagagctcga
ccacacaccatacagaatca
tcttcacttcaccatcacaa
ttggaagcacacagctacac
taccatatgcgctaatgtgt
aatcatctatgccagcagat

(continued)

Table 1
(continued)

ggacagatctggtcttacag
caaccatggctcttagcaaa
agacaggccacatttacatt
attgctctcaacctaatggt
tggctaaggcagtgatacat
agagatgggcagcaattcat

30. Tweezers.
31. Fluorescence microscope.

2.3 Detection of RSV DVGs—RNA-Seq Followed by ViReMa

1. ViReMa algorithm (https://sourceforge.net/projects/virema/).
2. Bowtie2 (https://sourceforge.net/projects/bowtie-bio/files/bowtie2/2.5.3/).
3. python3 (https://www.python.org/downloads/).
4. R (https://www.r-project.org/).
5. R studio (https://rstudio.com/products/rstudio/download/#download).
6. Human h38 reference genome (https://www.ncbi.nlm.nih.gov/datasets/genome/GCF_000001405.26/).
7. RSV full-length reference viral genome (from NCBI).
8. Sublime Text (https://www.sublimetext.com/3).

2.4 Enrichment of RSV DVGs

1. HEp2 cells (ATCC, CCL-23).
2. Human RSV strain A2 (ATCC, VR-1540) (*see* **Note 6**).
3. Mycoplasma removal agent.
4. Dulbecco's modified Eagle's medium (DMEM).
5. Opti-MEM™ I Reduced Serum Medium.
6. Trypsin–EDTA, 0.25% (wt/vol).
7. Fetal bovine serum (FBS) (*see* **Note 7).**
8. Gentamicin reagent solution (50 mg/mL).
9. Sodium pyruvate solution (100 mM).
10. L-glutamine (200 mM).
11. 1× Phosphate-buffered saline (PBS) (pH 7.4)
12. Dry ice.
13. Ethanol (used for quick-freeze, *see* **Note 8**).

14. Regular tissue culture medium (TCM): 500 mL DMEM supplemented with 10% FBS 50 mL, sodium pyruvate (100 mM) 5 mL, L-glutamine (200 mM) 5.5 mL, gentamicin (50 mg/mL) 500 mL. The entire TCM is passed the solution through a 0.22-μm filter.
15. Infection medium: 500 mL DMEM supplemented with 2% FBS 10 mL, sodium pyruvate (100 mM) 5 mL, L-glutamine (200 mM) 5.5 mL, gentamicin (50 mg/mL) 500 mL. The entire TCM is passed the solution through a 0.22-μm filter.
16. Sterile polypropylene conical 15- and 50-mL tubes.
17. Disposable cell scraper.
18. Sterile, aerosol-resistant micropipette tips (1–1000 mL).
19. Cotton-plugged, sterile serological pipettes (1–25 mL capacity).
20. Straight-neck polystyrene tissue culture flasks with vented caps, 75 cm^2 and 225 cm^2.
21. Polystyrene tissue culture plates with lids 12 and 96.
22. Cryotubes for storage viruses at −80 °C (internal thread).
23. Vacuum-driven Stericup® 500 mL Millipore ExpressPLUS 0.22 μm PES.
24. Class II biological safety cabinets.
25. Approved BSL-2 or enhanced BSL-2 laboratory facilities.
26. Personal protective equipment (PPE).
27. Spray bottles for 70% (vol/vol) ethanol.
28. Pipette controller.
29. Vortex mixer.
30. Water bath, 37 °C.
31. Refrigerated table-top centrifuge (for 15 mL and 50 mL conical tubes).
32. Freezers and refrigerators: −80 °C, −20 °C, and 4 °C.
33. Tissue culture incubator (37 °C, 5–7% CO_2).
34. Cell counter.
35. Cell culture microscope.

2.5 Purification of RSV DVGs

1. Ultra-Clear™ centrifuge tube, 14 × 89 mm.
2. Ultra-Clear™ centrifuge tube, 25 × 89 mm.
3. UltraPure™ EDTA.
4. Hanks' balanced salt solution (HBSS).
5. Cell culture microscope.
6. D-sucrose.

7. PNE buffer: PBS 50 mL, EDTA (0.5 M) 200 μL.
8. 20% sucrose: 10 g D-sucrose dissolved in a final volume of 50 mL of PNE buffer
9. 60% sucrose: 30 g D-sucrose dissolved in a final volume of 50 mL of PNE buffer
10. Optima™ L-90 K ultracentrifuge.
11. SW 32 Ti rotor package, swinging bucket (fit for 25 × 89 mm tube).
12. SW 41 Ti rotor package, swinging bucket (fit for 14 × 89 mm tube).
13. Gradient Master (BioComp Instruments Inc.)

2.6 Production of RSV Particles Packaging Specific DVGs

1. BSRT7 cells (BHK cells constitutively expressing the T7 polymerase).
2. Plasmids: Hamster-Codon-optimized pCDNA3.1-Phosphoprotein-IRES-Fusion and pCDNA3.1-Nucleoprotein-IRES-Matrix, pCDNA3.1-L; cbDVG plasmid encoding the specific cbDVG sequence under the T7 promoter (*see* **Note 9**).
3. Infection medium as Subheading 2.4.
4. TCM as Subheading 2.4, TCM used for BSRT7 cell growth contained additional 500 mg/mL geneticin.
5. PBS.
6. Opti-MEM™ I Reduced Serum Medium.
7. Lipofectamine 2000.
8. Nuclease-free water.
9. Polyclonal antirespiratory syncytial virus antibody.
10. RNaseA/T1 cocktail.
11. Turbo DNase I.
12. TRIzol-LS.
13. 0.45-mm filter
14. 10 mL luer-lok syringe
15. Ultra-Clear™ centrifuge tube, 14 × 89 mm.
16. SuperScript III First-Strand Synthesis System.
17. Power SYBR Green PCR Master Mixture.
18. Primers: RSV-small; RSV DI1; 15292F (5′-ATCGGAGGTTTACTTAGTCCT-3′); 15760F (5′- GTTATACAACTTTCATAATGAA -3′); cbDVGF (5′-TACAAATCTTGTTTGCTGTAATTGG-3′).
19. SW 41 Ti rotor package, swinging bucket (fit for 14 × 89 mm tube).
20. Viia7 Applied Biosystems Lightcycler.

3 Methods

3.1 Detection of RSV DVGs—RT-PCR

1. Total RNA is extracted from TRIzol according to manufacturer's instruction.
2. Extracted total RNA (1–2 μg) in 4 μL of nuclease-free H_2O is used in reverse transcription reaction (*see* **Note 10**).
3. Add 0.5 μL of the RSV DI1 primer (50 μM stock) and 0.5 μL of dNTP (10 mM) to RNA.
4. Incubate the mixture (total 5 μL) for 10 min at 65 °C (*see* **Note 11**).
5. During the **step 4** incubation, prepare the master mix as follows (5 μL per sample) on ice: RT buffer (5×) 1 μL, $MgCl_2$ (25 mM) 2 μL, DTT (0.1 M) 1 μL, RNaseOUT (40 U/μL) 0.5 μL, SSIII (200 U/μL) 0.5 μL (*see* **Note 12**).
6. Once the incubation is complete, add 5 μL/sample of master mix to individual samples on ice.
7. Incubate samples at 50 °C for 50 min and then 85 °C for 5 min (*see* **Note 13**).
8. Briefly centrifuge the reaction and then add 0.5 μL of RNase H.
9. Incubate at 37 °C for 20 min (*see* **Note 14**).
10. Prepare the master mix for PCR as follows (23 μL per sample): H_2O 13.75 μL, PCR buffer (10×) 2.5 μL, $MgSO_4$ (50 mM) 2.5 μL, dNTP (10 mM) 2 μL, RSV-small primer (10 μM) 1 μL, RSV DI1 primer (10 μM) 1 μL, Taq (5 U/μL) 0.25 μL. Aliquot the master mix into individual PCR tubes, and then add 2 μL of cDNA from **step 8**. Always include a "H_2O" sample as the negative control (*see* **Notes 15 16**).
11. Run PCR program as follows: 95 °C 10 min, 95 °C 30 s, 55 °C 30 s, 72 °C 90 s, cycle number 33–35, 72 °C 5 min, 4 °C hold (*see* **Note 16**).
12. Upon completion, run the entire PCR reaction on an agarose gel (Fig. 4) and excise candidate cbDVG bands for further confirmation by Sanger sequencing (*see* **Note 17**).

3.2 Detection of RSV DVGs—RNA-FISH in In Vitro Infected Cells (See Note 18)

1. Before RNA-FISH, clean workspace and instruments with RNaseZAP, including gloves, pipettes, and bench surfaces. Always use filtered pipette tips.
2. Seed cells on coverslips and infect with RSV. At the time of examination, collect the coverslips and transfer them to a six-well plate, wash with 1 mL of 1× PBS twice, and fix cells with 1 mL of the fixation buffer (*see* **Note 19**).
3. Incubate at room temperature for 10 min.
4. Wash cells twice with 1 mL of 1× PBS.

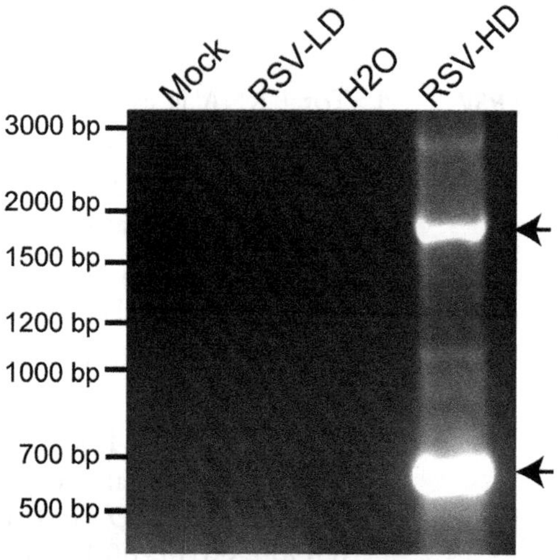

Fig. 4 Representative RSV cbDVG agarose gel image of DVG-specific RT-PCR. HEp2 cells were infected with RSV-LD (low DVG content) or HD (high DVG content) at an MOI of 1 $TCID_{50}$/cell, and DVGs were detected by DVG-specific RT-PCR at 24 h post-infection. DVGs RT-PCR used RSV-small and RSV DI1 primers. Bands highlighted by the black arrows were further confirmed by Sanger sequence as RSV cbDVGs. "Mock" indicated the addition of the same volume of infection media to cells without virus. An additional sample used H_20 as the template $_2$instead of cDNA in the PCR step, referred to as "H_20"

5. Permeabilize cells with 1 mL of 70% ethanol for at least 1 h at room temperature (*see* **Note 20**).
6. Aspirate the 70% ethanol off the coverslips.
7. Add 1 mL of wash buffer and incubate at room temperature for 2–5 min.
8. Prepare hybridization mix (per sample): 1 μL of 1:50 probe working solution from each set of probes diluted in total 50 μL of hybridization buffer.
9. Assemble the humidified hybridization chamber as shown in Fig. 5.
10. Add (Dot) 50 μL of the hybridization mix containing probes onto the clean parafilm in the chamber.
11. Gently transfer the coverslips from **step 7** with cells facing down onto the hybridization mix (50 μL/coverslip). Cover the humidified chamber with the lid. Incubate in the dark at 37 °C overnight.
12. The next day, transfer the coverslips, cells side up, to a fresh six-well plate containing 1 mL of wash buffer and incubate in the dark at 37 °C for 30 min.

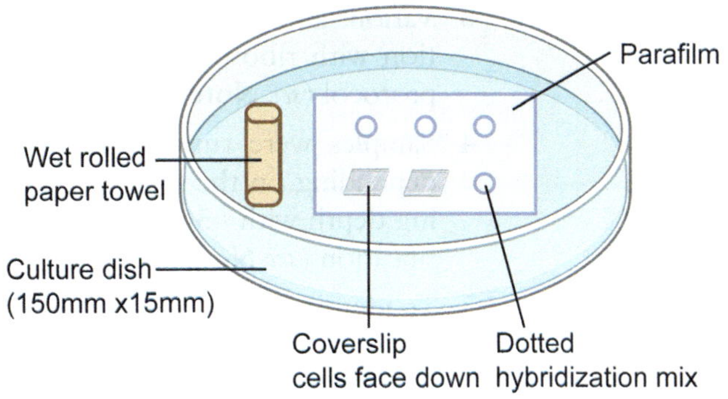

Fig. 5 Assembly of the hybridization chamber for RNA-FISH. This schematic illustrates the assembly of the hybridization chamber. Before assembly, the culture dish is cleaned with RNAseZAP. Cut a piece of parafilm to the proper size and place it in the culture dish with the clean side up. Dot the prepared hybridization mix on top of the parafilm (50 μL/dot/sample), and then place the coverslip on top of the hybridization mix as stated in Subheading 3.2 **step 10** with the cell side down. Lastly, roll a clean paper towel into a cylindrical shape, wet with nuclease-free water, and remove excessive H_20. Then place it to the side of the parafilm without touching the coverslip. Make sure no water is dripping before placement. The paper towel aims to humidify the chamber overnight, so that the coverslips do not dry out during hybridization. Put the lid on and keep the chamber at 37 °C incubator overnight in the dark

13. Wash with 1 mL of wash buffer consisting of Hoechst (1 μL into 10 mL washing buffer) to stain the nuclei. Incubate in the dark at 37 °C for 30 min (*see* **Note 21**).
14. Wash with 1 mL of 2X SSC. Incubate at room temperature in the dark for 2–5 min.
15. Mounting slides: Put a drop of ProLong Diamond Antifade onto slide, dry off the coverslip using aspirator and flip cells side down onto the mounting media. Make sure that there are no bubbles. The mounting media needs to "cure" overnight in the dark to achieve correct refractive index, followed by imaging (*see* **Note 22**).

3.3 Detection of RSV DVGs—RNA-seq Followed by ViReMa

Here, we use deep sequencing of RSV viral stocks as an example. This pipeline can be also used for deep sequencing of RNA from infected cells and clinical nasal samples. You can also use other algorithms for DVG identification. VODKA2 [25] is specialized for cbDVG identification.

1. Total RNAs were extracted from virus stocks using TRIzol LS reagent.
2. RNA quality was assessed on an Agilent Tapestation 2100 (Agilent Technologies) prior to cDNA library preparation.

3. Various amount of total RNAs can be used for library preparation with ribosomal RNA depletion using Illumina standard protocol (*see* **Note 23**).
4. Samples were run on Illumina NextSeq 500 or Nova seq depending on the sample numbers. We recommend sequencing depth with ~50 million reads/sample for RSV DVG identification (*see* **Note 24**).
5. To identify RSV DVGs, we used the ViReMa algorithm, which can identify both deletion and cbDVGs. Install ViReMa, bowtie2, python3, R, R studio according to individual instruction on the websites provided in Subheading 2 (*see* **Note 25**).
6. Raw fastaq.gz files from deep sequencing are first aligned to host reference genomes (hg38index) via bowtie2 using the following commands. If the deep sequencing is single-ended reads, use "bowtie2 -x /path_for_hg38index -U /path_for_sample.fastq.gz -p 8 --un-gz. /sample_unmapped.fq.gz --al-gz. /sample_mapped.fq.gz". If the deep sequencing is paired-end reads, use "bowtie2 -x /path_for_hg38index -1 /path_for_sample_R1.fastq.gz -2 /path_for_sample_R2.fastq.gz -p 8 --un-conc-gz. /sample.unmapped.fq.gz --al-conc-gz. /sample.mapped.fq.gz" (*see* **Note 26**).
7. As viral reads should not be aligned to human genomes, run ViReMa using unmapped output files from **step 6**, and align those files to RSV reference genome to identify DVGs as follows. "python3 /ViReMa_0.21/ViReMa.py /RSV_reference.fasta /sample_unmapped.fq.gz /sample_recombinations.bam --MicroInDel_Length 5 --Aligner_Directory /software/bowtie/0.12.9 -BED --Output_Dir /sample" (*see* **Note 27**).
8. Upon completion, output files should look like images in Fig. 6. The relevant output file for our analyses is the "Virus_Fusions.BEDPE" file within the "BED_Files" folder. When opening this file in Sublime Text, it should contain 10 columns. We explained each column in Fig. 7 and highlighted three columns (in red squares) we used for downstream analyses in

Fig. 6 Representative image of ViReMa output files

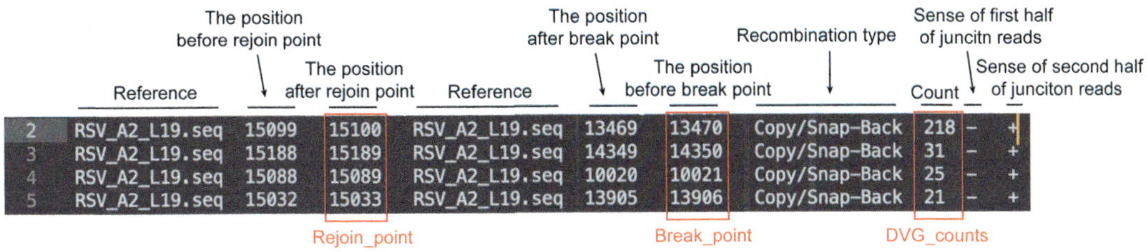

Fig. 7 Detailed indication of individual columns in the "Virus_Fusions.BEDPE" file. A screenshot of the output file, named "Virus_Fusions.BEDPE," is shown with explanations for each column. For our downstream analysis, three columns (highlighted with red squares) are selected. Among them, the first column indicates the rejoin point (Y in Fig. 1) of the identified DVG, the second column indicates the break point (W' in Fig. 1), and the last column indicates the read count (abundance) of this particular DVG. Note that this algorithm can identify both senses of DVGs. If the sense of DVGs is opposite from DVGs in this example, it is possible that the first column could indicate the break point, whereas the second column indicates the rejoin point. In terms of - sense cbDVGs, the genomic position for their break points should always be smaller than their rejoin points

R environment, including positions for break point and rejoin point and counts for individual identified DVGs.

9. To obtain viral reads that fully aligned to the viral reference genome, unmapped files are additionally aligned to RSV reference genome via bowtie2 using the following commands. Single-end: "bowtie2 -x /path_for_RSV_reference.fasta -U / path_for_sample.unmapped.fq.gz -p 8 --al-gz. /sample_mapped_bt.fq.gz". Paired-end: "bowtie2 -x / path_for_RSV_reference.fasta -1 /path_for_ sample.unmapped.1.fq.gz -2 / path_for_ sample.unmapped.2.fq.gz -p 8 --al-conc-gz. /sample.mapped.fq.gz". Make sure to set up the log file, which records the total number of reads aligned to the virus reference genome (refer to as total viral reads). The output bam files can then be used to visualize in IGV tool to identify potential genomic mutations if interested.
10. We then compute the DVG junction frequency for a given sample by calculating the ratio of total DVG counts (obtained from **step 8**, the sum of counts for all identified DVGs) over total viral reads. This number is important when comparing DVG abundance across various infections or different RNA-seq datasets.
11. Using results in "Virus_Fusions.BEDPE," DVGs with various break and rejoin points are ranked according to their counts (Fig. 7). The top 3–5 most abundant DVG species are selected for qPCR verification. Based on the breakpoint of a specific DVG, design primers using the same strategy as shown in Fig. 2 and keep the amplicon between two primers within 300 bp (*see* **Note 28**).
12. The distribution of DVG junctions can be visualized and compared using R. As R script is highly personalized, here we

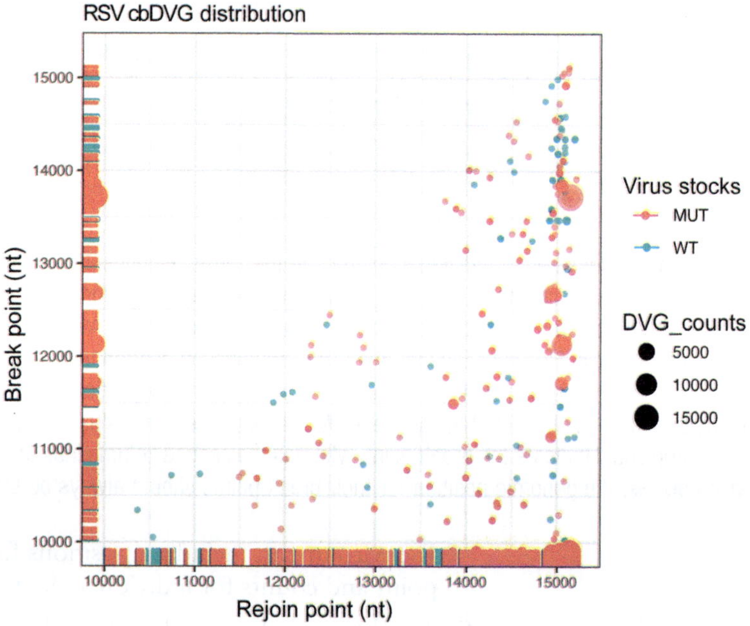

Fig. 8 Representative scatterplot showing the distribution of RSV cbDVGs identified by ViReMa. This scatterplot is graphed using the R scripts shown right after the the scatter plot. Note that the input of the cbDVGs, virus stocks, and color of dots can be all self-defined

provide one example of our scripts to visualize break and rejoin points of all identified DVGs in a scatterplot (Fig. 8) (*see* **Note 29**). Our scripts can serve as your starting point, but they should not be regarded as the definitive or exclusive scripts for your analyses.

```
" library(ggplot2, warn.conflicts = FALSE)
library(dplyr, warn.conflicts = FALSE)
library(stringr, warn.conflicts = FALSE)
library(readr)
library(plotly)
library(openxlsx)
#Input files
DVG_ViReMa <- read.csv("DVG_ViReMa_final.csv")
HD_WT <- subset(DVG_ViReMa,virus=="WT" & DVG_content=="HD")
 HD_MUT  <-  subset(DVG_ViReMa,virus=="10U"  &  DVG_con-
tent=="HD")
#Calculate the total DVG read counts
summary<- DVG_ViReMa %>%
group_by(ID) %>%
summarise(count = n(), Total_DVG = sum(counts))
write_csv(summary, "DVG_summary_ViReMa.csv")
```

```
#graphing DVG distribution
dvgcount_All <- ggplot(DVG_ViReMa_Sep, aes(y = Break_point, x
= Rejoin_point,
size = DVG_counts, color = virus)) +
geom_point(aes(alpha = 0.5)) +
labs(y = "Break point (nt)", x = "Rejoin point (nt)", title=
"DVG distribution",
color = "Virus stocks") +
geom_rug(aes(color = virus)) +
theme_bw() +
xlim(10000,15223)+ ylim(10000,15223)+
theme(legend.position = "right", axis.text = element_text
(size = 8), axis.title = element_text(size = 16)) +
theme(aspect.ratio = 1) +
guides(alpha = "none")
dvgcount_All"
```

3.4 Enrichment of RSV DVGs (See Note 30)

1. To maintain RSV stocks with low DVG content, referred to as RSV-LD, the parental RSV stock is passaged at an MOI of 0.001. We refer to the RSV stock enriched with DVGs as RSV-HD. To produce RSV-HD stocks, infect HEp2 cells with RSV-LD at an MOI of 4–10 $TCID_{50}$/cell according to the following steps. Infections to produce RSV-LD and RSV-HD are identical, differing only in their MOIs.
2. One day before infection, seed 3×10^6 HEp2 cells in T75 flasks.
3. The next day, aspirate the medium and wash the cells with OPTI-MEM twice. Dilute RSV-LD viral stock at an MOI of 4–10 $TCID_{50}$/cell in 1 mL total of infection medium and overlay on top of the washed cells.
4. Gently shake the flasks every 20–30 min. After 2 h of incubation in a tissue culture incubator, add 9 mL of infection medium to each flask.
5. Harvest the virus 2 days after infection by scraping the cells and collecting them together with the culture supernatants (*see* **Note 31**).
6. Centrifuge the pooled cells and supernatants for 5 min at $280 \times g$ in a table-top centrifuge at 4 °C to separate the supernatant and cells (save them in a separate sterile conical tube on ice). Leave <1 mL of supernatant in the tube with the pelleted cells.
7. Resuspend the cell pellet with the residual supernatant and quick-freeze the mixture in dry ice/ethanol, followed by thawing in a 37 °C water bath. Repeat this process at least three

times and vortex after each thaw to release more virions associated with cell membrane.

8. Centrifuge the mixture from **step 7** for 10 min at 280 × g in a table-top centrifuge to eliminate cell debris. Pool the supernatants from the freeze-thawed mixture with the previously saved supernatants (from **step 6**), vortex to mix well, aliquot (500 μL/vial) in cryotubes, and quick-freeze the virus in dry ice/ethanol. This concludes the first round of passage.
9. Titrate the virus via $TCID_{50}$ assay.
10. Repeat **steps 2–9** for two additional rounds of passaging (*see* **Note 32**).
11. To verify whether DVGs are enriched, RNAs from RSV-LD and RSV-HD stocks are extracted by TRIzol LS and cbDVG-specific RT-PCR is performed (*see* Subheading 3.1). If DVGs are enriched, you should observe one or more strong DVG bands (RSV-HD lane vs RSV-LD lane in Fig. 4). Importantly, all candidate DVG bands must be verified by Sanger sequencing. Additionally, we check the IFN stimulation of RSV-HD stock. When DVGs are sufficiently enriched in the RSV stock, types I and III IFN responses should be rapidly induced post-infection [5]. To do this, HEp2 or A549 cells are infected with RSV-LD and RSV-HD at an MOI of 1 $TCID_{50}$/cell. Cells are collected at 6–10 h post-infection using TRIzol. Total RNAs are then extracted from infected cells, followed by qPCR examination of types I and III IFN-related gene expression, such as IFNB1, IL-29, ISG56, and RSV G (*see* Fig. 9). For the unbiased characterization of cbDVGs within the RSV-HD stock, RNA deep sequencing can be applied (*see* Subheading 3.4).

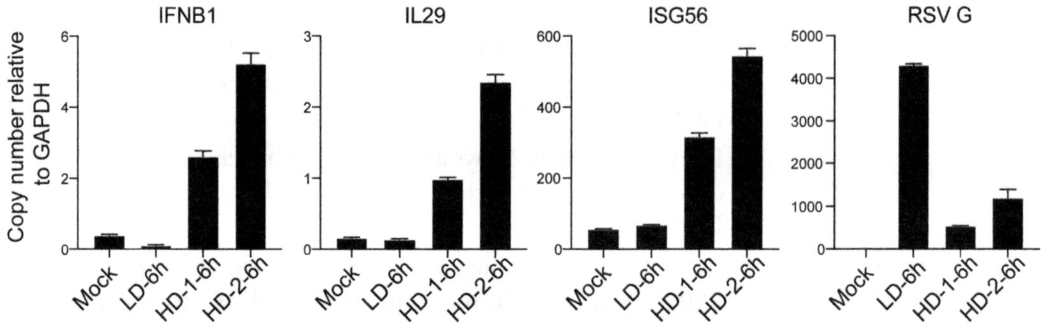

Fig. 9 Induction of type I/III IFN responses by RSV-HD infections. HEp2 cells were infected with one RSV-LD and two different RSV-HD (RSV-HD-1 and RSV-HD-2) virus stocks at an MOI of 1 $TCID_{50}$/cell. Cells were harvested at 6 h post-infection via TRIzol. Total RNA was extracted, followed by qRT-PCR examining the indicated host genes and RSV G. Gene expressions are quantified as their copy numbers relative to the housekeeping gene GAPDH. As shown in the figure, an RSV-HD stock enriched with a sufficient quantity of cbDVGs induces a much stronger IFN response, but relatively less RSV G expression compared to RSV-LD at 6 h post-infection

3.5 Purification of RSV DVGs (See Note 30)

1. To purify DVG containing defective viral particles, start with 8–10 T225 flasks to produce large quantities of RSV-HD viruses as follows.
2. One day before infection, seed 10^7 HEp2 cells in one T-225 flask and infect with RSV-HD at an MOI of 10 $TCID_{50}$/cell. Use an inoculum of 4 mL per T-225 flask and follow the infection procedure described above. Add 26 mL of infection medium after 2 h inoculation and return to the tissue culture incubator.
3. Two days post-infection, or when more than half of the cells have detached, harvest viruses as stated in the above section.
4. Combine all final supernatants and ultracentrifuge for 2.5 h at ~27,000× *g (15,000 rpm)* at 4 °C using an SW 32 Ti rotor and Ultra-Clear™ 25 × 89 mm centrifuge tube (*see* **Note 33**). This step is to concentrate the RSV-HD stock.
5. After centrifugation, lightly wash with 1 mL of HBSS twice without disturbing the pellet for each tube.
6. Suspend the pellet in a total of 200 μL of infection medium / T225 flask and combine all ultracentrifuged RSV-HD (*see* **Note 34**).
7. To purify defective viral particles, make 100 mL of PNE buffer fresh in the biosafety hood, shake and mix well.
8. Prepare 20% and 60% sucrose solutions fresh using PNE buffer as indicated in Subheading 2.5, shake and mix well.
9. Set out six tubes of Ultra-Clear™ 14 × 89 mm centrifuge tubes in the biosafety hood.
10. Fill each tube with 5 mL of 20% sucrose solution. Carefully and slowly pipette 5 mL of 60% sucrose solution starting at the bottom of the tube and slowly lift the 20% sucrose layer (*see* **Note 35**).
11. Carefully load the tubes containing sucrose solution on the Gradient Master to generate 20–60% sucrose gradient. Do this in the biosafety hood to maintain sterility.
12. Carefully overlay 1.5 mL of ultracentrifuged RSV-HD from **step 6** on top of each sucrose gradient (*see* **Note 36**).
13. Ultracentrifuge the gradients layered with virus for 2 h at ~106,000× *g (25,000 rpm)* at 4 °C using an SW 41 Ti rotor.
14. Carefully remove the top pink layer after centrifugation.
15. After centrifugation, several cloudy layers should be observed. Extract the cloudy layers from top to bottom using a P1000 micropipette. Transfer and combine the same fraction from all six tubes to a clean Ultra-Clear 14 × 89 mm centrifuge tube (one fraction, one tube) (*see* **Note 37**).

16. Fill up each tube to 2–3 mm from the top with sterile PBS (1×) and ultracentrifuge for 2 h at 4 °C at ~106,000× *g (25,000 rpm)* using SW 41 Ti rotor.
17. Suspend the pellet from each fraction in 1 mL of infection medium, aliquot 100 μL per vial, snap freeze, labeled with fraction numbers, and store at −80 °C.
18. To characterize the amount of DVGs and standard virus in each collected fraction, thaw one aliquot of each fraction. Use 20 μL to perform regular $TCID_{50}$ to determine the amount of infectious viral particles contained in the fraction. The remaining 80 μL are lysed in TRIzol for total RNA extraction, followed by DVG-specific RT-PCR (*see* Subheading 3.1) to determine DVG content. If the dominant DVG species is identified previously from RSV-HD stocks via deep sequencing, the level of dominant DVG species in each fraction should be examined via qPCR. The ratio of DVGs over infectious virus titer should be calculated, and the fraction with the most DVGs, but the least infectious virus, should be selected for further functional studies (*see* **Note 38**).

3.6 Production of RSV Particles Packaging Specific cbDVGs

1. To generate RSV defective virus particles (DVPs) packaging specific cbDVGs, seed 3 × 10^6 BSRT7 cells per 10 cm dish in regular TCM one day prior to transfection. At least two dishes are recommended for sufficient DVP generation (*see* **Note 39**).
2. Transfect pCDNA3.1-Np-IRES-M (4.8 μg), pCDNA3.1-P-IRES-F (4.8 μg), cbDVG expressing plasmid (9.6 μg), and pcDNA3.1-L (2.4 μg) using lipofectamine 2000 (DNA:lipofectamine 2000 ratio of 1:3) for two 10-cm dishes of BSRT7 cells according to the manufactural protocol. As a negative control, cbDVG plasmid (9.6 μg), pCDNA3.1-L (2.4 μg), and pcDNA3.1 empty vector (9.6 μg) were transfected similarly (*see* **Notes 40** and **41**).
3. Place transfected BSRT7 cells at 37 °C for 2 h.
4. Add OPTI-MEM to each 10 cm dish to reach a total volume of 7.2 mL. Place cells at 37 °C for 18 h (*see* **Note 42**).
5. The next day, wash the cells twice with 10 mL PBS and add 10 mL infection medium without antibiotics to the dishes and return cells to 37 °C for 4 h.
6. Twenty-four hours post-transfection, collect supernatants from the cells with a 10 mL luer-lok syringe and filter using a 0.45-mm filter into a 15-mL conical tube, quick-freeze, and place at −80 °C. Add 10 mL infection medium without antibiotics to each dish and place cells at 37 °C.
7. Repeat **step 6** for 48- and 72-h post-transfection. Immunofluorescent detection of viral inclusion bodies and filamentous protrusions from transfected cells can be observed between 24 and 72 h post-transfection (Fig. 10) (*see* **Note 43**).

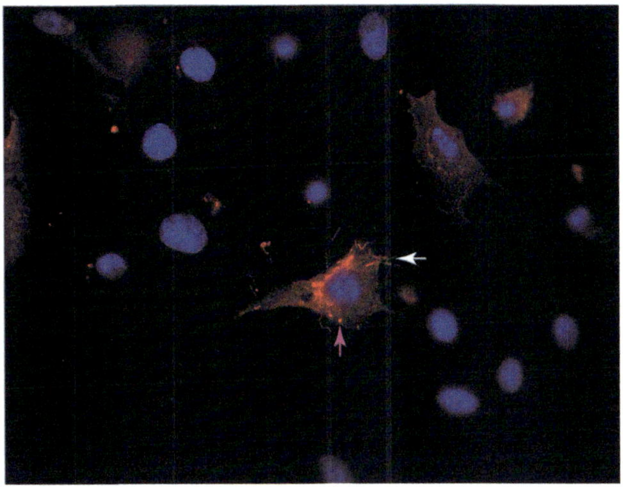

Fig. 10 Representative immunofluorescent image of transfected cells producing virus-like particles. Cells were transfected with 400 ng of pCDNA3.1-Phosphopreotein-IRES-Fusion, 400 ng pCDNA3.1-Nucleoprotein-IRES-Matrix, 800 ng cbDVG expressing plasmid and 200 ng pCDNA3.1-L. At 24 h post-transfection, cells were fixed and stained with goat anti-respiratory syncytial virus polyclonal antibody (1:500) and Hoescht (1:10,000). Images were taken at 630× magnification. Note the formation of viral inclusion bodies (magenta arrow) within the cytoplasm and filamentous protrusions (white arrow) extending away from the cell, indicating the formation of virus particles

8. Thaw supernatants collected at all three time points on ice and centrifuge at 2000 RCF for 20 min at 4 °C. Combine the clarified supernatants in a 50-mL conical tube.
9. Ultracentrifuge supernatants at 106,000 RCF with a 2 mL 20% sucrose cushion for 2 h at 4 °C using a SW 41 Ti rotor (*see* **Note 44**).
10. Decant the supernatant and resuspend the pellet in 600 μL infection medium. Aliquot 75 μL of purified DVPs per vial, immediately quick-freeze and place at −80 °C. A second round of sucrose purification is highly recommend to increase purity.
11. To characterize purified DVPs, thaw one aliquot on ice. Add 2 μL RNaseA/T1 cocktail and place at 37 °C for 30 min.
12. Extract total RNA using TRIzol-LS according to the manufacturer's instructions (*see* **Note 45**) and resuspend the RNA in 17 μL nuclease-free H_2O.
13. Treat the total RNA with DNase I (2 μL 10× turbo DNase buffer, 1 μL turbo DNase I to each RNA sample). Incubate at 37 °C for 30 min.
14. Purify the DNase treated RNA using phenol–chloroform extraction and allow RNAs to precipitate overnight at −20 °C.
15. Resuspend the RNA pellet in 10–20 μL nuclease-free water and measure RNA concentration via NanoDrop.

16. Reverse transcribe 100 ng of RNAs with RSV DI1 or 15292F to detect the + and − sense cbDVGs, respectively, using the SSIII reverse transcription kit as illustrated in Subheading 3.1.
17. For qPCR, RSV-small and cbDVGF were used for + sense cbDVG quantification. RSV-small and 15760F were used for −sense cbDVG quantification. cDNAs from **step 16** were diluted 1:8 in nuclease-free H_2O. Additionally, to generate standard curves of these two primer sets, make 1:10 serial dilutions of the cbDVG expressing plasmid, used for the generation of DVPs. qPCR reactions (10 μL /reaction) were performed in triplicate in a Viia7 Applied Biosystems Lightcycler as follows: diluted template (cDNA or plasmid) 4 μL, PowerTrack SYBR Green PCR Master Mixture (Applied Biosystems) 5 μL, the above primer pairs (5 μM) 0.5 μL/primer. Standard curves are shown in Fig. 11a, b, and the quantifications of purified DVPs via qPCR based on the standard curve are shown in Fig. 11c, d (*see* **Note 46**).

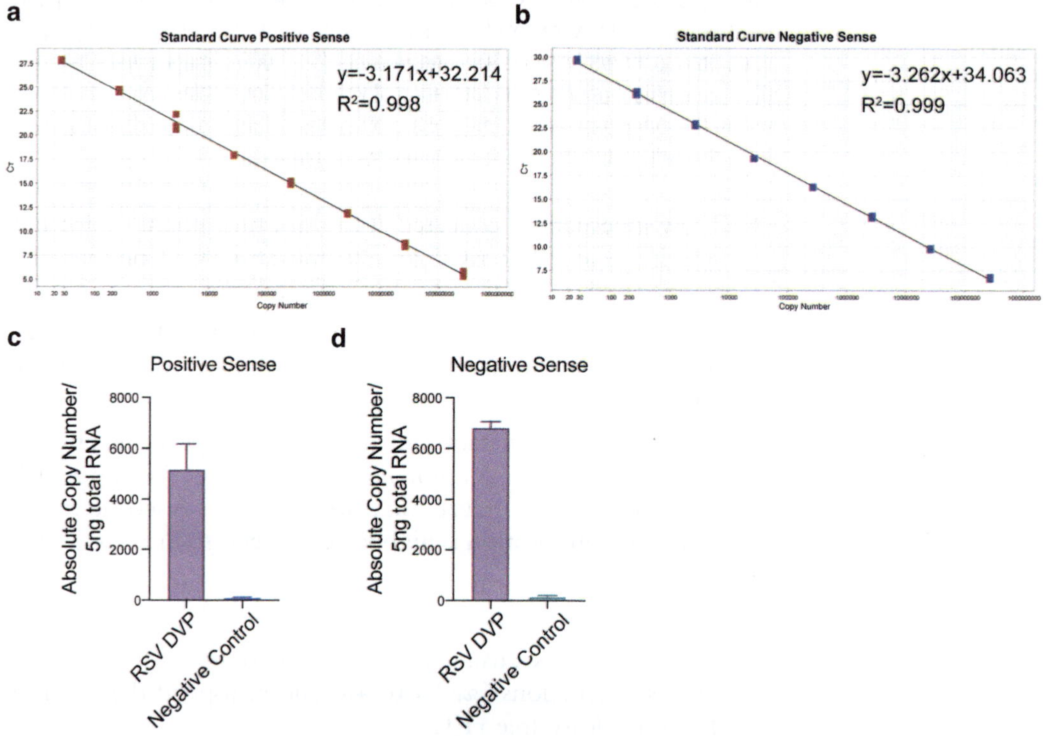

Fig. 11 Quantification of RSV DVP packaging specific cbDVG species. (**A** and **B**) Standard curves for primer sets detecting +sense (**A**) and −sense (**B**) cbDVGs using the serially diluted cbDVG plasmid. An R^2 value greater than 0.98 is necessary to ensure the reliability of the standard curve. (**C** and **D**) The quantification of the packaged +sense (**C**) and −sense (**D**) cbDVGs in the RSV DVPs and negative control conditions (from **step 2** in Subheading 3.6). Absolute copy number is calculated by employing standard curves to convert the Ct values of the RSV DVP and negative control conditions (using cDNA from 5 ng of RNA) into copy numbers

4 Notes

1. Primers can be self-designed based on the principle illustrated in Fig. 2. The primers listed here preferentially detect DVGs with rejoin points close to the 3′ end of viral antigenomes (trailer end) and break points before RT primer location. For RT reaction, primers are diluted in 50 μM working stocks and for PCR reaction, primers are diluted in 10 μM working stocks.
2. This buffer must be freshly prepared.
3. Mix by rotation and dissolve for 30 min.
4. Aliquot 500 μL per tube and store at −20 °C. The buffer is viable for years.
5. All probes need to be prepared and stored in the dark. 1:50–1: 100 dilution is typically used for in vitro infections. More concentrated probes, such as 1:10 dilution are needed for the hybridization using in vivo and ex vivo tissues.
6. This protocol can be adapted to enrich DVGs from other strains of RSV.
7. FBS should be heat-inactivated at 56 °C for 30 min, aliquoted, and stored at −20 °C. Thaw them before use.
8. This ethanol is used to quick-freeze viruses. We typically use 190 proof instead of 200 proof.
9. All plasmids used in this section are expressed under the T7 promoter and cbDVG expressing plasmids contain a 5′ hammerhead ribozyme sequence and a 3′ hepatitis delta virus ribozyme sequence.
10. Total amount of RNAs can be reduced to as low as 100 ng if you have low RNA yield. All steps involving RNA must be performed on ice using filter tips.
11. For all the incubation steps, we use the thermocycler.
12. All reagents are contained in the SuperScript® III First-Strand Synthesis kit.
13. After this step, DVG cDNAs are generated. We found that DVG cDNAs easily contaminate other materials, especially samples that are expected to have a high DVG content. Therefore, to avoid contamination, we recommend having a designated area and pipettes for the steps after this point. A 10% bleach will eliminate nucleic acids, including DVGs. We use 10% bleach to clean the designated area and pipettes before and after the experiments. Furthermore, we recommend adding cDNA samples in the following order: samples with no DVGs, less DVGs, high DVGs.

14. This is a good step to stop. cDNAs can be stored at −20 °C long-term.
15. We always include one negative control, where cDNAs are replaced by H_2O, to check the potential contamination by DVGs. This sample is always added as the last one of this step and should not contain any bands on agarose gel. If any bands are observed, even the faint ones, it is an indication of contamination of your reagents. In case of identifying contamination resources, we recommend checking in the following order based on our experience: H_2O, primers, $MgSO_4$ solution, other PCR reagents, RT reagents. It is more common to notice contaminations in PCR steps than RT steps. Therefore, it is critical to change gloves and clean pipettes and bench area after PCR steps. Do not touch primer stocks while performing PCR reaction, as this is the most common route to contaminate primer stocks with DVGs.
16. The amount of cDNA added in PCR reaction can be varied from 1 to 4 μL. Depending on the length of DVGs, the elongation time of PCR can be altered. The Taq polymerase we used in this protocol has a speed of 1 kb/min and working temperature of 72 °C. If other Taq polymerase is used, such as high-fidelity polymerase, the elongation time and temperature need to be changed correspondingly. The annealing temperature is specific to primers used.
17. Bands must be further confirmed by Sanger sequencing. Usually, DVG bands (indicated by black arrows in Fig. 4) are cut, purified by gel extraction, and ligated into pGEM-T-easy vector (other vectors also work). Selected colonies are sent for Sanger sequencing. Confirmed DVG sequencing must contain identifiable break and rejoin points of DVGs.
18. This protocol is adapted from the previous publication [22] and contains two sets of probes. Related images can refer to this publication. It is helpful to initially conduct RNA-FISH on infected cells using one set of probes before combining the two sets.
19. When washing cells, do not disturb the cells on the coverslips.
20. Alternative: fix overnight at 4 °C. Cells can be stored at 2–8 °C in 70% ethanol up to a week before hybridization.
21. Hoechst can be replaced with 5 ng/mL DAPI.
22. Alternatively, GLOX antifade solution may be used to mount the coverslips and image right away. In this case, seal the coverslips with nail polish, and allow to dry. Store slides on ice. Proceed directly to imaging.
23. We have used 75–450 ng of total RNAs for successful library preparations using Illumina TruSeq Stranded Total RNA LT

kit with Ribo-Zero Gold. For RSV clinical samples, frequently, total RNAs extracted are low yield and highly degraded. In this case, Takara SMARTer Stranded Total RNA-seq kit v2-Pico Input Mammalian was used to prepare cDNA libraries according to the manufacturer's instructions.

24. We recommend higher sequencing depth for clinical samples compared to in vitro infections and viral stocks.

25. We recommend running alignment jobs (**steps 6–9** in Subheading 3.3) in high-performance computer clusters, referred to as the "server," instead of a regular local computer. Once DVGs are identified in a spreadsheet, the data can be further analyzed and graphed via the R environment on local computers. In most cases, commonly used packages, such as python, bowtie2, and R have already installed in the "server." Installation of ViReMa can be slightly different depending on the different servers used.

26. In this command, put human reference genomes index after -x, the input fastq.gz files after -U for single-end reads. For paired-end, R1 file is put after −1 and R2 file is put after −2. After alignment, the aligned reads will be in the mapped file, and the unaligned reads will be in the unmapped file. For pair-end reads, since there are two input files, there will be two mapped and two unmapped files at the end of the alignment.

27. This step takes the longest time. These alignment outputs can be in BAM, SAM, or text file format and give you the reads of all types of virus–virus and virus–host recombination identified by ViReMa. In terms of RSV cbDVGs, we focused on the "Virus_Fusion" file. In the commend, the —MicroInDel_-Length option sets a threshold length for microindel identification. The —Aligner_Directory gives the path to where Bowtie is installed. The -BED option creates an output folder of BED files appropriately titled BED_Files within the output folder you provide. You need to provide a path after the —Output_Dir option to an output folder that ViReMa will automatically create. Before you save and run this script, make sure all of the code after python3 is on the same line. Importantly, always check on the updates of the algorithm, try to use the most updated version, and match all the requirements.

28. We often observe the same DVGs are identified as different break and rejoin points in ViReMa output. Two reasons can cause this. One is the sense, in which break and rejoin points are switched between two DVG species. We usually focus on one sense and flip the break and rejoin points of DVGs on the other sense to overcome this. The other reason is more complicated. Sometimes, the nucleotides immediately upstream of the break point are identical to the nucleotides immediately

downstream of the rejoin point, which leads to the ambiguity of the exact junction sites by the algorithm [34]. To identify them, break and rejoin points of two DVGs should be close, and the difference between break and rejoin positions of two DVGs should be the same. Unfortunately, we can only identify those incidences of DVGs manually at the top of the lists so far.

29. Package ggplot2 in R is a powerful tool for visualization. You can visualize DVG break point, rejoin point separately or in combination, and you can graph them in many other types of graphs using different parameters via this package.

30. Protocols for producing RSV-LD and RSV-HD and purification of RSV DVGs were previously published in bio-protocol [13]. Related images can refer to this publication.

31. We typically harvest viruses when roughly 50–70% of cells have detached from the plate. This is true for both RSV-LD and RSV-HD preparations. Due to the initial input differences, RSV-LD is collected about 5–6 days post-infection, and RSV-HD is collected at 2 days post-infection.

32. We observed DVG enrichment after three rounds of high MOI passaging. However, if DVGs are not observed or not enriched enough in the verification step, repeat **steps 2–9** and passage the virus for additional rounds.

33. Ultra-Clear™ 25 × 89 mm centrifuge tube can hold roughly 38 mL of medium. You should observe a brown pellet at the bottom of each tube after ultracentrifugation.

34. You can safely stop at this step. However, concentrated RSV-HD must be aliquoted into 1.5 mL/vial, quick-frozen in dry ice/ethanol for storage, and stored at −80 °C.

35. Caution: slow action is critical for this step. A sharp separation between 20% and 60% sucrose solutions should be observed. All steps need to be performed in the biosafety hood and kept sterile.

36. At this step, use a P1000 micropipette to add viruses slowly. The tubes should be filled to 2–3 mm from the top to avoid collapsing during ultracentrifugation. Do not disturb the sucrose gradient interface when adding the concentrated viruses.

37. Previously, three fractions were collected. The pellet at the bottom of the tubes contains standard viruses. All of the cloudy layers as well as the pellet should be collected to proceed to **step 18** for further characterization. Details refer to the publication [13]. This step can also be performed using automated fractionation machine. All fractions will be used in **step 18** to choose the factions containing most DVG particles and least infectious virus particles.

38. From previous experiments, the uppermost fraction contained the least infectious viral particles but also less DVPs. The fraction closest to the bottom of the tube contained most standard viral particles. The middle fraction contained abundant DVG particles with minimal amounts of infectious viruses and thus was preferred for our studies. It is important to note that the number of fractions and the ratio of DVGs to infectious viruses within each fraction may vary based on the species and the quantity of DVGs enriched in different RSV-HD stocks. Therefore, we recommend collecting all layers for characterization and selecting the appropriate fraction based on the specific results obtained. Total amount of viral particles can also be estimated by measuring the total protein amount in each fraction via BCA or Branford assay [27]. However, this estimation is less accurate compared to qPCR or deep sequencing.
39. This protocol describes the conditions for one RSV-DVP generating condition and one negative control condition, each using two 10 cm dishes of BSRT7 cells.
40. We prefer using OPTI-MEM throughout the transfection and changing media after overnight incubation. To reduce cost, PEI can be used instead of lipofectamine 2000. We have used a DNA:PEI ratio of 1:5 and increased the number of 10 cm dishes used for transfection, resulting in good DVP yield. Keep in mind that if lipofectamine 2000 is used, the incubation time for DNA and lipofectamine is 20 min, whereas if PEI is used, the incubation time is 10 min.
41. At this step, a minimum OPTI-MEM volume is recommended to increase the transfection efficiency. We typically use 4 mL/dish at this step.
42. This step ensures that sufficient OPTI-MEM is added to sustain the cells during the overnight incubation. For one 10 cm dish, we recommend 7–8 mL of OPTI-MEM for overnight incubation.
43. With this plasmid construction approach, cells with detectable viral inclusion bodies have been co-transfected with both plasmids required for virus particle generation.
44. This step is to purify and concentrate DVPs. Begin by adding the supernatants to the ultracentrifuge tube, then slowly add 20% sucrose from the bottom using a 1-mL serological pipette to lift the supernatants. It is critical to maintain the sharp interface between sucrose and supernatants.
45. We recommend overnight precipitation at −20 °C at this step for RNA extraction.

46. Primer pairs used here are specific to the cbDVG examined and cannot be used universally for all RSV cbDVGs. Appropriate primers should be designed according to your own cbDVG of interest based on the primer design strategy illustrated in Fig. 2. For better quantification of DVPs, Droplet Digital PCR (ddPCR) is recommended.

References

1. Huang AS (1973) Defective interfering viruses. Ann Rev Microbiol 27:101–117. https://doi.org/10.1146/annurev.mi.27.100173.000533
2. De BK, Nayak DP (1980) Defective interfering influenza viruses and host cells: establishment and maintenance of persistent influenza virus infection in MDBK and HeLa cells. J Virol 36(3):847–859. https://doi.org/10.1128/jvi.36.3.847-859.1980
3. Kajigaya S, Arakawa H, Kuge S, Koi T, Imura N, Nomoto A (1985) Isolation and characterization of defective-interfering particles of poliovirus Sabin 1 strain. Virology 142(2):307–316. https://doi.org/10.1016/0042-6822(85)90339-3
4. Baum A, Sachidanandam R, Garcia-Sastre A (2010) Preference of RIG-I for short viral RNA molecules in infected cells revealed by next-generation sequencing. Proc Natl Acad Sci USA 107(37):16303–16308. https://doi.org/10.1073/pnas.1005077107
5. Sun Y, Jain D, Koziol-White CJ, Genoyer E, Gilbert M, Tapia K, Panettieri RA Jr, Hodinka RL, López CB (2015) Immunostimulatory defective viral genomes from respiratory syncytial virus promote a strong innate antiviral response during infection in mice and humans. PLoS Pathog 11(9):e1005122. https://doi.org/10.1371/journal.ppat.1005122
6. Chaturvedi S, Vasen G, Pablo M, Chen X, Beutler N, Kumar A, Tanner E, Illouz S, Rahgoshay D, Burnett J, Holguin L, Chen PY, Ndjamen B, Ott M, Rodick R, Rogers T, Smith DM, Weinberger LS (2021) Identification of a therapeutic interfering particle-a single-dose SARS-CoV-2 antiviral intervention with a high barrier to resistance. Cell 184(25):6022–6036 e6018. https://doi.org/10.1016/j.cell.2021.11.004
7. Perrault J, Leavitt RW (1978) Inverted complementary terminal sequences in single-stranded RNAs and snap-back RNAs from vesicular stomatitis defective interfering particles. J Gen Virol 38(1):35–50. https://doi.org/10.1099/0022-1317-38-1-35
8. Re GG, Gupta KC, Kingsbury DW (1983) Genomic and copy-back 3′ termini in Sendai virus defective interfering RNA species. J Virol 45(2):659–664. https://doi.org/10.1128/jvi.45.2.659-664.1983
9. Kolakofsky D (1976) Isolation and characterization of Sendai virus DI-RNAs. Cell 8(4):547–555. https://doi.org/10.1016/0092-8674(76)90223-3
10. Lazzarini RA, Keene JD, Schubert M (1981) The origins of defective interfering particles of the negative-strand RNA viruses. Cell 26(2 Pt 2):145–154. https://doi.org/10.1016/0092-8674(81)90298-1
11. Perrault J (1981) Origin and replication of defective interfering particles. Curr Top Microbiol Immunol 93:151–207. https://doi.org/10.1007/978-3-642-68123-3_7
12. Felt SA, Sun Y, Jozwik A, Paras A, Habibi MS, Nickle D, Anderson L, Achouri E, Feemster KA, Cárdenas AM, Turi KN, Chang M, Hartert TV, Sengupta S, Chiu C, López CB (2021) Detection of respiratory syncytial virus defective genomes in nasal secretions is associated with distinct clinical outcomes. Nat Microbiol 6(5):672–681. https://doi.org/10.1038/s41564-021-00882-3
13. Sun Y, Lopez CB (2016) Preparation of respiratory syncytial virus with high or low content of defective viral particles and their purification from viral stocks. Bio Protoc 6(10). https://doi.org/10.21769/BioProtoc.1820
14. Sun Y, Lopez CB (2016) Respiratory syncytial virus infection in mice and detection of viral genomes in the lung using RT-qPCR. Bio Protoc 6(10). https://doi.org/10.21769/BioProtoc.1819
15. Tapia K, Kim WK, Sun Y, Mercado-Lopez X, Dunay E, Wise M, Adu M, Lopez CB (2013)

Defective viral genomes arising in vivo provide critical danger signals for the triggering of lung antiviral immunity. PLoS Pathog 9(10): e1003703. https://doi.org/10.1371/journal.ppat.1003703

16. Calain P, Curran J, Kolakofsky D, Roux L (1992) Molecular cloning of natural paramyxovirus copy-back defective interfering RNAs and their expression from DNA. Virology 191(1): 62–71. https://doi.org/10.1016/0042-6822(92)90166-m
17. Linder A, Bothe V, Linder N, Schwarzlmueller P, Dahlstrom F, Bartenhagen C, Dugas M, Pandey D, Thorn-Seshold J, Boehmer DFR, Koenig LM, Kobold S, Schnurr M, Raedler J, Spielmann G, Karimzadeh H, Schmidt A, Endres S, Rothenfusser S (2021) Defective interfering genomes and the full-length viral genome trigger RIG-I after infection with vesicular stomatitis virus in a replication dependent manner. Front Immunol 12:595390. https://doi.org/10.3389/fimmu.2021.595390
18. Mura M, Combredet C, Najburg V, Sanchez David RY, Tangy F, Komarova AV (2017) Nonencapsidated 5′ copy-back defective interfering genomes produced by recombinant measles viruses are recognized by RIG-I and LGP2 but not MDA5. J Virol 91(20). https://doi.org/10.1128/JVI.00643-17
19. Santak M, Markusic M, Balija ML, Kopac SK, Jug R, Orvell C, Tomac J, Forcic D (2015) Accumulation of defective interfering viral particles in only a few passages in Vero cells attenuates mumps virus neurovirulence. Microbes Infect 17(3):228–236. https://doi.org/10.1016/j.micinf.2014.11.006
20. Calain P, Monroe MC, Nichol ST (1999) Ebola virus defective interfering particles and persistent infection. Virology 262(1): 114–128. https://doi.org/10.1006/viro.1999.9915
21. Johnson RI, Boczkowska B, Alfson K, Weary T, Menzie H, Delgado J, Rodriguez G, Carrion R Jr, Griffiths A (2021) Identification and characterization of defective viral genomes in Ebola virus-infected rhesus macaques. J Virol 95(17): e0071421. https://doi.org/10.1128/JVI.00714-21
22. Gonzalez Aparicio LJ, Yang Y, Hackbart M, Lopez CB (2023) Copy-back viral genomes induce a cellular stress response that interferes with viral protein expression without affecting antiviral immunity. PLoS Biol 21(11): e3002381. https://doi.org/10.1371/journal.pbio.3002381
23. Xu J, Sun Y, Li Y, Ruthel G, Weiss SR, Raj A, Beiting D, Lopez CB (2017) Replication defective viral genomes exploit a cellular pro-survival mechanism to establish paramyxovirus persistence. Nat Commun 8(1):799. https://doi.org/10.1038/s41467-017-00909-6
24. Sun Y, Kim EJ, Felt SA, Taylor LJ, Agarwal D, Grant GR, Lopez CB (2019) Correction: a specific sequence in the genome of respiratory syncytial virus regulates the generation of copy-back defective viral genomes. PLoS Pathog 15(10):e1008099. https://doi.org/10.1371/journal.ppat.1008099
25. Achouri E, Felt SA, Hackbart M, Rivera-Espinal NS, Lopez CB (2023) VODKA2: a fast and accurate method to detect non-standard viral genomes from large RNA-seq datasets. bioRxiv. doi:https://doi.org/10.1101/2023.04.25.537842
26. Sotcheff S, Zhou Y, Yeung J, Sun Y, Johnson JE, Torbett BE, Routh AL (2023) ViReMa: a virus recombination mapper of next-generation sequencing data characterizes diverse recombinant viral nucleic acids. Gigascience 12. https://doi.org/10.1093/gigascience/giad009
27. Olmo-Uceda MJ, Munoz-Sanchez JC, Lasso-Giraldo W, Arnau V, Diaz-Villanueva W, Elena SF (2022) DVGfinder: a metasearch tool for identifying defective viral genomes in RNA-seq data. Viruses 14(5). https://doi.org/10.3390/v14051114
28. Beauclair G, Mura M, Combredet C, Tangy F, Jouvenet N, Komarova AV (2018) DI-tector: defective interfering viral genomes' detector for next-generation sequencing data. RNA 24(10):1285–1296. https://doi.org/10.1261/rna.066910.118
29. Von Magnus P (1947) Studies on interferenee in experimental influenza. I. Biological Observations Arkiv for Kemi, Mineralogi och Geologi (7)
30. Huang AS, Baltimore D (1970) Defective viral particles and viral disease processes. Nature 226(5243):325–327. https://doi.org/10.1038/226325a0
31. Walpita P, Johns LM, Tandon R, Moore ML (2015) Mammalian cell-derived respiratory syncytial virus-like particles protect the lower

as well as the upper respiratory tract. PLoS One 10(7):e0130755. https://doi.org/10.1371/journal.pone.0130755

32. Meshram CD, Baviskar PS, Ognibene CM, Oomens AGP (2016) The respiratory syncytial virus phosphoprotein, matrix protein, and fusion protein carboxy-terminal domain drive efficient filamentous virus-like particle formation. J Virol 90(23):10612–10628. https://doi.org/10.1128/JVI.01193-16

33. Shaikh FY, Crowe JE Jr (2013) Molecular mechanisms driving respiratory syncytial virus assembly. Future Microbiol 8(1):123–131. https://doi.org/10.2217/fmb.12.132

34. Routh A, Johnson JE (2014) Discovery of functional genomic motifs in viruses with ViReMa-a virus recombination mapper-for analysis of next-generation sequencing data. Nucleic Acids Res 42(2):e11. https://doi.org/10.1093/nar/gkt916

Chapter 4

A Rapid Yeast-Based System for Modifying Respiratory Syncytial Virus

Mahesh K. C., Ilada Thongpan, Namita Bhattarai, Phylip Chen, Michelle Chamblee, Cheng Chih Hsu, Mark E. Peeples, and Jianrong Li

Abstract

The addition of marker protein genes to respiratory syncytial virus (RSV) has enabled studies of the spread of RSV in different types of cell cultures and quantification of viral replication in those cultures. Genetic deletion of individual RSV genes from RSV genome has been used to determine their importance in virus infection and the differences between infection of cultured cells lines and of primary well-differentiated human bronchial epithelial (HBE) cultures. Modifications of individual viral proteins can identify the importance of a particular glycosylation, cleavage, or antigenic sites or reveal sites with these functions. However, the standard recombinant systems for the RSV genome, based on natural and inserted restriction sites, have been difficult to use, slow to accomplish, and frequently not successful. Here, we describe a yeast-based cDNA assembly system that streamlines both the construction of a cDNA clone of an RSV strain as well as any modification of an RSV cDNA, thereby enabling the rapid generation of an RSV mutant virus.

Key words Respiratory syncytial virus (RSV), Polymerase chain reaction (PCR), Copy DNA (cDNA), Yeast-based cDNA assembly, Homologous recombination, T7 polymerase, Virus rescue from cDNA

1 Introduction

In the 40 years between the discovery of human respiratory syncytial virus (RSV) in 1956 [1] and the development of an RSV reverse genetics system in 1995 [2], the study of RSV and similar viruses was limited to descriptions of the viral proteins in infected cells, virions, and mRNAs. The only viral genetic studies were comparisons of RSV strains or conditional mutant viruses such as temperature-sensitive mutants [3, 4]. In this chapter, we briefly describe systems that have been established since then to genetically modify RSV, enabling modifications of its genomic elements,

K. C. Mahesh and Ilada Thongpan contributed equally to this chapter. Mark E. Peeples and Jianrong Li are the co-corresponding authors for this chapter.

Bo Liang and Rachel Fearns (eds.), *Human Respiratory Syncytial Virus: Methods and Protocols*, Methods in Molecular Biology, vol. 2948, https://doi.org/10.1007/978-1-0716-4666-3_4,

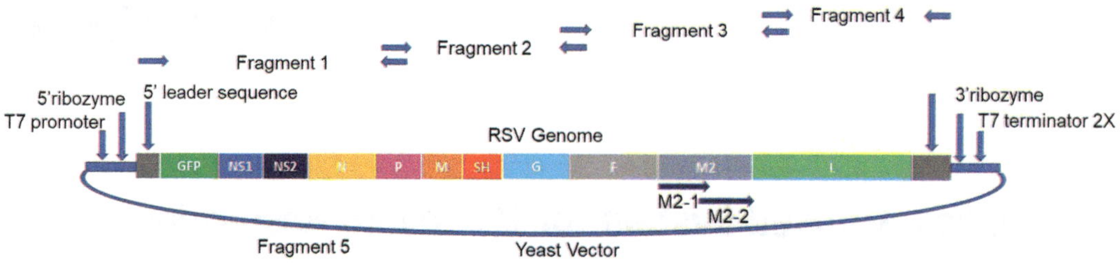

Fig. 1 Schematic diagram (not to scale) of the recombinant green fluorescent protein-expressing (rgRSV) genome in the yeast vector

deletion of individual genes and the proteins they express, and insertion of foreign genes and the proteins they express, primarily to quantify and trace virus spread in cultured cells. We focus on a relatively new, rapid method for cloning and modifying RSV genomes and rescuing modified RSV viruses that they encode.

The first full-length RSV genomic cDNA (Fig. 1) was constructed by Peter Collins [2] in the Laboratory of Infectious Diseases of the National Institute of Allergy and Infectious Diseases (NIAID), on the NIH campus in Bethesda, MD, USA. Before arriving at NIH, he had cloned and sequenced most of the RSV genes at the University of North Carolina-Chapel Hill, working with Gail Wertz. The complete RSV cDNA was constructed in a low copy plasmid pBR322 [2] behind a T7 bacteriophage promotor and transcribed as the antigenome full-length positive-sense RNA, beginning with the RSV leader (Le) sequence, followed by the 10 RSV gene units, and ending with the trailer (Tr) sequence [5, 6]. Each gene unit was preceded by a gene start (GS) sequence, followed by a gene end (GE) sequence, and separated from the neighboring gene by an intergenic (IG) sequence of variable length, depending on the gene. The other nine genes of the virus had a similar gene organization. Following the final, large (L), gene unit that encodes the viral polymerase and the Tr, a hammerhead or hepatitis delta virus (HDV) ribozyme was added to cleave and release itself the RNA transcript, and a T7 terminator to halt the T7 polymerase transcription.

The first complete genomic RSV cDNA sequence was derived from the A2 strain of RSV and was 15,222 bp in length (Genbank Accession: M74568.1 GI: 333959). The Collins group used this cDNA to produce many variants of RSV, deleting genes to determine which of its 11 genes were essential, and to study their functions. They mutated the L polymerase gene and other genes to attenuate the virus for the purpose of developing a live attenuated RSV vaccine [2, 7–17].

Modified versions of RSV rescued from this cDNA which express green [18] or red fluorescent proteins [19] can be obtained from BEI Resources. Others expressing cystic fibrosis transmembrane conductance regulator (CFTR), or with restriction sites to insert foreign genes at different positions in the viral genome, or lacking particular viral genes can generally be obtained from the senior authors of the publications in which they were first described [18, 20–23]. As with any unique biological, a Material Transfer Agreement (MTA) from the originator's institution is required to obtain the cDNA or the virus. Because the original RSV A2 cDNA was generated at the NIH, an MTA from the NIH is required for any full-length cDNA clone generated from it.

RSV strains expressing additional genes whose products color the cell in which they are replicating have been useful for identifying infected cells, quantifying viral infection and spread, and detecting neutralizing antibodies or antiviral drug inhibition of virus infection or production. An RSV with an inserted CFTR, the largest exogenous gene to be expressed from RSV, 4.5 kb, restores CFTR activity to primary well-differentiated human bronchial epithelial (HBE) cultures derived from persons with cystic fibrosis [24].

RSVs expressing an enzyme such as luciferase enable quantification of viral replication enzymatically. We generated an RSV replicon, a virus lacking its G and F protein genes, which produces luciferase in addition to green fluorescent protein (GFP). This replicon was able to replicate in Vero, BHK, or HeLa cells without causing obvious damage or slowing cell growth [25]. Cells containing the RSV replicon have been used to screen libraries of chemical compounds to identify those that were able to inhibit the RSV polymerase, and therefore luciferase production.

In addition, viral genes have been deleted to determine their importance for infection of cell lines vs. HBE cultures that represent the in vivo target cells for RSV. Deleting the G gene proved that the G protein was essential for infection of HBE cultures, more specifically the ciliated epithelial cells, but was not essential for infection of cell lines [26, 27] leading to the conclusion that the receptors on these two cell types were also different.

Recombinant, infectious RSV virus is "rescued" as infectious virus by co-transfecting Vero, HEp-2, or HeLa cell lines with both a plasmid expressing the RSV genome as a positive-sense RNA and plasmids expressing each of the four RSV nucleocapsid-associated proteins, N, P, M2-1, and L. Transcription of all five of these plasmids is accomplished by the bacteriophage T7 polymerase, a single-molecule DNA-dependent RNA polymerase, expressed from Modified Vaccinia Ankara (MVA-T7), an attenuated vaccinia virus. Vaccinia viruses are the only DNA viruses that replicate in the cytoplasm and can therefore provide the T7 polymerase in the location where RSV also replicates, the cytoplasm. The T7 polymerase transcribes each of these plasmids to produce the antigenomic RNA and mRNAs for the four RSV proteins needed for RSV to replicate, N, P, L and M2-1. The result is the rescue of a self-replicating recombinant RSV. Cell lines that constitutively express T7 polymerase have also been used to rescue RSV [9, 28, 29].

The N protein encapsidates the positive-sense RNA "antigenome," and the P, M2-1, and L proteins associate with the nucleocapsid. Together they transcribe the antigenome into the negative-sense "genome" which is also encapsidated by the N protein. The genome is then transcribed into the 10 RSV mRNAs which are translated into the 11 RSV proteins that support RSV replication and transcription (N, P, M2-1, and L) and enable virion formation (M, F, G, and SH) at the plasma membrane. The three other RSV proteins, NS1, NS2, and M2-2, are also produced, inhibiting the host interferon response (NS1 and NS2) and controlling the balance between the production of genome and antigenome (M2-2).

A second approach for producing recombinant RSV, based on the Bacterial Artificial Chromosome (BAC) system, was developed by Martin Moore [30]. This BAC system has been used to produce a bioluminescent RSV for studies of RSV infection in mice for use in testing potential antiviral drugs and vaccines in vivo [31]. It has also been used to produce an RSV vaccine candidate that incorporates many deoptimized codons to attenuate the virus. One of these candidates is currently in a Phase 1C clinical trial.

A New, Efficient, Yeast-Based System for Modifying the RSV Genome The traditional, restriction enzyme-based method originally used to modify the RSV cDNAs has been very useful but is inefficient and requires that restriction enzyme sites first be positioned at the insertion site or a long synthetic DNA between two restriction sites. To overcome these problems, we have modified a yeast homologous recombination system that is more efficient, quicker, and less expensive [32].

2 Materials

Yeasts have a very active homologous recombination system that only requires the 30-nucleotide (nt) homologous overhangs on DNA fragments and at both ends of the pYES2 vector to connect them. All that is required is knowledge of the complete sequence of the viral genome and made-to-order oligonucleotide primers that provide the 30-nt overhangs.

2.1 Materials Required for the Yeast Homologous Recombination System

1. pYES2 vector (Invitrogen, Cat. No. V825-20).
2. INVSc1, *Saccharomyces cerevisiae* yeast strain (Invitrogen, Cat. No. C81000).
3. Yeast extract peptone dextrose (YPD) broth.
4. 1 M lithium acetate (LiAc), (Cat. No. A8509)
5. 0.1 M LiAc
6. ssDNA 5 mg/mL (Abcam, Cat. No. Ab229278).
7. PEG3350 (50%) (Cat. No. 39713-100ML-F).
8. SD/Ura$^-$ agar plate.
9. SD/Ura$^-$ broth.
10. Digestion buffer (Cat. No. D2001-1-15).
11. Zymolyase (Cat. No. E1004-A).
12. Bacterial miniprep kit.
13. 10 beta/DH10B/Top10 competent cells.
14. Endotoxin-free (EndoFree) Plasmid Maxi Kit (QIAGEN, Cat. No. 12362).

2.2 Materials Required for Rescuing Recombinant and Mutant RSV Virus

1. DMEM with L-Glutamine, 4.5 g/L Glucose, and Sodium Pyruvate (Fisher Scientific, Cat. No. MT10013CV).
2. Opti-MEM™ I Reduced Serum Medium (1×) liquid (Gibco, Cat. No. 31985062).
3. Cytiva HyClone™ Phosphate Buffered Saline (PBS) (Fisher Scientific, Cat. No. SH3025601).
4. Penicillin–Streptomycin (10,000 U/mL) (Gibco, Cat. No. 15140122).
5. Fetal Bovine Serum (FBS), Premium, heat-inactivated (Fisher Scientific, Cat. No. A5670801).
6. TransIT-HeLaMONSTER® Transfection Kit (Mirus Bio, Cat. No. MIR 2900).
7. Full-length cDNA plasmid encoding the full RSV genome.
8. RSV helper plasmids encoding RSV N, P, M2-1, and L genes.
9. Recombinant modified vaccinia virus Ankara expressing the T7 RNA polymerase (MVA-T7).
10. HeLa cells (human origin) (ATCC CCL-2).
11. Non-coated 6-well cell culture plates.
12. Filtered pipette tips.

3 Methods

3.1 Assembly of RSV cDNA Using the Yeast Homologous Recombination System

Full-length genomic cDNA can be assembled into the pYES2 vector (Fig. 1) by mixing DNA fragments of any size generated by PCR using primers that provide homologous overhangs with the neighboring fragments (*see* **Note 1**). We have found that fragments with a maximum size of 4–5 kb are optimal for reducing/avoiding mutations that can be introduced by PCR while using the minimal number of fragments (*see* **Note 2**). Mutations are incorporated in the primers used to amplify adjacent fragments [32, 33]. These PCR fragments can be produced from an RSV cDNA clone template or RT-PCR fragments from an RSV genomic RNA. Foreign genes can be added to the final cDNA clone in a similar manner.

The pYES2 vector contains a yeast replication origin, T7 RNA polymerase promoter, 5′ hammerhead ribozyme, 3′ hepatitis delta virus ribozyme, and a 2× T7 terminator [24, 32, 34]. The pYES2 vector is amplified such that its ends overlap with the terminal fragments of the RSV genome insert, with the fragment in the correct orientation relative to the T7 promotor to produce a positive-sense (anti-genomic) RNA. When mixed with the single fragment of the yeast vector, the fragments assemble into a plasmid containing a complete RSV cDNA (Fig. 2).

Steps:

1. Amplify 4–5 kb RSV genomic DNA fragments and the yeast vector with primers that include 30-nt overlaps with the adjacent fragments.
2. Grow INVSc1, *Saccharomyces cerevisiae* yeast strain in YPD broth (1 μL/mL YPD broth) overnight at 30 °C with rotation (225 rpm). Use 1 mL of YPD broth per sample transformation in yeast.
3. Check the optical density (OD_{620}) of the yeast suspension after 12–14 h and dilute to ~0.6.
4. Spin down 1 mL of yeast at 5000 rpm in a microcentrifuge tube for 5 min for each reaction. Discard the supernatant.

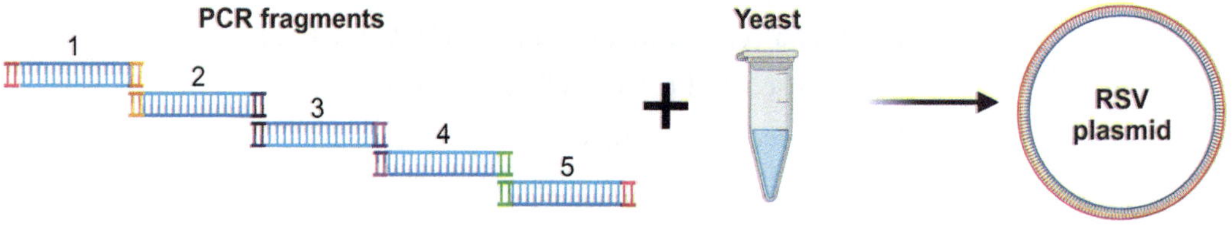

Fig. 2 Schematic diagram of the assembly of viral genome PCR products (fragments 1–4) and the yeast plasmid vector (fragment 5) that are transformed into yeast, where they recombine to form the complete RSV cDNA in the yeast plasmid

5. Resuspend the pellet in 1 mL dH_2O for washing, spin down, and discard the supernatant.
6. Resuspend the pellet in 500 μL of 0.1 M LiAc, spin down, and discard the supernatant.
7. Resuspend the pellet in 400 μL of 0.1 M LiAc and incubate in a 30 °C water bath for 5 min, then pellet the cells and discard the supernatant.
8. Resuspend the yeast pellet with the mixture described in Table 1, and vortex (Level 5–6) for 30 s. Prepare the mixture of DNA fragments (~100 ng/fragment) in a separate tube. Add the DNA to the yeast and mix.
9. Incubate the mixture at 30 °C for 30 min, with occasional agitation.
10. Incubate the mixture at 42 °C for 10 min.
11. Spin down the mixture for 2–3 min and discard the supernatant. Resuspend the yeast pellet in 100 μL of ddH_2O and spread on an SD/Ura^- agar plate, invert, and incubate at 30 °C for 2–3 days.
12. Pick yeast colonies and perform colony PCR with primer pairs to amplify the segments. Send the amplified fragments or the plasmid after extraction for sequencing.
13. Grow the correct colony in 5 mL SD/Ura^- broth, at 30 °C, shaking at 225 rpm overnight. The broth may need to be kept in the shaker for up to 2 days for larger plasmids.
14. Extract the plasmid DNA from the yeast.
    (a) Spin down the yeast cells (4000 rpm for 10 min) and discard the supernatant.
    (b) Resuspend the yeast pellet in 150 μL of Digestion buffer containing 2 μL of Zymolyase and incubate at 37 °C for 15–60 min.

Table 1
Composition of yeast homologous recombination system master mix

Reagent	Volume (μL per reaction)	Note
1 M LiAc	36	
ssDNA 5 mg/mL	20	Preheat at 100 °C for 5 min and keep on ice.
PEG3350 (50%) (Cat. No. 39713-100ML-F)	240	Dissolve in water and filter using 0.45-μm filter
dH_2O	59	
Total volume/reaction	355	

(c) Add 100 μL of lysis buffer (from any bacterial miniprep kit) and mix. Incubate the mixture for 2–5 min at room temperature.

(d) Add 200 μL of neutralizing buffer and mix.

(e) Centrifuge at 13,000 rpm for 10 min.

(f) Transfer the supernatant to a spin column and centrifuge for 30–60 s. Discard the flow-through.

(g) Wash the spin column by adding 500 μL of PB buffer, centrifuge for 30–60 s, and discard the flow-through.

(h) Wash the spin column with 750 μL of washing buffer (PE), centrifuge for 30–60 s, and discard the flow-through.

(i) Centrifuge again to remove any residual wash buffer.

(j) Place the spin column in a clean 1.7 mL microcentrifuge tube and elute DNA by adding 30 μL of dH_2O or elution buffer. The DNA concentration should be 10–30 ng/μL.

15. Transform 12–15 μL of this plasmid into 10 beta/DH10B/Top10 competent cells using the standard transformation protocol, plate the bacteria on an Amp^+ LB plate, and incubate at 30 °C for 2 days.
16. Pick bacterial colonies and perform colony PCR with primer pairs to amplify the segments. Send the amplified fragments or the plasmid after extraction along with the sequencing primers.
17. Grow the correct colony in 5 mL Amp^+ LB broth, at 30 °C, shaking at 225 rpm overnight. The broth may need to be kept in the shaker for up to 2 days for larger plasmids.
18. Extract the plasmid DNA as described above (skip **step b**, breakdown of fungal wall, for bacterial culture).
19. Prepare a maxiprep of a plasmid with the correct sequence to use for virus rescue.

3.2 Rescuing Modified RSV

A schematic diagram of the RSV reverse genetic process is shown in Fig. 3.

3.2.1 The Day Prior to Transfection (Day − 1): Split HeLa Cells

1. Wash, trypsinize, and count HeLa cells either manually with a hemocytometer or using an automated cell counter.
2. Plate 2 mL of HeLa cells (1×10^6 cells per mL DMEM supplemented with 10% FBS without antibiotics in each well of a 6-well plate. Allow the cells to grow overnight at 37 °C and 5% CO_2, or until cells reach ≥80% confluence (*see* **Notes 3–5**).

3.2.2 The Next Day (Day 0): Transfect HeLa Cells

1. Wash the cells once with 1 mL of 1× PBS and discard the PBS.
2. Wash once more with 1 mL of Opti-MEM and discard the medium.

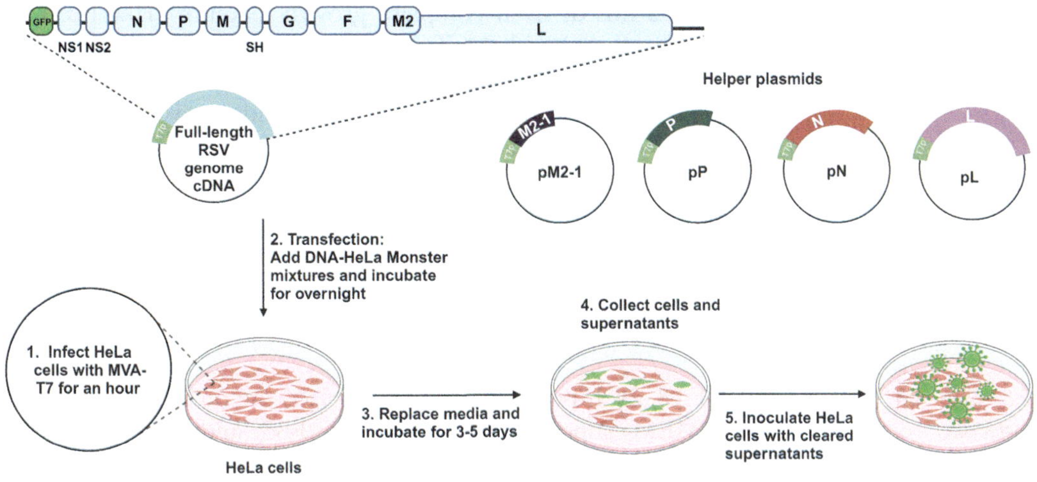

Fig. 3 Schematic diagram of the RSV rescue procedure

3. Inoculate cells with MVA-T7 at an MOI of 1 in 300–500 μL of Opti-MEM. Apply carefully dropwise to avoid disrupting the cell monolayer. Rock the plates gently to evenly distribute the inoculum.
4. Incubate the cells for 1 h at 37 °C in 5% CO_2.
5. In the meantime, prepare TransIT-HeLa DNA complexes. Warm the TransIT-HeLa reagent to room temperature and gently vortex before use.
6. Mix full-length RSV cDNA (0.6 μg) with helper plasmids that express the N, P, M2-1, and L protein at a ratio of 4:3:1:1 [N (0.2 μg); P (0.15 μg); M2-1 (0.05 μg); and L (0.05 μg)] in a 1.5-mL microcentrifuge tube (*see* **Notes 6–9**). After mixing all five plasmid DNAs, add 250 μL of Opti-MEM (*see* **Note 5**). Gently pipette to mix completely.
7. Add 7.5 μL TransIT-HeLa reagent to the diluted DNA mixture. Pipette gently to mix.
8. Incubate at room temperature for 15–30 min to allow complexes to form.
9. Warm MONSTER reagent to room temperature and vortex gently before using.
10. During TransIT-HeLa DNA complex formation, dilute MONSTER reagent by adding 5–45 μL sterile water and mix.
11. One hour after inoculating cells with MVA-T7, remove the inoculum and wash the cells once with 1 mL of PBS.
12. Wash once more with 1 mL of Opti-MEM.

13. Add 2 mL of fresh complete growth medium (10% FBS) to each well and add the TransIT-HeLa DNA complexes dropwise to different areas of the wells. Gently rock the culture vessel back-and-forth and from side-to-side to evenly distribute the TransIT-HeLa DNA complexes (*see* **Note 10**).
14. Add the prepared diluted MONSTER reagent dropwise directly to the cells. Gently rock the culture vessel back-and-forth and from side-to-side to evenly distribute the diluted MONSTER reagent.
15. Incubate overnight (16–24 h) at 33 °C with 5% CO_2.

3.2.3 Days Post-transfection

1. At 18–24 h post-transfection, replace the medium with fresh DMEM supplemented with 2% FBS and 1% penicillin/streptomycin.
2. Incubate for 3–5 days at 33 °C in 5% CO_2, check cells daily for fluorescence (if a fluorescence gene is expressed from the rescued RSV) or cytopathic effects, particularly syncytia formation.
3. Check cells daily for cytopathology. When about 50% of the cells have died, add fresh cells. If the cells keep dying, remove them with a cell scraper and collect them in a sterile tube. Vortex to detach virions from the cell surface and centrifuge the mixture at 3000 rpm for 5 min to remove cell debris. Collect the supernatant and discard the cell pellet.
4. Inoculate fresh cells with diluted and centrifuged supernatants (dilution factor depends on virus production), and keep the remainder of the virus stock frozen for future use. Snap freeze virus-containing vials on dry ice for 1 h before storing them at −80 °C.
5. Several passages of the virus may be required to reach high titers. Use of plates with increasing sizes from 6-well plates to a 100-mm dish in each successive passage is suggested.

4 Notes

1. *Primer length:* approximately 30 nt with some Gs and/or Cs at the 3′ end (25–35 nt) support good amplification with low false priming.
2. *Number of PCR fragments:* 4–5 fragments of ~4 kb are enough to copy the RSV genome, unless the genome has additional inserts. More fragments (at least up to 9) can be added.

3. *Cell density at transfection:* Cells should be approximately 80% confluent, because they will continue to multiply over the days needed for RSV to begin to replicate. Split the cells 18–24 h before transfection to ensure that they are actively dividing and attain the optimum cell density at the time of transfection. Cell viability may be increased by splitting cells at higher or lower densities, depending on the cell type.
4. *Cell contamination.* Mycoplasma contamination can alter cell morphology and reduce transfection efficiency. Check cells for mycoplasma contamination. Use a fresh frozen cell stock or treat with an appropriate antibiotic to eliminate mycoplasma before attempting to rescue RSV from cDNA.
5. *Serum and antibiotics inhibit transfection complex formation.* Prepare transfection complexes in serum-free growth medium, such as Opti-MEM Reduced Serum Medium. Once transfection complexes are formed, they can be directly added to cells. The following day, the medium can be changed to complete growth medium containing serum and 0.1–1× antibiotics where the presence of serum in growth medium in cell cultures improves transfection efficiency and reduces cytotoxicity.
6. *DNA purity.* Transfect DNA that is sterile, highly pure, and contaminant-free. EndoFree plasmid DNA preps with an A260/280 absorbance ratio of 1.8–2.0 are desirable. Miniprep kit-prepared DNA may contain high endotoxin levels and is not recommended. To ensure that DNA preparation is endotoxin-free, we recommend using an EndoFree Plasmid Maxi Kit, which increases transfection efficiency by 50%.
7. *Precipitate formation.* Excess DNA may cause precipitation during complex formation, thereby reducing transfection efficiency. When transfecting larger amounts of DNA, add twice the volume of serum-free medium during complex formation to reduce precipitation.
8. *DNA mutations.* Always sequence the final product after RT-PCR, PCR amplification, or any genetic manipulation of the genome. Verify that no unexpected mutations have been introduced, and gene inserts or gene replacements are in the correct region and orientation.
9. *Use a positive control to verify transfection efficiency.* Use the TransIT-HeLaMONSTER Transfection Kit to deliver a positive control plasmid encoding luciferase, beta-galactosidase, or green fluorescent protein (GFP).
10. Avoid rotating or swirling the transfected culture plate, as this might lead to an uneven distribution of transfection complexes across the cells and reduced efficiency.

Acknowledgments

The authors thank the College of Veterinary Medicine at The Ohio State University for providing bridge funds for developing yeast-based cDNA assembly system for large RNA viruses. The Ohio State University has filed a patent application for yeast-based cDNA assembly system (Tech. Ref. T2024-205), for which J. Li is the inventor.

References

1. Blount RE Jr, Morris JA, Savage RE (1956) Recovery of cytopathogenic agent from chimpanzees with coryza. Proc Soc Exp Biol Med 92(3):544–549
2. Collins PL et al (1995) Production of infectious human respiratory syncytial virus from cloned cDNA confirms an essential role for the transcription elongation factor from the 5′ proximal open reading frame of the M2 mRNA in gene expression and provides a capability for vaccine development. Proc Natl Acad Sci USA 92(25):11563–11567
3. Horowitz NH (1950) Biochemical genetics of Neurospora. Adv Genet 3:33–71
4. Fried M (1965) Cell-transforming ability of a temperature-sensitive mutant of polyoma virus. Proc Natl Acad Sci USA 53(3):486–491
5. Collins PL, Wertz GW (1983) cDNA cloning and transcriptional mapping of nine polyadenylylated RNAs encoded by the genome of human respiratory syncytial virus. Proc Natl Acad Sci USA 80(11):3208–3212
6. Collins PL et al (1986) Nucleotide sequences for the gene junctions of human respiratory syncytial virus reveal distinctive features of intergenic structure and gene order. Proc Natl Acad Sci USA 83(13):4594–4598
7. McFarland EJ et al (2020) Live-attenuated respiratory syncytial virus vaccine with M2-2 deletion and with small hydrophobic noncoding region is highly immunogenic in children. J Infect Dis 221(12):2050–2059
8. Karron RA et al (2015) A gene deletion that up-regulates viral gene expression yields an attenuated RSV vaccine with improved antibody responses in children. Sci Transl Med 7(312):312ra175
9. Luongo C et al (2013) Respiratory syncytial virus modified by deletions of the NS2 gene and amino acid S1313 of the L polymerase protein is a temperature-sensitive, live-attenuated vaccine candidate that is phenotypically stable at physiological temperature. J Virol 87(4):1985–1996
10. Whitehead SS et al (1999) Recombinant respiratory syncytial virus bearing a deletion of either the NS2 or SH gene is attenuated in chimpanzees. J Virol 73(4):3438–3442
11. Teng MN, Collins PL (1999) Altered growth characteristics of recombinant respiratory syncytial viruses which do not produce NS2 protein. J Virol 73(1):466–473
12. Teng MN et al (2000) Recombinant respiratory syncytial virus that does not express the NS1 or M2-2 protein is highly attenuated and immunogenic in chimpanzees. J Virol 74(19): 9317–9321
13. Harker J et al (2007) Virally delivered cytokines alter the immune response to future lung infections. J Virol 81(23):13105–13111
14. Fearns R, Collins PL (1999) Role of the M2-1 transcription antitermination protein of respiratory syncytial virus in sequential transcription. J Virol 73(7):5852–5864
15. Bukreyev A et al (2000) Effect of coexpression of interleukin-2 by recombinant respiratory syncytial virus on virus replication, immunogenicity, and production of other cytokines. J Virol 74(15):7151–7157
16. Alamares-Sapuay J et al (2024) Mutations in the F protein of the live-attenuated respiratory syncytial virus vaccine candidate DeltaNS2/Delta1313/I1314L increase the stability of infectivity and content of prefusion F protein. PLoS One 19(4):e0301773
17. Bukreyev A, Yang L, Collins PL (2012) The secreted G protein of human respiratory syncytial virus antagonizes antibody-mediated restriction of replication involving macrophages and complement. J Virol 86(19): 10880–10884
18. Hallak LK et al (2000) Iduronic acid-containing glycosaminoglycans on target cells are required for efficient respiratory syncytial virus infection. Virology 271(2):264–275
19. Guerrero-Plata A et al (2006) Differential response of dendritic cells to human

metapneumovirus and respiratory syncytial virus. Am J Respir Cell Mol Biol 34(3): 320–329
20. Rameix-Welti MA et al (2014) Visualizing the replication of respiratory syncytial virus in cells and in living mice. Nat Commun 5:5104
21. Cheng X et al (2002) Expression of beta-galactosidase by recombinant respiratory syncytial viruses for microneutralization assay. J Virol Methods 105(2):287–296
22. Hotard AL et al (2012) A stabilized respiratory syncytial virus reverse genetics system amenable to recombination-mediated mutagenesis. Virology 434(1):129–136
23. Fuentes S et al (2013) Development of a simple, rapid, sensitive, high-throughput luciferase reporter based microneutralization test for measurement of virus neutralizing antibodies following respiratory syncytial virus vaccination and infection. Vaccine 31(37):3987–3994
24. Kwilas AR et al (2010) Respiratory syncytial virus engineered to express the cystic fibrosis transmembrane conductance regulator corrects the bioelectric phenotype of human cystic fibrosis airway epithelium in vitro. J Virol 84(15):7770–7781
25. Malykhina O et al (2011) A respiratory syncytial virus replicon that is noncytotoxic and capable of long-term foreign gene expression. J Virol 85(10):4792–4801
26. Teng MN, Whitehead SS, Collins PL (2001) Contribution of the respiratory syncytial virus G glycoprotein and its secreted and membrane-bound forms to virus replication in vitro and in vivo. Virology 289(2):283–296
27. Techaarpornkul S, Barretto N, Peeples ME (2001) Functional analysis of recombinant respiratory syncytial virus deletion mutants lacking the small hydrophobic and/or attachment glycoprotein gene. J Virol 75(15): 6825–6834
28. Buchholz Ursula J, Finke S, Conzelmann K-K (1999) Generation of bovine respiratory syncytial virus (BRSV) from cDNA: BRSV NS2 is not essential for virus replication in tissue culture, and the human RSV leader region acts as a functional BRSV genome promoter. J Virol 73(1):251–259
29. Lo Mindy S, Brazas Robert M, Holtzman Michael J (2005) Respiratory syncytial virus nonstructural proteins NS1 and NS2 mediate inhibition of Stat2 expression and alpha/beta interferon responsiveness. J Virol 79(14): 9315–9319
30. Stobart CC et al (2017) BAC-based recovery of recombinant respiratory syncytial virus (RSV). Methods Mol Biol 1602:111–124
31. Fuentes S et al (2017) Development of bioluminescence imaging of respiratory syncytial virus (RSV) in virus-infected live mice and its use for evaluation of therapeutics and vaccines. Vaccine 35(4):694–702
32. Lu Y et al (2020) Porcine epidemic diarrhea virus deficient in RNA cap guanine-N-7 methylation is attenuated and induces higher type I and III interferon responses. J Virol 94(16): e00447-20
33. Joska TM et al (2014) A universal cloning method based on yeast homologous recombination that is simple, efficient, and versatile. J Microbiol Methods 100:46–51
34. Wang Y et al (2018) Enhancement of safety and immunogenicity of the Chinese Hu191 measles virus vaccine by alteration of the S-adenosylmethionine (SAM) binding site in the large polymerase protein. Virology 518: 210–220

Chapter 5

Respiratory Syncytial Virus Fusion Activity Syncytia Assay

Kearstin Edmonds and Rebecca Dutch

Abstract

The methods detailed here measure the membrane fusion activity of viral fusion proteins and are optimized for the respiratory syncytial virus (RSV). RSV utilizes a class I fusion protein (F) to facilitate fusion between a viral and cellular membrane to allow viral entry into host cells. This process is carried out with the assistance of the glycoprotein (G), which attaches to the host-cell surface, allowing F to interact with the target cell. The F protein then undergoes conformational changes, which drive membrane fusion, thus creating a path for viral entry. The fusogenic activity of viral fusion proteins can be quantified by observing and measuring syncytia formation via wide-field microscopy. Syncytia are multinucleated cell masses formed by cell–cell membrane fusion, which in this case are driven by viral fusion proteins expressed on the host-cell surface.

Key words Respiratory syncytial virus (RSV), Fusion protein, Fusogenic activity, Syncytia, Wide-field microscopy

1 Introduction

1.1 Respiratory Syncytial Virus Fusion and Syncytia

Fusion between the viral membrane and host-cell membrane is a crucial step in the infection cycle of enveloped viruses, including respiratory syncytial virus (RSV), which is a member of the *Pneumoviridae* family. This family of viruses utilizes class I fusion proteins to access and infect target host cells [1, 2]. The fusion (F) protein contains regions that are structurally conserved across the *Pneumoviridae* and closely related *Paramyxoviridae* families [3, 4]. The RSV F protein is initially synthesized as an immature precursor known as F_0 [2] before undergoing proteolytic cleavage at two sites by furin-like proteases, resulting in disulfide-linked F_1 and F_2 subunits [2, 5]. In addition, three F monomers fold into a trimeric form, which is the active form of F [4, 5]. Mature F is then trafficked to the surface of the host cell and incorporated into new viral particles during the budding process (Fig. 1A) [6]. In order to undergo fusion, the F protein must be in proximity to a new host cell. Here, the glycoprotein (G) aids in attachment to the target cell

Bo Liang and Rachel Fearns (eds.), *Human Respiratory Syncytial Virus: Methods and Protocols*, Methods in Molecular Biology, vol. 2948, https://doi.org/10.1007/978-1-0716-4666-3_5,

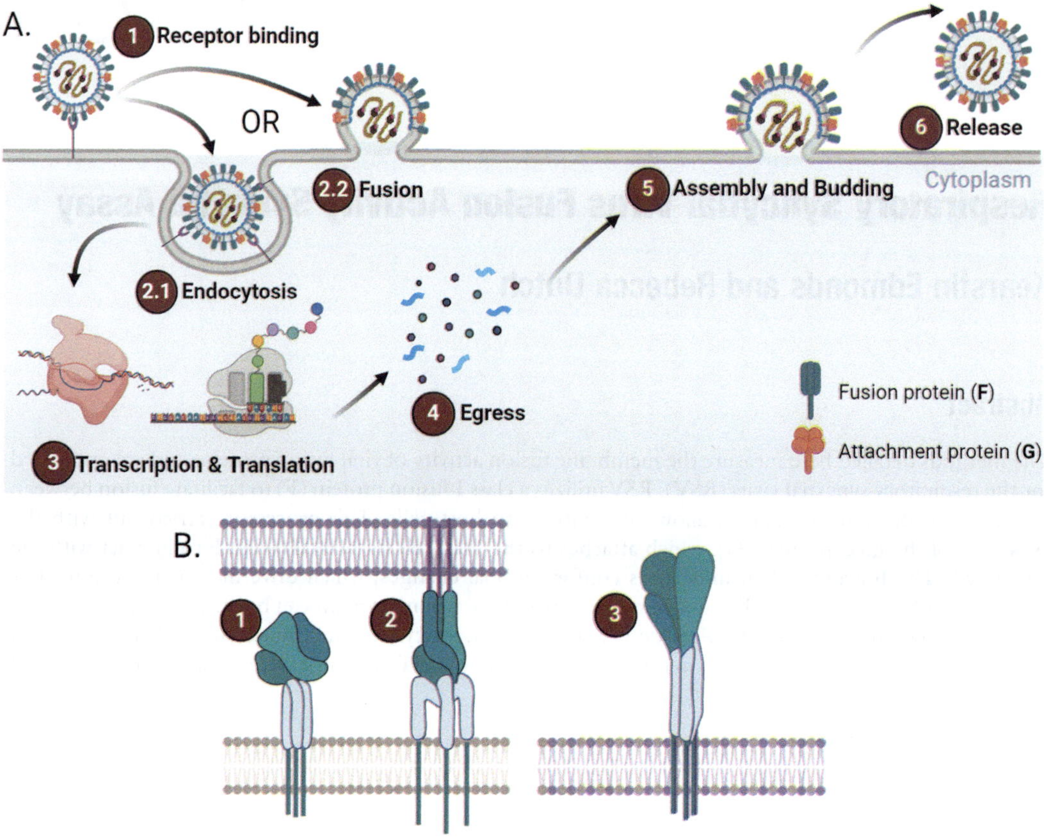

Fig. 1 RSV infection cycle. (**A**) The respiratory syncytial virus (RSV) viral glyco- and fusion proteins will bind (1) to target host-cell receptors to facilitate entry. RSV is thought to enter cells via endocytosis and escape (2. A) or direct fusion with the host membrane (2.B). The viral genome will undergo replication, transcription, and translation (3) before egress (4) toward the cell membrane, where viral particle assembly will occur (5). Lastly, the viral particle will fully release (6) from the host cell to restart the process with a new target cell. (**B**) RSV F undergoes conformational changes to drive membrane fusion. Nascent F undergoes proteolytic processing, resulting in a trimer of disulfide-bonded F1 and F2 subunits embedded in the virus or host-cell membranes in a pre-fusion F form (1). Initiation of F conformational changes results in a pre-hairpin intermediate, which inserts a portion of F into the target membrane (2), and further conformational changes result in the final post-fusion conformation and membrane merger (3). (Made with Biorender.com)

by binding to host-cell surface components such as glycosaminoglycans (GAGs) or CX3CR1 [2, 7–9]. RSV G attachment is not required for F to facilitate fusion in mammalian cell culture, but the presence of RSV G aids in the fusion efficiency in cell culture and is essential in vivo [10–13]. F interaction with target cells via IGF1R, EGFR, or nucleolin may also provide the needed interaction with the host cell [14–18]. F-promoted fusion may occur on the cell surface or in the endosome, depending on the cell type and strain [5, 12, 19–21]. Upon activation, the pre-fusion F trimer undergoes a series of conformational changes that lead to the post-fusion F trimer conformation, whose energetically stable structure is hypothesized to drive fusion (Fig. 1B) [2]. Due to the important role of the F protein in host infection and the fact that F is a key immunological target, it is often a target of vaccine development [13, 22, 23]. However, some aspects of RSV F-mediated membrane fusion and cell entry remain to be elucidated [5]. With this, it is important to have standard methods to assay the fusogenic activity of RSV F.

A commonly used method to investigate RSV F fusion activity is the observation and quantitation of syncytium formation [24–27]. Syncytia are formed through fusion events between multiple cells as viral infection spreads, in RSV and some other viruses whose fusion proteins are expressed on the cellular surface after infection (Fig. 2) [26]. Syncytium formation can also be observed when fusion proteins such as RSV F are transiently expressed in cells via transfection, allowing for analysis of F fusogenic activity outside the context of infection. Very low levels of multinucleated cells are observed in nontransfected or noninfected cells. Therefore, comparisons to negative controls such as empty vector or uninfected cells must always be conducted [24–27]. Through the use of wide-field microscopy on transfected or infected cells, it is possible to quantify the level of syncytia formation occurring in different samples for comparison [27]. Thus, assessment of syncytia formation can be used to characterize RSV F protein fusogenic activity and examine the importance of other viral or cellular proteins in the process. The assay is relatively straightforward and works well for

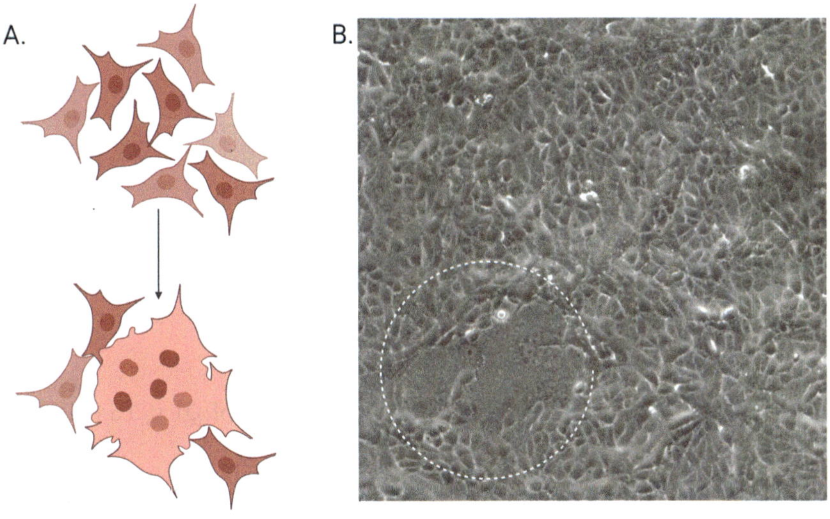

Fig. 2 Syncytia formation. (**A**) During the course of infection, several cells may undergo RSV F-promoted membrane fusion events, resulting in multinucleated giant cells. These cells are referred to as a syncytium. (**B**) Image collected from wild-type RSV fusion protein transfection, 48 h post-transfection, on a 20× objective with Axiovert 200 M. Circled in white dots is a syncytia containing multiple nuclei from fused cells. (Figure made with Biorender.com)

multiple viral strains of RSV. The methods required to perform a syncytia assay using cells transiently transfected to express RSV F are outlined here.

1.2 Alternative Approach

The syncytia assay described here is the most straightforward method for measuring RSV F protein-mediated fusion, and it works consistently for F proteins from multiple strains. Alternative methods for measuring cell–cell fusion or virus–cell fusion have been developed for many virus fusion protein systems. Several reports have utilized a quantitative luciferase reporter gene method to study RSV F fusion [28, 29]. This assay works by expressing the RSV F protein in a T7-polymerase producing stable effector cell type. Subsequently, target cells that express the receptor for the viral fusion protein and are transfected with a T7-dependent luciferase reporter plasmid are overlaid onto the F-expressing cells. When the F-expressing cells undergo fusion with the target cells, the T7-dependent luciferase reporter plasmid gains access to the T7-polymerase, and the luciferase reporter is produced. Firefly luciferase production correlates to the level of cell–cell fusion. When an injecting luminometer is used to supply luciferin to each sample, an enzymatic reaction will occur, which results in measurable relative light units (RLUs). This method has been validated for use in other related viruses, like human metapneumovirus and Hendra virus [30–34]. However, while this technique has been utilized in a few publications with RSV F [29], challenges with the use of this assay for RSV F from different strains and with the consistency of the results have been observed in our group and others, suggesting that initial analysis with syncytia formation may be more straightforward.

2 Materials

1. Vero E6 cells (ATCC, CCL-81).
2. DMEM: 4.5 g/L glucose, L-glutamine, sodium pyruvate, supplemented (Corning, 10-013-CV) with 10% heat-inactivated FBS (Corning, 35-010-CV).
3. PBS 1×.
4. Trypsin: 0.5% EDTA (Life Technologies, 15400-054).
5. Sterile clear 6-well plates.
6. Bright-field microscope with camera.
7. Lipo3000 (Life Technologies, L3000001).
8. OptiMem (Life Technologies, 31985-088).
9. Plasmid with RSV F gene appropriate for transient transfection.

10. Empty vector plasmid.
11. Image analysis software, ImageJ is detailed here.

3 Methods

3.1 Prepare Cells for Transfection

1. Have confluent Vero E6 cells in T75 flasks and DMEM media (see items 1–11).
2. Determine the number of 6-well plates needed (*see* **Note 1**).
3. Wash with PBS 8 mL and aspirate.
4. Lift with Trypsin: 5 min at 37 °C.
5. Add 8 mL of DMEM to neutralize.
6. Seed Vero E6 cells into 6-well plates to be 80–85% confluent the following day (*see* **Note 1**).
7. Place the cells in a sterile incubator at 37 °C and 5% CO_2 for 24 h.

3.2 Transfection to Facilitate Expression of the Fusion Protein

1. In a microtube, dilute 4 μL/well of Lipo3000 into 100 μL/well OptiMem, to create the Lipo3000 reagent mix.
2. Next, dilute 2 μg of RSV F plasmid DNA (*see* **Note 2**) into 100 μL OptiMem per sample, followed by 4 μL of the P3000 additive. Repeat this with the empty vector negative control and any other controls or samples needed.
3. Add 100 μL of the Lipo3000 reagent mix to each DNA mixture tube and thoroughly combine via pipetting.
4. Allow the mixtures to combine for 15 min before adding to the wells in a dropwise fashion.
5. Place the cells for 24 h in a sterile incubator at 37 °C and 5% CO_2 (*see* **Note 3**).

3.3 Preparation of Samples and Imaging

1. Wash cells twice with PBS at 24 hpt (hours post-transfection). This step minimizes the amount of cell debris that can interfere with image collection.
2. Replenish media.
3. Carefully carry to bright-field microscope with camera, one plate at a time. Plates not being imaged are left in and returned to the incubator.
4. Image plates (*see* **Note 4**).
5. After imaging, place the cells back in 37 °C and 5% CO_2, for an additional 24 h.
6. At 48hpt (*see* **Note 5**), wash the cells again, then repeat the imaging process.

3.4 Analysis Method I

An image analysis software such as ImageJ, or a similar program, can be utilized to quantify the images. The step-by-step method supplied here is as performed.

1. Open 10× magnification photos in ImageJ (1.54d).
2. Draw an ROI (region of interest) around a syncytia area (*see* **Note 6**). Make sure to fully encompass each nucleus and cell border included in the syncytia region without extending past them.
3. Right click the area and add it to the ROI manager.
4. Repeat this for all syncytia areas in the image.
5. Open the ROI manager (analyze>tools) and select each corresponding syncytium number (automatically assigned), followed by the measure button. This will result in area measurements that can be used to quantify total syncytia area in each image (Fig. 3).
6. Save the file as a new TIFF, to avoid altering the original image, with an indicator that it has been edited, like "_analysis". This will allow access to the image later to see the exact locations of each ROI drawn, as well as the ability to recalculate values (*see* **Note 7**).

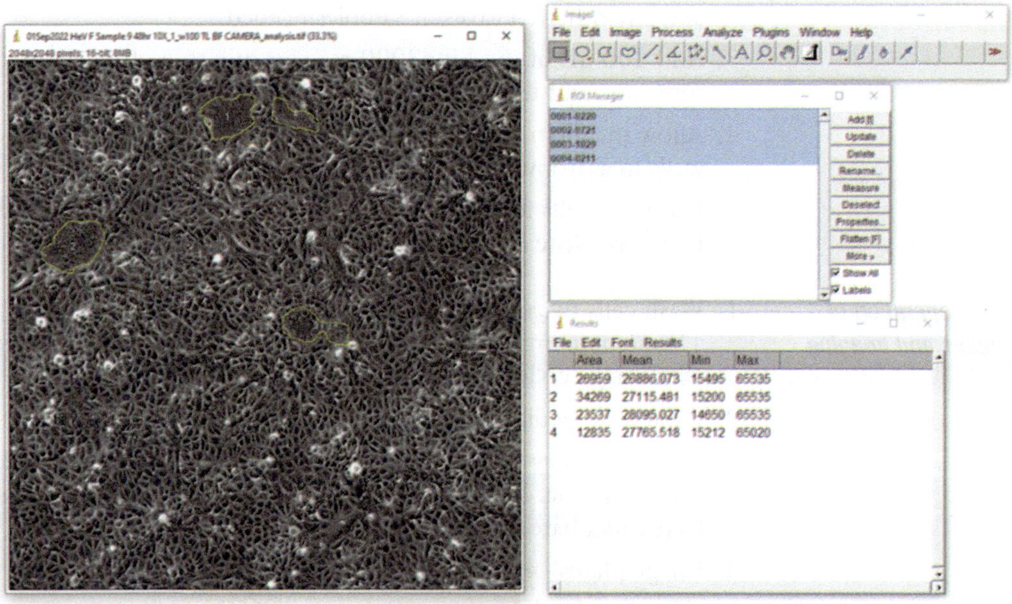

Fig. 3 ImageJ analysis. Image collected at 10×, on an Axiovert 200 M, 48 h post-transfection with an RSV F mutant in Vero cells. Regions of interest (ROIs) were drawn around each area of syncytia (yellow) in the image. Once ROIs are added to the ROI manager tool, they can be highlighted and measured. Resulting measurements will be added to the results box and include: area, mean, minimum, and maximum. These measurements can be further analyzed to yield comparative quantitative data

7. The data obtained through ImageJ data can then be analyzed to calculate total syncytial volume.
8. Average the area of the replicate images from each sample, i.e., five images collected on the wild-type (WT) sample are averaged together.
9. Subtract the empty vector negative control average from the rest of the sample averages.
10. Normalize data to the WT sample by dividing each sample's value by the WT sample value. The resulting data represents syncytia formation area compared to WT (*see* **Note 8**).
11. Once the experiment is completed in triplicate, a standard ANOVA can be used to analyze statistical significance.
12. Represent data with graph of choice, bar graph is common.

3.5 Analysis Method II

1. Using the 20× magnification images (*see* **Note 9**).
2. If necessary, increase the contrast of images to better differentiate nuclei from cell mass (*see* **Note 10**).
3. Nuclei are counted for the entire image and then counted only in syncytia formations. These values can then be further analyzed with the following formula: 1 − (total nuclei − nuclei in syncytia + number of syncytia) / total nuclei. This analysis results in a fusion index.
4. Statistically analyze using simple techniques like ANOVA (*see* **Note 11**).
5. Represent data with graph of choice, bar graph is common.

4 Notes

1. The number of plates used will be determined by the number of variable RSV F samples and controls. Include sufficient wells for technical duplicates and negative controls. It is also recommended to include a positive control, which could be an RSV F or other viral fusion protein previously shown to promote syncytia formation. Other cell types can be used if desired, but it is important to ensure the cell type chosen is validated for RSV fusion activity.
2. The proper amount of the expression plasmid to use can depend on the cell type and expression plasmid. If a new expression construct is used, a titration of the amount should be performed. A plasmid expressing the RSV G protein can also be added into the transfection with the F protein expression plasmid if desired. The total DNA concentration should remain at 2 μg per well by reducing the amount of F plasmid added.

Within an individual experiment, the total amount of F plasmid must be kept constant between wells.

3. The syncytia assay can also be performed with infected cells, allowing comparison of wild-type and recombinant viruses with mutations in the F protein. In this case, either appropriate BSL imaging facilities should be used or plates should be fixed prior to analysis.
4. If possible, it is best to have predetermined set locations for image collection to avoid bias. This can be accomplished with a multipoint image acquisition setting, built into many types of acquisition software, or by predetermining XY coordinates for each image. If image acquisition through automated multipoint, or manual XY coordinates, is not possible, an alternative approach would be imaging with the same pattern on each well. For example, starting on the left healthy cell boundaries and moving three frames to the right between each image collection. This approach is less controlled for bias but is still preferred to well-to-well individual selection of images to collect. When choosing the locations for image collection, an empty plate, or the first negative control sample, can be used to estimate boundaries for each well. In 6-well plates, the cells near the edges of the well are often more dispersed and less representative of the sample. Due to this fact, it is important to collect images within a healthy cell border, which can be determined by morphology and general dispersion. The majority of the well will be inside this healthy cell border. It is recommended to take a minimum of five images per well on both a 10× objective and a 20× objective, per image location. This provides two available methods of analysis, which are outlined in this protocol. When adjusting the microscope for image collection, aim for strong contrast between nuclei and cell mass. The contrast can also be appropriately enhanced post-acquisition in either the acquisition software or photoshop, so long as the same adjustments are made to all images identically.
5. Timepoints other than 24 h or 48 h can be used, as desired.
6. When using the area analysis method for syncytia, be sure to be consistent with the parameters established for drawing ROIs.
7. To access previously drawn ROIs in ImageJ, pull up the analysis TIFF saved post-edit, and choose "image" at the top tool bar, followed by "overlay" and "to ROI manager."
8. Using analysis method I, the syncytia area can also be divided by the total image area to yield a percent of total cells that underwent fusion.
9. Analysis method II is more time consuming, but still robust. The 20× images can be used in both analysis methods I and II,

but the 10× images are preferable in method I only because it is difficult to count nuclei in a wider field of view.

10. This can be done in Adobe Photoshop, or another image editing software with value controls, but it is important that all images undergo the same measurable level of contrast change. This is to prevent misrepresentation of images compared to one another.
11. There are multiple methods of appropriate statistical analysis not listed here, ANOVA analysis is one option.

References

1. Schibli DJ, Weissenhorn W (2004) Class I and class II viral fusion protein structures reveal similar principles in membrane fusion. Mol Membr Biol 21:361–371. https://doi.org/10.1080/09687860400017784
2. Mclellan JS, Ray WC, Peeples ME (2013) Structure and function of RSV surface glycoproteins. Curr Top Microbiol Immunol 372:83–104. https://doi.org/10.1007/978-3-642-38919-1_4
3. Lambert DM, Barney S, Lambert AL, Guthrie K, Medinas R, Davis DE, Bucy T, Erickson J, Merutka G, Petreway SR (1996) Peptides from conserved regions of paramyxovirus fusion (F) proteins are potent inhibitors of viral fusion. Med Sci 93:2186–2191
4. McLellan JS, Yang Y, Graham BS, Kwong PD (2011) Structure of respiratory syncytial virus fusion glycoprotein in the postfusion conformation reveals preservation of neutralizing epitopes. J Virol 85:7788–7796. https://doi.org/10.1128/jvi.00555-11
5. Çadena-Cruz C, Villarreal Camacho JL, De Ávila-Arias M, Hurtado-Gomez L, Rodriguez A, San-Juan-Vergara H (2023) Respiratory syncytial virus entry mechanism in host cells: a general overview. Mol Microbiol 120:341–350. https://doi.org/10.1111/mmi.15133
6. Gilman MSA, Furmanova-Hollenstein P, Pascual GB, van't Wout A, Langedijk JPM, McLellan JS (2019) Transient opening of trimeric prefusion RSV F proteins. Nat Commun 10:1–13. https://doi.org/10.1038/s41467-019-09807-5
7. Hallak LK, Spillmann D, Collins PL, Peeples ME (2000) Glycosaminoglycan sulfation requirements for respiratory syncytial virus infection. J Virol 74:10508–10513
8. Green G, Johnson SM, Costello H, Brakel K, Harder O, Oomens AG, Peeples ME, Moulton HM, Niewiesk S (2021) CX3CR1 is a receptor for human respiratory syncytial virus in cotton rats. J Virol 95:1–12. https://doi.org/10.1128/jvi.00010-21
9. Anderson CS, Chu CY, Wang Q, Mereness JA, Ren Y, Donlon K, Bhattacharya S, Misra RS, Walsh EE, Pryhuber GS, Mariani TJ (2020) CX3CR1 as a respiratory syncytial virus receptor in pediatric human lung. Pediatr Res 87:862–867. https://doi.org/10.1038/s41390-019-0677-0
10. Chirkova T, Lin S, Oomens AGP, Gaston KA, Boyoglu-Barnum S, Meng J, Stobart CC, Cotton CU, Hartert TV, Moore ML, Ziady AG, Anderson LJ (2015) CX3CR1 is an important surface molecule for respiratory syncytial virus infection in human airway epithelial cells. J Gen Virol 96:2543–2556. https://doi.org/10.1099/vir.0.000218
11. Heminway BR, Yu Y, Tanaka Y, Perrine KG, Gustafson E, Bernstein JM, Galinski MS (1994) Analysis of respiratory syncytial virus F, G, and SH proteins in cell fusion. Virology 200:801–805. https://doi.org/10.1006/viro.1994.1245
12. Cox RG, Williams JV (2013) Breaking in: human metapneumovirus fusion and entry. Viruses 5:192–210. https://doi.org/10.3390/v5010192
13. Biacchesi S, Pham QN, Skiadopoulos MH, Murphy BR, Collins PL, Buchholz UJ (2005) Infection of nonhuman primates with recombinant human metapneumovirus lacking the SH, G, or M2-2 protein categorizes each as a nonessential accessory protein and identifies vaccine candidates. J Virol 79:12608–12613. https://doi.org/10.1128/jvi.79.19.12608-12613.2005
14. Griffiths CD, Bilawchuk LM, McDonough JE, Jamieson KC, Elawar F, Cen Y, Duan W, Lin C, Song H, Casanova JL, Ogg S, Jensen LD, Thienpont B, Kumar A, Hobman TC, Proud D, Moraes TJ, Marchant DJ (2020)

IGF1R is an entry receptor for respiratory syncytial virus. Nature 583:615–619. https://doi.org/10.1038/s41586-020-2369-7

15. Mastrangelo P, Chin AA, Tan S, Jeon AH, Ackerley CA, Siu KK, Lee JE, Hegele RG (2021) Identification of rsv fusion protein interaction domains on the virus receptor, nucleolin. Viruses 13:192–210. https://doi.org/10.3390/v13020261
16. Tayyari F, Marchant D, Moraes TJ, Duan W, Mastrangelo P, Hegele RG (2011) Identification of nucleolin as a cellular receptor for human respiratory syncytial virus. Nat Med 17:1132–1135. https://doi.org/10.1038/nm.2444
17. Hayes RS, Oraby AK, Camargo C, Marchant DJ, Sagan SM (2024) Mapping respiratory syncytial virus fusion protein interactions with the receptor IGF1R and the impact of alanine-scanning mutagenesis on viral infection. J Gen Virol 105:1–11. https://doi.org/10.1099/jgv.0.001951
18. Currier MG, Lee S, Stobart CC, Hotard AL, Villenave R, Meng J, Pretto CD, Shields MD, Nguyen MT, Todd SO, Chi MH, Hammonds J, Krumm SA, Spearman P, Plemper RK, Sakamoto K, Peebles RS, Power UF, Moore ML (2016) EGFR interacts with the fusion protein of respiratory syncytial virus strain 2–20 and mediates infection and mucin expression. PLoS Pathog 12:1–22. https://doi.org/10.1371/journal.ppat.1005622
19. Krzyzaniak MA, Zumstein MT, Gerez JA, Picotti P, Helenius A (2013) Host cell entry of respiratory syncytial virus involves macropinocytosis followed by proteolytic activation of the F protein. PLoS Pathog 9:1–19. https://doi.org/10.1371/journal.ppat.1003309
20. Kolokoltsov AA, Deniger D, Fleming EH, Roberts NJ, Karpilow JM, Davey RA (2007) Small interfering RNA profiling reveals key role of clathrin-mediated endocytosis and early endosome formation for infection by respiratory syncytial virus. J Virol 81:7786–7800. https://doi.org/10.1128/jvi.02780-06
21. McLaughlin JM, Khan F, Schmitt HJ, Agosti Y, Jodar L, Simoes EAF, Swerdlow DL (2022) Respiratory syncytial virus-associated hospitalization rates among US infants: a systematic review and meta-analysis. J Infect Dis 225:1100–1111. https://doi.org/10.1093/infdis/jiaa752
22. Lee SH, Chu KB, Kim MJ, Mao J, Eom GD, Yoon KW, Ahmed MA, Quan FS (2023) Virus-like particle vaccine expressing the respiratory syncytial virus pre-fusion and G proteins confers protection against RSV challenge infection. Pharmaceutics 15:1–16. https://doi.org/10.3390/pharmaceutics15030782
23. Krarup A, Truan D, Furmanova-Hollenstein P, Bogaert L, Bouchier P, Bisschop IJM, Widjojoatmodjo MN, Zahn R, Schuitemaker H, McLellan JS, Langedijk JPM (2015) A highly stable prefusion RSV F vaccine derived from structural analysis of the fusion mechanism. Nat Commun 6:1–12. https://doi.org/10.1038/ncomms9143
24. Gower TL, Pastey MK, Peeples ME, Collins PL, McCurdy LH, Hart TK, Guth A, Johnson TR, Graham BS (2005) RhoA signaling is required for respiratory syncytial virus-induced syncytium formation and filamentous virion morphology. J Virol 79:5326–5336. https://doi.org/10.1128/jvi.79.9.5326-5336.2005
25. Tian J, Huang K, Krishnan S, Svabek C, Rowe DC, Brewah Y, Sanjuan M, Patera AC, Kolbeck R, Herbst R, Sims GP (2013) RAGE inhibits human respiratory syncytial virus syncytium formation by interfering with F-protein function. J Gen Virol 94:1691–1700. https://doi.org/10.1099/vir.0.049254-0
26. Jessie B, Dobrovolny HM (2021) The role of syncytia during viral infections. J Theor Biol 525:1–9. https://doi.org/10.1016/j.jtbi.2021.110749
27. Bitko V, Oldenburg A, Garmon NE, Barik S (2003) Profilin is required for viral morphogenesis, syncytium formation, and cell-specific stress fiber induction by respiratory syncytial virus. BMC Microbiol 3:1–12
28. Hotard AL, Lee S, Currier MG, Crowe JE, Sakamoto K, Newcomb DC, Peebles RS, Plemper RK, Moore ML (2015) Identification of residues in the human respiratory syncytial virus fusion protein that modulate fusion activity and pathogenesis. J Virol 89:512–522. https://doi.org/10.1128/jvi.02472-14
29. Battles MB, Langedijk JP, Furmanova-Hollenstein P, Chaiwatpongsakorn S, Costello HM, Kwanten L, Vranckx L, Vink P, Jaensch S, Jonckers THM, Koul A, Arnoult E, Peeples ME, Roymans D, McLellan JS (2016) Molecular mechanism of respiratory syncytial virus fusion inhibitors. Nat Chem Biol 12:87–93. https://doi.org/10.1038/nchembio.1982
30. Russell CJ, Kantor KL, Jardetzky TS, Lamb RA (2003) A dual-functional paramyxovirus F protein regulatory switch segment: activation and membrane fusion. J Cell Biol 163:363–374. https://doi.org/10.1083/jcb.200305130
31. Porotto M, Doctor L, Carta P, Fornabaio M, Greengard O, Kellogg GE, Moscona A (2006) Inhibition of hendra virus fusion. J Virol 80:9837–9849. https://doi.org/10.1128/jvi.00736-06
32. Barrett CT, Webb SR, Dutch RE (2019) A hydrophobic target: using the paramyxovirus fusion protein transmembrane domain to

modulate fusion protein stability. J Virol:1–15. https://doi.org/10.1128/JVI.00863-19

33. Chang A, Masante C, Buchholz UJ, Dutch RE (2012) Human metapneumovirus (HMPV) binding and infection are mediated by interactions between the HMPV fusion protein and heparan sulfate. J Virol 86:3230–3243. https://doi.org/10.1128/jvi.06706-11

34. Masante C, El Najjar F, Chang A, Jones A, Moncman CL, Dutch RE (2014) The human metapneumovirus small hydrophobic protein has properties consistent with those of a viroporin and can modulate viral fusogenic activity. J Virol 88:6423–6433. https://doi.org/10.1128/jvi.02848-13

Chapter 6

Visualization of Respiratory Syncytial Virus RNA Synthesis Sites by Ethynyl Uridine Labeling

Risso-Ballester J., Rincheval V., and Marie-Anne Rameix-Welti

Abstract

Modified cell-permeable nucleosides such as 5-bromouridine 5′-triphosphate (BrUTP) or 5-ethynyluridine (5-EU), together with the powerful click chemistry, have been widely used to specifically label newly synthetized RNA in cells and tissues in a simple two-step approach. The use of cellular transcription inhibitors has established metabolic RNA labeling as an optimal approach for the precise visualization of viral RNA transcripts and the microscopic analysis of their distribution in infected cells. The labeling of nascent viral RNA of respiratory syncytial virus (RSV) and other *Mononegavirales (MNV)* (e.g., rabies virus (RABV), Ebola virus (EBOV), or human metapneumovirus (HMPV)) demonstrated the presence of newly synthesized viral RNAs in virally induced cytoplasmic inclusions called inclusion bodies (IBs). This was the crucial result to confirm that viral RNA synthesis occurs in these IBs, which could thus be renamed viral factories. Here, we describe a method for the unbiased detection of newly synthetized viral RNA in RSV-infected cells via metabolic labeling with 5-EU and subsequent detection using highly specific alkyne-azide "click" chemistry.

Key words 5-Ethynyluridine (EU), Respiratory syncytial virus (RSV), Inclusion bodies (IBs), RNA synthesis, Nascent viral RNA and click chemistry

1 Introduction

Metabolic labeling methods have long been used to detect and track the fate of newly synthesized molecules (DNA, RNA, proteins, etc.). These methods were mostly based on radioactive metabolites, which raised safety issues and resulted in cumbersome and expensive manipulations. Fluorescent labeling of metabolites using click chemistry offers a powerful alternative method to label newly synthesized molecules to study their localization and dynamics. Here, we describe the fluorescent metabolic labeling of newly synthesized RSV RNAs. RSV is a single-stranded RNA virus, which replicates in the cytoplasm of infected cells and induces the formation of spherical cytoplasmic granules commonly referred to as inclusion bodies (IBs). IBs were identified as a hallmark of

Bo Liang and Rachel Fearns (eds.), *Human Respiratory Syncytial Virus: Methods and Protocols*, Methods in Molecular Biology, vol. 2948, https://doi.org/10.1007/978-1-0716-4666-3_6,

infection for other RNA viruses in the *MNV* order, including vesicular stomatitis virus (VSV) or parainfluenza virus (PIV). IBs concentrate all viral components and cellular proteins required for viral genome replication and transcription and were thought to host viral RNA synthesis steps. Metabolic RNA staining experiments revealed the critical role of IBs in viral RNA synthesis, leading to their designation as viral factories [1–6].

In these assays, cells were incubated with uridine analogs during a specific period to label newly synthetized viral RNA. Radioactive 5-bromouridine 5′-triphosphate (BrUTP) was first successfully used to demonstrate viral RNA synthesis in IBs [1, 4, 6]. However, BrUTP—based methods are expensive, laborious, and incompatible with some cell lines. Here we used another uridine analog, the 5-ethynyluridine (5-EU) which contains an alkyne group. This potent, cell-permeable nucleoside is rapidly taken up by living cells, converted to nucleotide phosphates, and then incorporated into RNA transcripts generated by RNA polymerases I, II, and III [7]. Interestingly, Salic and Jao previously showed that EU does not incorporate into DNA [7], which strongly demonstrates its specificity as a transcriptional label. Although here we focus on the metabolic staining of viral RNA in fixed cells, 5-EU nascent-RNA labeling is also compatible with live-cell imaging.

To specifically stain newly synthetized viral RNA, cellular transcription is inhibited during the 5-EU incorporation step. Actinomycin D, a potent inhibitor of the host DNA-dependent RNA polymerases, has been widely used to prevent cellular transcription and in consequence, to decrease undesired background signal. The general flow chart of the protocol is shown in Fig. 1.

After EU treatment, infected cells are fixed, permeabilized and subjected to click chemistry by using a fluorescent azide. The click chemistry approach is based on the chemoselective ligation between an azide and an alkyne where the modified RNA is detected with a corresponding azide-coupled dye [7] (Fig. 2). The azide-labeled RNA transcripts are finally visualized by fluorescence microscopy.

Metabolic RNA labeling experiments showed for the first time that IBs are the main sites of viral RNA synthesis. 5-EU signal corresponding to newly synthetized viral RNAs was only visible in IBs in cells infected with RSV (Figs. 3 and 4), while in the presence of AZ27 [8, 9], a specific inhibitor of RSV RNA polymerase, IBs were negative for 5-EU. Significantly, efficient labeling of IBs was observed when reducing the labeling period from one hour up to 15 min (Fig. 3), which is short compared to the average time expected for RSV mRNA synthesis. Indeed, based on the slow elongation rate reported for measles polymerase (~ 3 nucleotides/sec), transcription of N mRNA for example would take about 8 min [10].

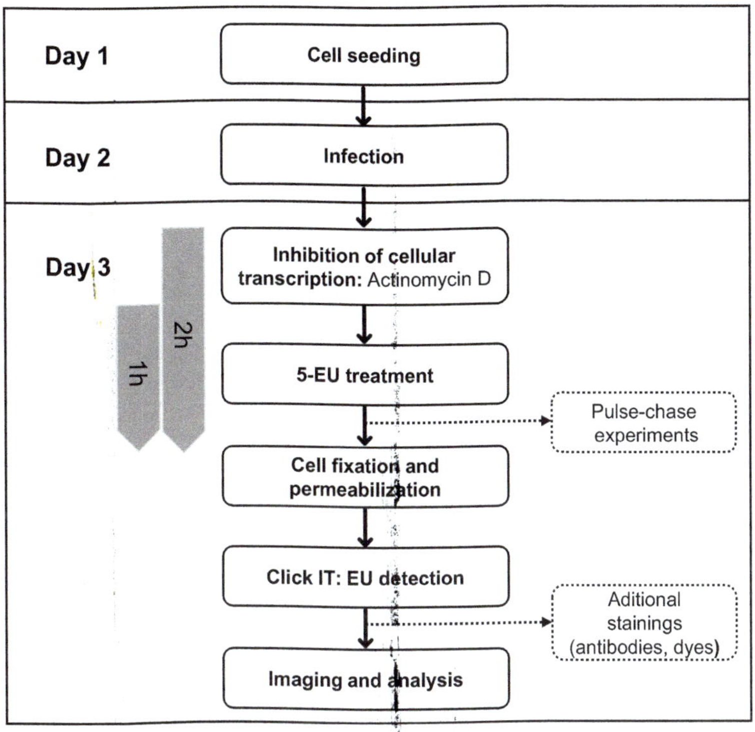

Fig. 1 Workflow diagram for the detection of newly synthetized viral RNA in RSV-infected cells via metabolic labelling with 5-EU and subsequent click chemistry detection

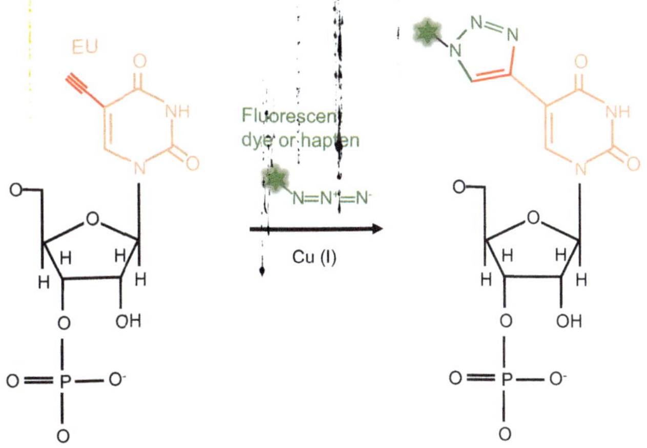

Fig. 2 Click reaction between EU and azide-modified dye or hapten. During the first step of the click reaction, an alkyne(red)-containing nucleoside, 5-EU (orange), is incorporated into RNA during active RNA synthesis. Subsequently, this alkyne will react with an azide linked to a fluorescent dye, biotin, or other detection reagent (green) to form a stable triazole ring in the presence of Cu (I)

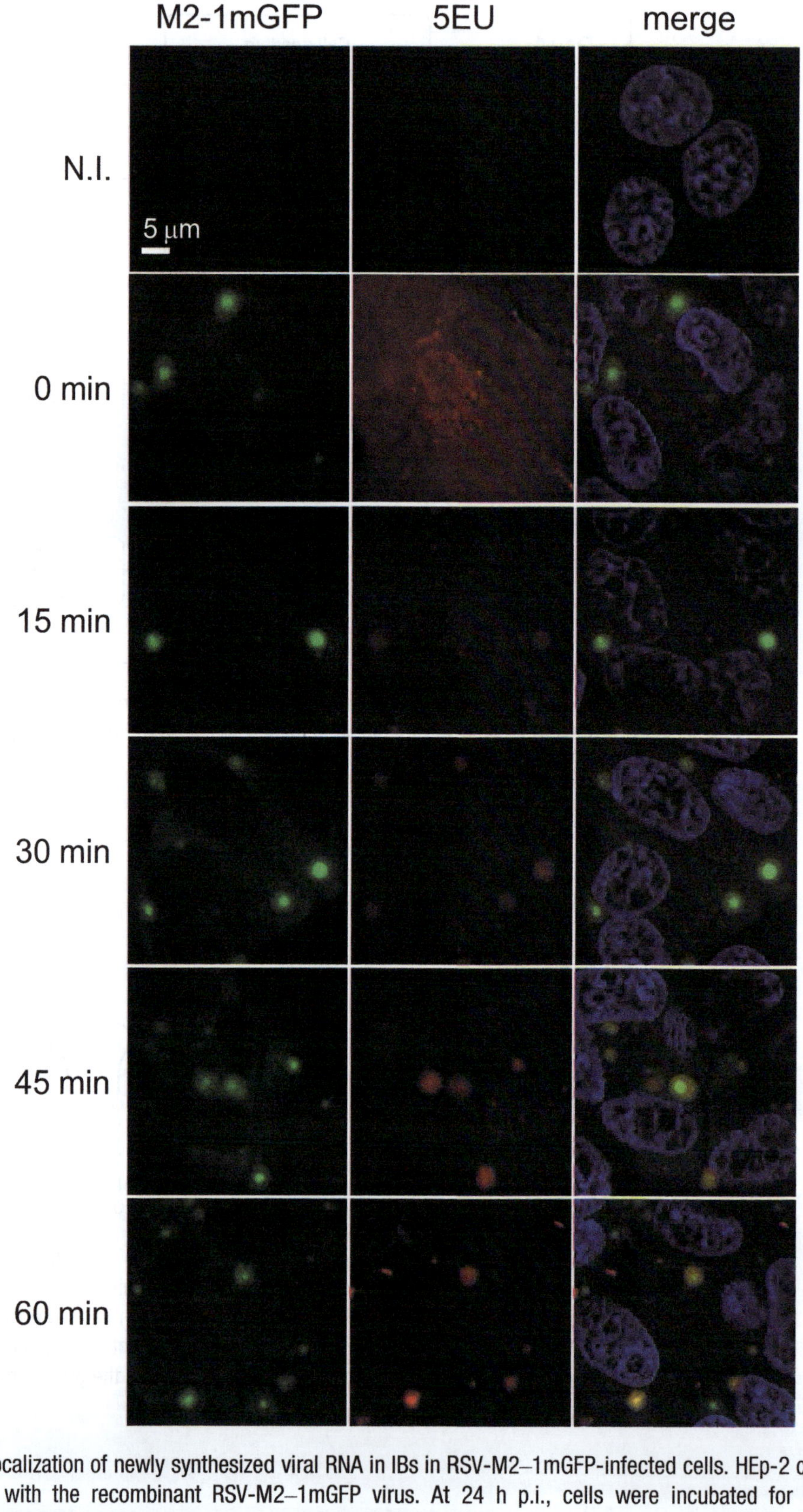

Fig. 3 Localization of newly synthesized viral RNA in IBs in RSV-M2–1mGFP-infected cells. HEp-2 cells were infected with the recombinant RSV-M2–1mGFP virus. At 24 h p.i., cells were incubated for 1 h with actinomycin D to inhibit cellular transcription, and then 5-EU was added for the indicated time before cell

Pulse-chase experiments with 5-EU were also used to investigate the fate of newly synthetized viral mRNAs in RSV IBs. After incubation with the uridine analog for the pulse, infected cells are incubated with non-marked uridine analog for several time periods (chase), enabling the RNA synthesis to proceed. The location of the newly synthetized viral RNAs is then analyzed at specific times after the chase period [5] (Fig. 4). In the absence of chase (Fig. 4a, 0 h), newly synthesized viral RNA was distributed throughout IBs and concentrated in IBAGs (inclusion bodies associated granules) were M2–1mGFP also accumulates. Infection was continued in the absence of 5EU (chase) to analyze the location of viral RNAs 3 or 6 h later. Under these conditions, concentration of the 5EU signal in IBAGs was no longer observed (Fig. 4a, 3h, and 6h). Intensity profiles shown in Fig. 4b reveal that, after 3 or 6 h of chase period, 5EU and M2–1mGFP fluorescence peaks no longer colocalized as they did directly after the pulse. This is consistent with ancient viral mRNAs being replaced by newly and unlabeled viral mRNAs. These observations reveal that viral mRNAs accumulate into IBAGs together with M2–1 shortly after synthesis, but then reach a different compartment. Using a similar pulse-chase approach, Heinrich et al. also showed that in the presence of nocodazole, viral RNAs were accumulated at IBs edges, which suggested the involvement of microtubule network in the traffic of viral transcripts from IBs to the cytosol [1].

2 Materials

2.1 Infection

1. RSV stock.
2. HEp-2 cells (ATCC CCL-23).
3. MEM-1% PS: Minimum Essential Medium (MEM) containing 1% Penicillin-Streptomycin (PS) (v/v). Store at 4 °C.
4. MEM-2% FBS-1% PS: MEM containing 2% fetal bovine serum (FBS) (v/v) and 1% PS (v/v). Store at 4 °C.
5. MEM-10% FBS-1% PS: MEM containing 10% FBS (v/v) and 1% PS (v/v). Store at 4 °C.
6. 24-well plates and glass 12-mm diameter coverslips.

1. Deionized water (dH_2O).
2. Dimethylsulfoxide (DMSO).

Fig. 3 (continued) fixation. The 5-EU incorporated in newly synthesized RNAs was detected using Alexa Fluor 647-azide (red), and nuclei were stained with Hoechst (merge). The M2–1mGFP protein is visualized through its spontaneous green fluorescence. Images were taken under a Leica SP8 confocal microscope. N.I.: Noninfected cells

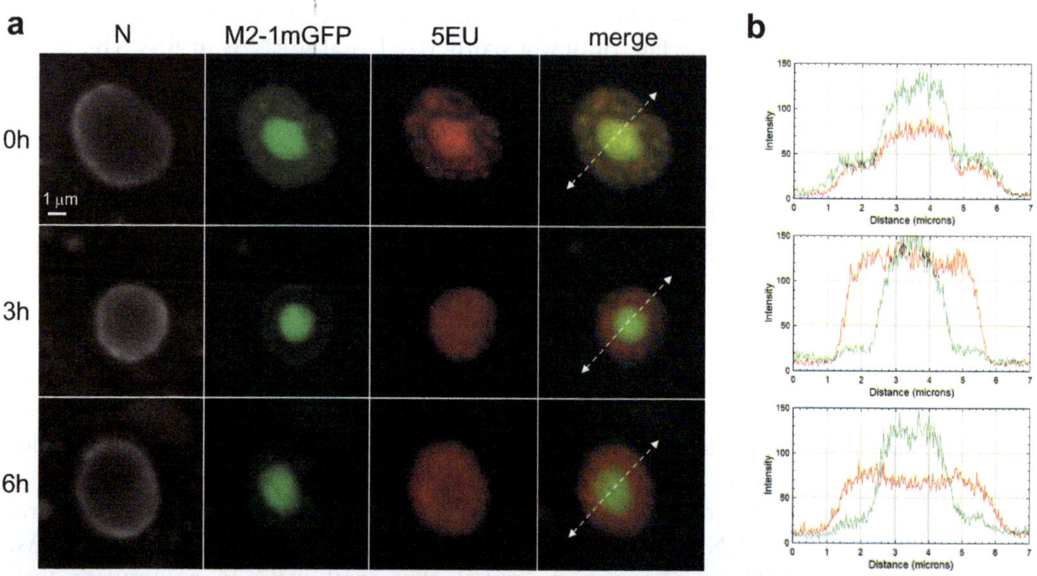

Fig. 4 Transient colocalization of newly synthesized viral RNA with M2–1 in RSV-M2–1mGFP-infected cells. (**a**) At 24 h p.i., HEp-2 cells were incubated for 1 h with actinomycin D to inhibit cellular transcription, and 5EU was added for one more hour (pulse). Cells were rinsed and fixed immediately (0 h) or after 3 or 6 h (chase). The 5-EU incorporated in newly synthesized RNAs was detected using Alexa Fluor 647 azide (red), and cells were stained with an anti-N antibody (gray). The M2–1mGFP protein is visualized through its spontaneous green fluorescence. M2–1mGFP and 5EU-associated fluorescence were merged to compare their spatial distribution. The white lines indicate the track of a line intensity profile across the IBs (**b**)

2.2 Inhibition of Cellular Transcription and 5-EU Incubation

3. 1 mg/mL stock solution of actinomycin D in DMSO. Store at −20 °C.
4. 100 mM stock solution of 5-EU in dH_2O. Store at −20 °C.

2.3 Fixation and Permeabilization

1. Washing solution: Phosphate-buffered saline (PBS), pH 7.2–7.6.
2. 32% reagent-grade formaldehyde.
3. Fixative solution: PBS containing 4% reagent-grade formaldehyde (v/v). Store at 4 °C.
4. Triton X-100 is suitable for immunofluorescence. Store at 4 °C.
5. 1% Triton X-100 (suitable for immunofluorescence) solution in PBS. Store at 4 °C.
6. Permeabilization solution: PBS containing 0.1% Triton X-100 (v/v). Store at 4 °C.
7. Laboratory tweezers.

2.4 2.4 5EU Labeling

1. Click-iT Cell Reaction Buffer Kit (Invitrogen C10329, C10330). Store all components at 4 °C, protected from light.
2. Alexa Fluor 647 azide (Invitrogen A10277) in DMSO. Store at −20 °C, protected from light.
3. 12-well plates.

2.5 Antibody Labeling

1. Desired antibodies to localize RSV IBs.
2. PBS.
3. Bovine serum albumin (BSA) lyophilized powder.
4. Diluent solution: PBS containing 1% BSA (w/v). Store at 4 °C.
   Store at 4 °C, and if long-term storage is desired, store at −20 °C.
5. Mounting media, such as ProLong Diamond antifade reagent (Invitrogen P36961).
   Store at 4 °C, and if long-term storage is desired, store at −20 °C.
6. Glass slides.
7. Moistened absorbent paper.
8. 150 mm petri dish.
9. Parafilm.
10. 10 mg/mL stock solution Hoechst 33342 (Invitrogen H1399) in dH_2O (*optional*)

2.6 Preparation of Stock Solutions

1. 100 mM 5-EU stock solution: follow the manufacturer's instructions. Store at −20 °C.
2. 1 mg/mL Actinomycin D solution: dissolve 1 mg in 1 mL of DMSO. Store at −20 °C.
3. Diluent solution: dissolve 1 g BSA in 100 mL of PBS (*see* **Note 1**). Store at 4 °C.
4. 1% Triton X-100 solution: add 500 μL Triton X-100 in 50 mL of PBS. Store at 4 °C.
5. Permeabilization solution: add 500 μL 1% Triton X-100 solution in 5 mL of PBS (*see* **Note 2**).
6. Fixative solution: dilute 4 mL 32% reagent-grade formaldehyde into 28 mL PBS under an appropriate chemical hood. Store at 4 °C.
7. 10× stock solution of the Click-iT® reaction buffer additive: prepare it by dissolving Component E of Click-iT Cell Reaction Buffer Kit in 2 mL of dH_2O. Mix until fully dissolved and store at −20 °C.
8. Alexa Fluor 647 azide: Dissolve the content of one vial in 100 μL of DMSO. Store at −20 °C, protected from light.

9. 10 mg/mL stock solution Hoechst 33342: dissolve 10 mg in 1 mL of dH_2O. Store at 4 °C and if long-term storage is desired store at −20 °C.
10. 1× working solution of actinomycin D (20 μg/ml): dilute actinomycin 1 mg/mL stock solution in MEM-2% FBS-1% PS.
11. 2 × working solution of 5-EU (2 mM): dilute 5-EU 100 mM stock solution in MEM-2% FBS-1% PS.

3 Methods

This protocol was optimized for HEp-2 cells infected with RSV. Variations in cell type (growth and density), medium, virus, treatments, and other factors may affect the assay conditions. Optimal conditions depending on your cell type, virus, and experimental conditions need to be experimentally adjusted.

Unless different conditions are specified, all living cell incubations are performed at normal cell culture conditions: 37 °C and CO_2 5%.

3.1 Infection

1. Seed 1.5×10^5 HEp-2 cells per well in MEM-10% FBS-1% PS on glass coverslips in 24-well plates (*see* **Note 3**). Allow cells to recover before performing any desired treatment or infection, ideally 16–24 h.
2. The day after seeding, check cell confluence under the microscope and perform treatments before virus infection if desired.
3. Aspirate medium and infect cells by adding RSV at desired MOI (*see* **Note 4**).

   Dilute virus stock in MEM-1% PS and add 300 μL per well. Incubate cells for 2 h (*see* **Notes 5** and **6**).
4. After virus adsorption step, aspirate the inoculum and add 500 μL MEM-2% FBS-1% PS per well and incubate the cells at 37 °C 5% CO_2 for 12 to 24 h.

   Perform cell treatments if desired..

3.2 Inhibition of Cellular Transcript and 5-EU Incubation

1. Warm up 1× working solution of actinomycin D (20 μg/mL) in a water bath at 37 °C before use.
2. At the desired times postinfection (p.i.), aspirate the medium and add 250 μL of 1× working solution of actinomycin D per well to inhibit cellular transcription.
3. Incubate for 1 h (*see* **Note 7**).
4. Warm up the 2× working solution of 5-EU (2 mM) solution in a water bath at 37 °C before use.

5. After 1 h actinomycin D treatment, add 250 μL of 2× working solution of 5-EU per well (final concentration = 1 mM) and incubate for 1 h (different pulse duration may be used) (*see* **Note 8**).
6. *Perform pulse-chase experiments if desired. For that purpose, rinse once cells with prewarmed fresh media (*MEM-2% FBS-1% PS*) without 5-EU, put 500 μL of prewarmed fresh media and incubate for the appropriate chase time. Proceed to* Subheading 3.4 *immediately after chase.*

3.3 Fixation and Permeabilization

1. Immediately following 5-EU treatment (or the chase period), aspirate the media and add 250–300 μL of fixative solution to each well.
2. Incubate for 10–15 min at room temperature (RT) (*see* **Note 9**).
3. Wash three times with PBS.
4. Permeabilize cells by adding 300 μL of permeabilization solution per well.
5. Incubate 10 min at RT (*see* **Note 9**).
6. Wash three times with PBS.
7. Carefully transfer single coverslips into wells of a 12-well plate using laboratory tweezers and a syringe needle.
8. Perform 5-EU labeling.

3.4 5-EULabeling

1. Prepare a 1× working solution of Click-iT® reaction buffer additive by diluting the 10× solution in dH_2O.
2. Prepare Click-iT® reaction cocktail according to Table 1 (*see* **Notes 10** and **11**).
3. Remove washing solution and add 250 μL of Click-iT reaction cocktail per well.
4. Incubate at RT for 30 min, protected from light (*see* **Note 13**).

Table 1
Click-iT® reaction cocktail preparation

Component	12 coverslips (μL)
Click-iT® RNA reaction buffer	2568
$CuSO_4$	120
Alexa Fluor® azide (*see* **Note 12**)	11.1
Click-iT® reaction buffer additive	300

5. After incubation, remove the Click-iT® reaction cocktail and wash cells once with 500 μL of Click-iT reaction rinse buffer.
6. Wash once with PBS.

   It is very important to keep samples protected from light during subsequent steps.

3.5 Antibody Labeling (Optional)

1. Prepare an incubation chamber by placing a 10 × 10 cm piece of moistened absorbent paper in the bottom of a 150-mm petri dish and position a piece of Parafilm directly on the paper layer.
2. Prepare primary antibody at the appropriate concentration in diluent solution (*see* **Note 14**).
3. Apply a 25–30 μL drop of primary antibody solution per glass coverslip directly to the Parafilm.
4. Carefully remove each coverslip from the 12-well plate. Drain the excess PBS by touching the edge of the coverslip to a piece of absorbent paper, and drop the coverslip on the antibody drop, making sure that cells are directly in contact with the antibody solution.
5. Cover the petri dish and incubate the chamber for 1 h at RT, protected from the light (*see* **Note 14**).
6. Recover coverslips from the incubation chamber and place them again on the 12-well plate.
7. Wash three times with PBS.
8. Prepare a solution containing the secondary antibody in diluent solution (*see* **Note 15**).
9. Repeat **steps 6** and 7.
10. Mount coverslips with a drop of mounting medium (*see* **Note 16**).
11. Store coverslips in the dark at −20 °C (especially for long-term storage) or 4 °C until proceeding to image using a 63× oil immersion objective.

4 Notes

1. Check the presence of contamination in the diluent solution before use. This solution can be filtered using a 0.2-μm filter before use. This solution can be used for up to one month.
2. If the antibody labeling step is going to be realized, Triton X-100 must be diluted in the diluent solution.
3. This protocol was developed with HEp-2 cells plated at 50–60% confluence.
4. In our laboratory, an MOI = 1 is routinely used and sufficient to see 5-EU positive IBs at 24 h p.i.

5. During the adsorption step, larger plate formats need to be incubated in gentle agitation to avoid monolayer drying.
6. Adsorption step is performed in the absence of FBS, since it could affect viral entry step.
7. To avoid cell toxicity, do not exceed DMSO concentrations of 0.5% when treating live cells.
8. We strongly recommend testing a range of EU concentrations to determine the optimal concentration for your cell type and experimental conditions. The used concentration may differ depending upon the duration of the pulse.
9. The proposed fixation and permeabilization solutions and corresponding conditions proposed here may need to be optimized based on each antibody and cell type investigated.
10. If Click-iT® reaction buffer additive solution develops a brown color, it should be discarded.
11. Prepare Click-iT® reaction cocktail fresh. Use it immediately after preparation, since Click-iT® reaction buffer additive is susceptible to oxidation, which decreases cocktail's efficiency over time.
12. Despite being mostly recommended and included as a component in the Click-iT® RNA Imaging Kits, we highly recommend using the Alexa Fluor® 647 azide instead of Alexa Fluor® 594 azide. Although the specificity and reactivity of both dyes should be the same, Alexa Fluor 647 is one of the brightest dyes and showed better sensitivity than Alexa Fluor 594 in our hands.
13. Incubation can be performed in a laboratory oven to maintain the appropriate incubation temperature (from 20 °C to 25 °C).
14. The composition of the antibody's dilution solution may vary depending on the used antibody. Refer to the manufacturer's instructions and adapt the protocol to the specific conditions related to your antibody, including the incubation time and temperature. For instance, some antibodies require a longer incubation at lower temperature, such as overnight at 4 °C.
15. If nuclei staining is desired, we recommend adding Hoechst 33342 at 1 μg/mL to the secondary antibody mix. Phalloidin staining is not compatible with the Click-iT® detection reaction. To stain cytoskeleton, the use of other antibodies, such as anti-α-tubulin, is then recommended.
16. ProLong Diamond antifade reagent is recommended, but other equivalent reagents can be used. Click-iT® EU-stained cells are compatible with all methods of slide preparation, including wet mount or prepared mounting media.

References

1. Heinrich BS, Cureton DK, Rahmeh AA, Whelan SPJ (2010) Protein expression redirects vesicular stomatitis virus RNA synthesis to cytoplasmic inclusions. PLoS Pathog 6(6): e1000958
2. Hoenen T, Shabman RS, Groseth A, Herwig A, Weber M, Schudt G et al (2012) Inclusion bodies are a site of ebolavirus replication. J Virol 86(21):11779–11788
3. Jobe F, Simpson J, Hawes P, Guzman E, Bailey D (2020) Respiratory syncytial virus sequesters NF-κB subunit p65 to cytoplasmic inclusion bodies to inhibit innate immune signaling. J Virol 94(22)
4. Lahaye X, Vidy A, Pomier C, Obiang L, Harper F, Gaudin Y et al (2009, August 15) Functional characterization of Negri bodies (NBs) in rabies virus-infected cells: Evidence that NBs are sites of viral transcription and replication. J Virol [Internet] [cited 2023 Feb 6];83(16):7948–7958
5. Rincheval V, Lelek M, Gault E, Bouillier C, Sitterlin D, Blouquit-Laye S et al (2017) Functional organization of cytoplasmic inclusion bodies in cells infected by respiratory syncytial virus. Nat Commun 8(1):563
6. Ringel M, Heiner A, Behner L, Halwe S, Sauerhering L, Becker N et al (2019) Nipah virus induces two inclusion body populations: identification of novel inclusions at the plasma membrane. PLoS Pathog 15(4):e1007733
7. Jao CY, Salic A (2008) Exploring RNA transcription and turnover in vivo by using click chemistry. Proc Natl Acad Sci USA 105(41): 15779–15784
8. Tiong-Yip CL, Aschenbrenner L, Johnson KD, McLaughlin RE, Fan J, Challa S et al (2014) Characterization of a respiratory syncytial virus L protein inhibitor. Antimicrob Agents Chemother 58(7):3867–3873
9. Zheng X, Gao L, Wang L, Liang C, Wang B, Liu Y et al (2019) Discovery of Ziresovir as a potent, selective, and orally bioavailable respiratory syncytial virus fusion protein inhibitor. J Med Chem 62(13):6315–6329
10. Plumet S, Duprex WP, Gerlier D (2005) Dynamics of viral RNA synthesis during measles virus infection. J Virol 79(11):6900–6908

Chapter 7

Minigenome System to Study Respiratory Syncytial Virus Gene Expression

Michael N. Teng and Kim C. Tran

Abstract

Minigenome assays have been essential tools in the understanding of viral transcription and RNA replication for respiratory syncytial virus (RSV). Here, we describe the RSV minigenome assay for determining transcription by the viral polymerase in the absence of infection. We detail three different methods of detecting viral RNA synthesis: a firefly luciferase assay for rapid and sensitive measurement of RSV polymerase activity; a dual luciferase assay for assessing contributions of *cis*-acting signals for RSV transcription; and a real-time quantitative PCR method for determination of specific effects on the transcription of individual viral genes and the polar transcription gradient of RSV.

Key words RSV transcription, Minigenome, Polymerase, Luciferase assay, Real-time quantitative PCR assay

1 Introduction

The advent of reverse genetics systems over 20 years ago revolutionized the study of viral transcription for non-segmented negative-sense RNA viruses. Initial studies utilized viral genome analogs cloned from defective viral particles (DVPs). Expression of these defective viral genomes in virus-infected cells resulted in the production of new DVPs [1]. Alternatively, synthetic viral genome analogs (minigenomes) were constructed where a reporter gene (chloramphenicol acetyltransferase, CAT) whose transcription was directed by viral transcription start and stop sequences was cloned between the 3′ and 5′ ends of the viral RNA (vRNA) [2, 3]. Thus, these minigenomes contained the signals for viral transcription and RNA replication without encoding any viral proteins. The minigenomes were transcribed in vitro, transfected into virus-infected cells, and viral transcription measured by CAT activity. Further refinement of these minigenome systems used co-transfection of

Bo Liang and Rachel Fearns (eds.), *Human Respiratory Syncytial Virus: Methods and Protocols*, Methods in Molecular Biology, vol. 2948, https://doi.org/10.1007/978-1-0716-4666-3_7,

cells with a plasmid expressing the minigenome under the control of a T7 RNA polymerase promoter with plasmids encoding the various viral polymerase components, also under the control of a T7 RNA polymerase promoter. The T7 RNA polymerase was then provided in *trans* by simultaneous infection with recombinant vaccinia virus (VV-T7) [4]. This transfection/infection system allowed for the transient expression of high levels of viral proteins and minigenome, which then could be replicated and transcribed. Importantly, T7 RNA polymerase expression in the cytoplasm allowed for correct subcellular targeting of the transcribed minigenome, since NNS RNA viruses, except bornaviruses, replicate in the cytoplasm. Cytoplasmic transcription also obviated the potential problem of cryptic splicing of viral protein mRNAs and minigenomes.

The recovery of recombinant NNS RNA virus soon followed [5–7]. However, there were a number of problems resulting from the use of VV-T7, including contamination of newly produced recombinant virus with the vaccinia virus and homologous recombination between transfected plasmids. Therefore, a cell line stably transfected to express T7 RNA polymerase was developed [8]. This cell line, BSR-T7/5, could then be transfected with the T7 plasmids to produce recombinant virus.

The use of the minigenome system for respiratory syncytial virus (RSV) was essential for identifying the M2–1 protein as an essential component of the viral polymerase [9]. In addition, the minigenomes have been further developed to include multiple transcription units and different reporter genes (e.g., green fluorescent protein, firefly luciferase) [10–12]. Currently, we use three different methods to detect viral transcription in minigenome assays. First, when examining the effect of *trans*-acting factors, we evaluate transcription by simple luciferase assay using a minigenome that encodes firefly luciferase as the only gene [12]. This assay shows global effects on RSV polymerase activity but does not distinguish between effects on viral transcription and RNA replication. For more detailed analysis of viral transcription and RNA replication, including the effect of *cis*-acting sequences, we use a dicistronic minigenome that encodes *Renilla* luciferase as the first transcription unit followed by firefly luciferase as the second transcription unit. Comparison of ratios of firefly:*Renilla* luciferase provides a relative measure of transcription efficiency. Finally, we utilize a real-time quantitative PCR (qPCR) assay to detect viral RNA products from a modified minigenome assay using the full-length RSV antigenome cDNA. The qPCR assay allows us to examine the transcription of individual RSV genes as well as global effects on RSV transcription and the polar transcription gradient. This assay relies on the use of expression plasmids encoding codon-optimized versions of the RSV polymerase genes, so that transcripts synthesized by the viral polymerase can be differentiated from those derived from T7 RNA polymerase transcription. In addition, the qPCR assay can be adapted for use in cells or tissues infected by RSV (subgroup A) [13].

2 Materials

2.1 Transfection

1. BSR-T7/5 cells (courtesy of Klaus Conzelmann)
2. Dulbecco's Modified Eagle Medium (DMEM). Store at 4 °C.
3. Complete medium (CM): DMEM containing 5% fetal bovine serum and MEM nonessential amino acids. Store at 4 °C.
4. Transfection reagent (Gene Juice). Store at 4 °C.
5. Plasmids: pcDNA3.1-N, pcDNA3.1-P, pcDNA3.1-M2-1, pcDNA3.1-L (courtesy of Martin Moore, Emory University); C2L and D53 (courtesy of Peter Collins, NIAID); phRL-TK. Store at −20 °C.
6. 24-well plates
7. Sterile 1.5 mL microfuge tubes

2.2 Luciferase Assay

1. Dual Luciferase Reporter Assay Kit. Store at −80 °C.
2. Phosphate-buffered saline (PBS): 10× (80 g NaCl, 2 g KCl, 11.5 g $Na_2HPO_4 \cdot 7H_2O$, 2 g KH_2PO_4 in 1 L ultrapure water). Dilute to 1× in ultrapure water before use.
3. 0.5-mL microfuge tubes
4. Black flat-bottom 96-well plate
5. Plate reading luminometer (SynergyMx, Biotek)
6. Distilled water
7. 70% ethanol in distilled water

2.3 Real-Time Quantitative PCR (qPCR) Assay

1. RNAzol (Molecular Research Center)
2. 1.5-mL microfuge tubes
3. Phosphate-buffered saline (PBS): 10× (80 g NaCl, 2 g KCl, 11.5 g $Na_2HPO_4 \cdot 7H_2O$, 2 g KH_2PO_4 in 1 L ultrapure water). Dilute to 1× in ultrapure water before use.
4. Isopropanol
5. Nuclease-free water
6. 75% ethanol in nuclease-free water
7. First-strand cDNA synthesis kit (iScript, Bio-Rad). Store at −20 °C.
8. 2× SYBRGreen QPCR kit (SensiFAST SYBR & Fluorescein kit, Bioline). Store at −20 °C.
9. 96-well PCR plate, 0.2-mL well (TempPlate Semi-Skirt)

10. Optical Film PCR plate cover (TempPlate RT)
11. Standard PCR machine (ABI 2720)
12. Real-time PCR machine

3 Methods

Perform all incubations at room temperature unless otherwise noted.

3.1 Transfection (see Note 1)

1. Plate BSR-T7/5 cells in 24-well clusters (~2 × 10 (*see* **Note 4**) cells/well) in CM. Seed enough wells for triplicate samples.
2. Incubate overnight at 37 °C/5% CO_2.
3. Mix support plasmids in a sterile 1.5-mL microfuge tube (150 ng pcDNA3.1-N, 100 ng pcDNA3.1-P, 50 ng pcDNA3.1-M2–1, 100 ng pcDNA3.1-L per sample) with either: (a) firefly luciferase-expressing minigenome (150 ng C2L) and *Renilla* luciferase transfection control (20 ng phRL-TK); (b) dicistronic *Renilla* and firefly luciferase minigenome (150 ng pFP24); or (c) full-length antigenome cDNA (300 ng D53) (*see* **Note 2**).
4. Make a master mix of transfection reagent (3 mL Gene Juice/1 mg DNA) in serum-free DMEM (250 mL/tube). Incubate 5 min.
5. Add 250 mL of transfection reagent master mix to each tube, mix by gentle agitation, and incubate 30 min (*see* **Note 3**).
6. Remove media from 24-well plate of BSR-T7/5 cells.
7. Add 83 mL of DNA/Gene Juice mix per well and incubate at 37 °C/5% CO_2 for 24 h (*see* **Note 4**).
8. Harvest cells for luciferase (Subheading 3.2) or qPCR (Subheadings 3.3, 3.4, and 3.5) assay (*see* **Note 5**).

3.2 Luciferase Assay (Dual Luciferase Reporter Assay System Method)

1. Dilute the 5× Passive Lysis Buffer (PLB) in distilled water to make a 1× solution. Make sure PLB is at room temperature before use.
2. Remove media from cells and wash with 1× PBS by rocking plate gently to remove detached cells and residual medium. Completely remove PBS before adding PLB.
3. Add 100 μL PLB to each well (24-well plate), swirl to cover cells. Rock plate for 30 min (*see* **Note 6**). Transfer lysate to 0.5-mL microfuge tube and keep on ice.
4. Spin samples at max speed (13,000 × g) for 1 min to clarify (*see* **Note 7**).

5. Prepare Luciferase Assay Reagent II (LAR II) by solubilizing the lyophilized Luciferase Assay Substrate in 10 mL of the Luciferase Assay Buffer II (*see* **Note 8**).
6. Preparation of Stop & Glo Reagent (prepare immediately prior to use). Make 1× Stop & Glo Reagent by mixing 1 volume of 50× Stop & Glo Substrate to 50 volumes of Stop & Glo Buffer in glass vial (100 μL reagent per assay) (*see* **Note 9**).
7. Program plate reader luminometer to perform a 2-s premeasurement delay, followed by a 10-s measurement period for each reporter assay. Program the injector system to dispense *50 μL* of *diluted* LAR II reagent per well and *50 μl* of *diluted* Stop & Glo Reagent per well.
8. Purge all storage liquid (distilled water or 70% ethanol wash solution) from injector. Prime empty injector system with LAR II (Injector #1) or Stop & Glo Reagent (Injector #2) and collect priming reagent in appropriate vials.
9. Dispense *30 μL* (for most luc assays from p24, We usually use 30 μL) of cell lysate samples into an appropriate well of black 96-well microplate (*see* **Note 10**). Place the microplate in plate reader (*see* **Note 11**).
10. Run Dual Luciferase Assay program. Make sure injection volume is set to *50 μL*, as the SynergyMx will inject 100 μL otherwise.
11. Download the data (Excel) and calculate firefly (FF) luciferase numbers relative to *Renilla* luciferase. Examples of the level of minigenome transcription detectable for the monocistronic minigenome, normalizing to mock-transfected cells, and for the dicistronic minigenome are shown in Figs. 1 and 2 (*see* **Note 12**).

3.2.1 General Injector Wash Protocol (see Note 13)

1. Purge Stop & Glo Reagent from injector lines by repeated priming/washing with distilled water (about 3 pump void volumes).
2. Prime injector system with at least 5 mL of 70% ethanol to completely replace the void volume and rinse the injector plumbing. It is preferable to soak in this wash solution for 30 min prior to rinsing with distilled water.
3. Rinse with enough distilled water (at least 3 pump void volumes) to thoroughly remove all traces of ethanol.

3.3 RNA Isolation (RNAzol Method) (see Note 14)

1. Aspirate medium from wells of transfected cells.
2. Wash cells with 1 mL ice-cold PBS and aspirate.
3. Add 250 mL of RNAzol and incubate 5 min at room temperature.

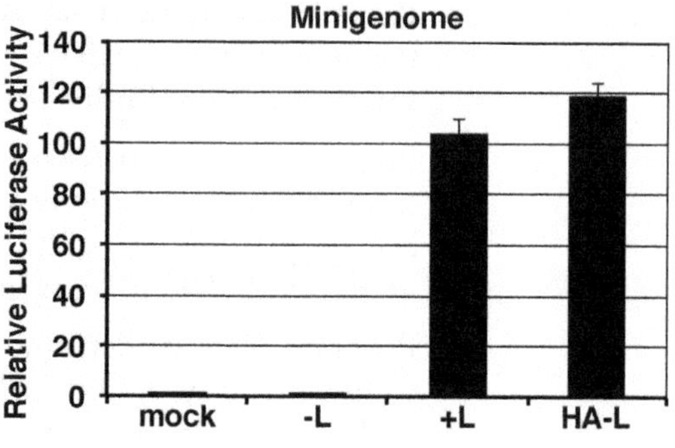

Fig. 1 Luciferase detection for minigenome assay. BSR-T7 cells were transfected with expression plasmids encoding N, P, M2–1, and *Renilla* luciferase (transfection control) plus a firefly luciferase-expressing RSV minigenome plasmid (C2L). Cells were co-transfected with empty vector (−L) or expression plasmids encoding wild-type L (+L) or HA-tagged L (+HA-L). Mock-transfected cells were used as a negative control. At 24 h post-transfection, lysates were harvested and subjected to Dual Luciferase Assay. Luciferase activities were normalized to those from mock-transfected cells. Shown are the averages ± SEM of triplicate samples

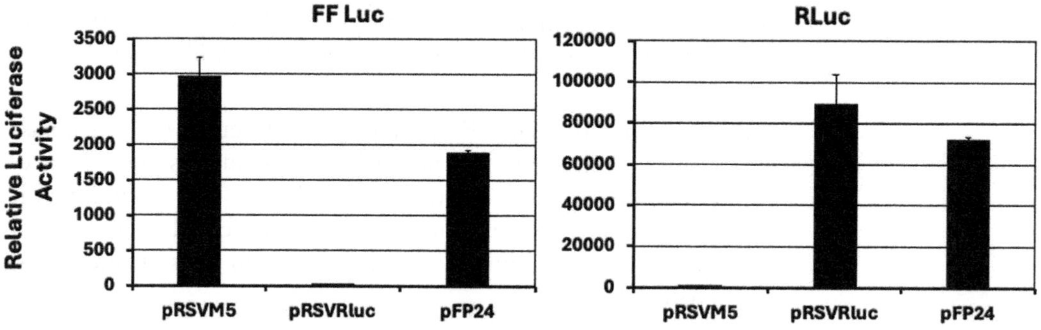

Fig. 2 Dual luciferase minigenome assay. BSR-T7 cells were transfected with expression plasmids encoding N, P, M2–1, and L plus RSV minigenome plasmids expressing firefly luciferase alone (pRSVM5), *Renilla* luciferase alone (pRSVRluc), or *Renilla* and firefly luciferase (FP24). At 24 h post-transfection, lysates were harvested and subjected to Dual Luciferase Assay. Luciferase activities were normalized to those from mock-transfected cells. Shown are the averages ± SEM of triplicate samples

4. Harvest the RNAzol mixture by carefully swirling the viscous liquid around the well and add to a 1.5-mL microfuge tube.
5. Add 100 mL nuclease-free water to each tube, mix by inversion 6–10 times, and incubate 15 min.
6. Centrifuge tubes for 10 min at 12,000× g (*see* **Note 15**).
7. Transfer the supernatant to a fresh microfuge tube, add 1 vol. of isopropanol, and mix by inversion. Incubate 15 min.
8. Centrifuge 12,000× g, 10 min.

9. Aspirate supernatant, add 250 mL 75% ethanol, and centrifuge 4000× g, 2 min.
10. Repeat **step 9**.
11. Resuspend pellet in 10 mL of nuclease-free H_2O.

3.4 First-Strand cDNA Synthesis (iScript Method) (see Note 16)

1. Reaction setup (*see* **Note 17**)

RNA template (500 ng)	× mL
5× reaction mix	4 mL
Reverse transcriptase	1 mL
RNase/DNase-free H_2O	× mL
Total volume	20 mL (0.2 mL microfuge tube)

2. Reaction protocol (performed in standard thermocycler)

5 min, 25 °C

60 min, 42 °C

5 min, 85 °C

Hold at 4 °C

3.5 Real-Time Quantitative PCR Protocol (see Note 18)

1. Reaction setup (*see* **Notes 19, 20** and **21**)

cDNA	2 mL
Forward primer (10 mM)	1.25 mL
Reverse primer (10 mM)	1.25 mL
RNase/DNase-free H_2O	9 mL
2× SYBRGreen mix	12.5 mL

2. Run protocol (*see* **Note 22**)
   95 °C, 5 min
   40 cycles

   95 °C, 5 s

   60 °C, 15 s
3. Download data and calculate fold increase using ΔΔC (t) method. An example of this analysis is shown in Fig. 3 (*see* **Note 23**).

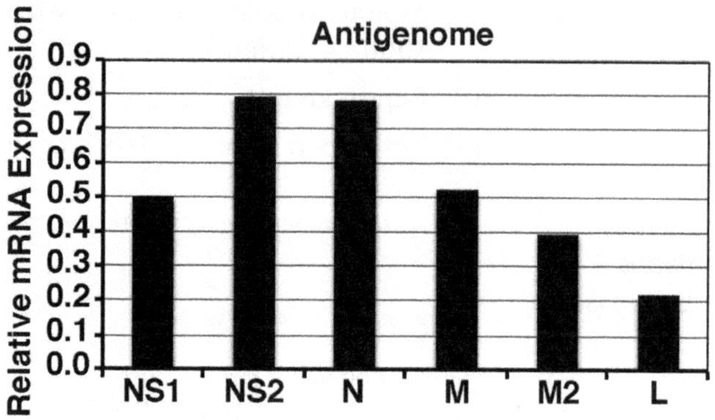

Fig. 3 qPCR assay for antigenome assay. BSR-T7 cells were transfected with expression plasmids encoding N, P, M2–1, and L plus a plasmid containing the full-length antigenome cDNA of RSV (D53). At 24 h post-transfection, total RNA was harvested, and first-strand cDNA was produced using random hexamer priming. qPCR was performed with primers for the indicated genes and SH-G (as a measure of genome/antigenome) and 18S rRNA. Transcript levels were calculated by the DDC(t) method. Shown are the averages of duplicate samples (*see* **Note 24**)

4 Notes

1. The transfection protocol has been optimized for 24-well plates. However, the protocol can readily be scaled up or down for larger or smaller plates.
2. For negative control, replace pcDNA3.1-L with an equal amount of empty pcDNA3.1.
3. Making a master mix of DNA/transfection reagent for each triplicate then dispensing 1/3 of the volume per well gives us the least amount of sample-to-sample variability.
4. Best results occur with transfection for 6–8 h followed by the addition of fresh media (800 mL CM/well).
5. Alternatively, an additional (fourth) well can be transfected to harvest to check for protein expression by immunoblotting.
6. We generally lyse for a minimum 30 min and up to 3 h.
7. Samples can be stored at −80 °C at this point for later analysis.
8. LAR II can be stored in 1 ml aliquots at −80 °C (good for up to 1 year). Make sure LAR II is at room temperature before use. Mix LAR II by inversion or gentle vortexing prior to use.
9. LAR II and Stop & Glo can be diluted to a working stock before use. We generally use a 1:5 dilution.
10. It is essential to use opaque microplates to prevent cross talk between samples.

11. The microplate can be washed with water then EtOH and reused 4–8 times.
12. With the dicistronic minigenome, the absolute luminescence for *Renilla* luciferase is usually 30- to 50-fold higher than that of firefly. We normalize the FF/*Renilla* ratio to the wild-type (FP24) plasmid.
13. It is essential to carefully clean the injection apparatus after each use.
14. Other RNA isolation methods can be used; however, we use RNAzol because of the lack of DNA contamination in the RNA sample. For other methods, treatment with DNase is recommended prior to cDNA synthesis to avoid spurious results due to contamination of the RNA by the transfected plasmids.
15. It is important to centrifuge specifically at 12,000× g. Higher centrifugation speeds will compress the pellet containing the DNA/protein contaminants.
16. We have had good success using a variety of first-strand cDNA synthesis kits, though the iScript kit works best in our hands.
17. We add the RNA to the tube and make up the volume to 5 μL with nuclease-free H_2O. Then, we make a master mix of the 5× reaction mix (4 μL/reaction), reverse transcriptase (1 μL/reaction), and nuclease-free H_2O (10 μL/reaction) and dispense 15 μL of master mix per tube.
18. The sequences of the forward and reverse primers for both RSV genes and intergenic regions are listed in Table 1.
19. We usually perform duplicates of each sample to minimize the effects of well-to-well variation.
20. We make a master mix of the forward and reverse primers, then dispense 2.5 μL per well. In addition, we make a master mix of the cDNA, H_2O, and 2× SYBRGreen reagent then dispense 22.5 μL per well.
21. We generally use human 18S rRNA primers as a housekeeping control. Although BSR-T7/5 cells are derived from the BHK-21 cell line, we see good C(t) values with these primers. We have not seen an improvement in inter-assay variation using multiple housekeeping controls.
22. The SH-G primer pair gives the most consistent C(t) values between assays for genome/antigenome. The L-tr primer pair gives C(t) values that are somewhat higher than the other two pairs.
23. Alternatively, a melt curve protocol (increasing temperature by 0.5 °C/15 s steps) can be added to the end of the run to ensure the relative specificity of the PCR.

Table 1
List of qPCR primers

RSV gene	Primer sequence (5′ to 3′)	Antigenome position	Fragment size
NS1	F: CACAACAATGCCAGTGCTACAA R: TTAGACCATTAGGTTGAGAGCAATGT	329–343 411–386	82
NS2	F: TGAATGCATAGTGAGAAAACTTGATG R: GGAAAGTGCCATATTTTGTGTTG	762–787 901–879	139
N	F: AAGGGATTTTTGCAGGATTGTTT R: CTCCCCACCGTAGCATTACTTG	1857–1880 1923–1908	66
P	F: TGACCAATGACAGATTAGAAGCT R: TCATTGTCACTATCATTCCCTTCC	2905–2927 3049–3026	144
M	F: ATGTGCTAATGTGTCCTTGGATGA R: TGATTTCACAGGGTGTGGTTACA	3530–3543 3597–3575	67
G	F: AGCCCACAGAAGAGCCAACCA R: TTCCTGTGGTGTTGGAGGTGAGT	5373–5393 5450–5428	77
F	F: TAAGCAGCTCCGTTATCACATCTC R: ATTGGATGCTGTACATTTAGTTTTGC	6865–6888 6938–6913	73
L	F: CACTCTACAAAACAAAAAGACACAATCA R: AGGATGCTGCATTGAACACATT	9026–9053 9097–9076	71
Le-NS1	F: TGCGTACAACAAACTTGCAT R: GCTGCCCATCTCTAACCAAG	12–31 107–88	95
SH-G	F: TTAACATCCCACCATGCAAA R: GCATTTGCCCCAATGTTATT	4575–4594 4683–4664	108
L-tr	F: CACAGGTAGTCTGTTATACAACTTTCA R: ACAGTGTTAGTGTATAGCTATGGGAAT	14959–14985 15045–15019	86
18S rRNA	F: GTAACCCGTTGAACCCCATT R: CCATCCAATCGGTAGTAGCG		

24. The fold increase for the viral RNAs is not as high as that seen in infection, likely due to the comparison with excess production of T7-transcribed antigenome that is not properly encapsidated.

Acknowledgments

The authors thank David Fitzhugh for help in developing the real-time quantitative PCR assay, Martin Moore (Emory University) for the optimized codon expression plasmids for the RSV proteins, Peter Collins (NIAID) for the minigenome and antigenome constructs, and Klaus Conzelmann (Ludwig-Maximillians-Universitat) for the BSR-T7/5 cells. This work was supported in part by a grant from the NIAID and funds from the Joy McCann Culverhouse

Endowment. The content is solely the responsibility of the authors and does not necessarily represent the official views of the National Institutes of Health.

References

1. Pattnaik AK, Ball LA, LeGrone AW, Wertz GW (1992) Infectious defective interfering particles of VSV from transcripts of a cDNA clone. Cell 69:1011–1020
2. Park KH, Huang T, Correia FF, Krystal M (1991) Rescue of a foreign gene by Sendai virus. Proc Natl Acad Sci USA 88:5537–5541
3. Collins PL, Mink MA, Stec DS (1991) Rescue of synthetic analogs of respiratory syncytial virus genomic RNA and effect of truncations and mutations on the expression of a foreign reporter gene. Proc Natl Acad Sci USA 88: 9663–9667
4. Calain P, Curran J, Kolakofsky D, Roux L (1992) Molecular cloning of natural paramyxovirus copy-back defective interfering RNAs and their expression from DNA. Virology 191:62–71
5. Schnell MJ, Mebatsion T, Conzelmann KK (1994) Infectious rabies viruses from cloned cDNA. EMBO J 13:4195–4203
6. Whelan SP, Ball LA, Barr JN, Wertz GT (1995) Efficient recovery of infectious vesicular stomatitis virus entirely from cDNA clones. Proc Natl Acad Sci USA 92:8388–8392
7. Collins PL, Hill MG, Camargo E, Grosfeld H, Chanock RM, Murphy BR (1995) Production of infectious human respiratory syncytial virus from cloned cDNA confirms an essential role for the transcription elongation factor from the 5′ proximal open reading frame of the M2 mRNA in gene expression and provides a capability for vaccine development. Proc Natl Acad Sci USA 92:11563–11567
8. Buchholz UJ, Finke S, Conzelmann KK (1999) Generation of bovine respiratory syncytial virus (BRSV) from cDNA: BRSV NS2 is not essential for virus replication in tissue culture, and the human RSV leader region acts as a functional BRSV genome promoter. J Virol 73: 251–259
9. Collins PL, Hill MG, Cristina J, Grosfeld H (1996) Transcription elongation factor of respiratory syncytial virus, a nonsegmented negative-strand RNA virus. Proc Natl Acad Sci USA 93:81–85
10. Kuo L, Fearns R, Collins PL (1996) The structurally diverse intergenic regions of respiratory syncytial virus do not modulate sequential transcription by a dicistronic minigenome. J Virol 70:6143–6150
11. Hardy RW, Harmon SB, Wertz GW (1999) Diverse gene junctions of respiratory syncytial virus modulate the efficiency of transcription termination and respond differently to M2-mediated antitermination. J Virol 73: 170–176
12. Teng MN, Collins PL (1998) Identification of the respiratory syncytial virus proteins required for formation and passage of helper-dependent infectious particles. J Virol 72:5707–5716
13. Piedra FA, Qiu X, Teng MN, Avadhanula V, Machado AA, Kim DK, Hixson J, Bahl J, Piedra PA (2020) Non-gradient and genotype-dependent patterns of RSV gene expression. PLoS One 15:e0227558

Chapter 8

Minigenome System to Study Respiratory Syncytial Virus Transcription Versus Genome Replication

Heesu Kim and Sarah L. Noton

Abstract

The respiratory syncytial virus (RSV) minigenome system is a powerful analytical tool that has enabled viral transcription and genome replication to be modeled in a cellular context without viral infection. The minigenome assay can be used to dissect the different steps in RSV transcription and genome replication, assess the effects of mutations in *cis*-acting RNA elements, analyze mutations in the polymerase and other proteins in the transcription–replication machinery, and assess the effect of antiviral compounds on RSV polymerase functions. Here, we describe the RSV minigenome assay and detail a northern blotting technique for detecting viral mRNA transcripts and genome replication products.

Key words RSV, Negative-sense RNA virus, Minigenome, Reporter gene, Genome replication, Transcription, Northern blotting

1 Introduction

The respiratory syncytial virus (RSV) minigenome assay is a valuable tool that has enabled the intracellular study of mechanisms of RSV transcription and genome replication in the absence of viral infection [1–7]. The assay can recapitulate many aspects of RSV transcription and replication, including RNA synthesis from an encapsidated RNA template, subgenomic mRNA synthesis, cap addition, and polyadenylation and generation of encapsidated RNA replication products. The assay can be used to examine the impact of mutations in key viral proteins and *cis*-acting RNA elements that are required for RSV transcription and genome replication that might otherwise be lethal for a recombinant viral rescue [6, 8–14]. The minigenome system can also be used to assess the effect of antiviral compounds on RSV transcription and genome replication, helping to decipher their mechanism of action [8, 15–18]. The RSV minigenome system is outlined in Fig. 1. In essence, to create the minigenome, the ten viral genes comprising the

Bo Liang and Rachel Fearns (eds.), *Human Respiratory Syncytial Virus: Methods and Protocols*, Methods in Molecular Biology, vol. 2948, https://doi.org/10.1007/978-1-0716-4666-3_8,

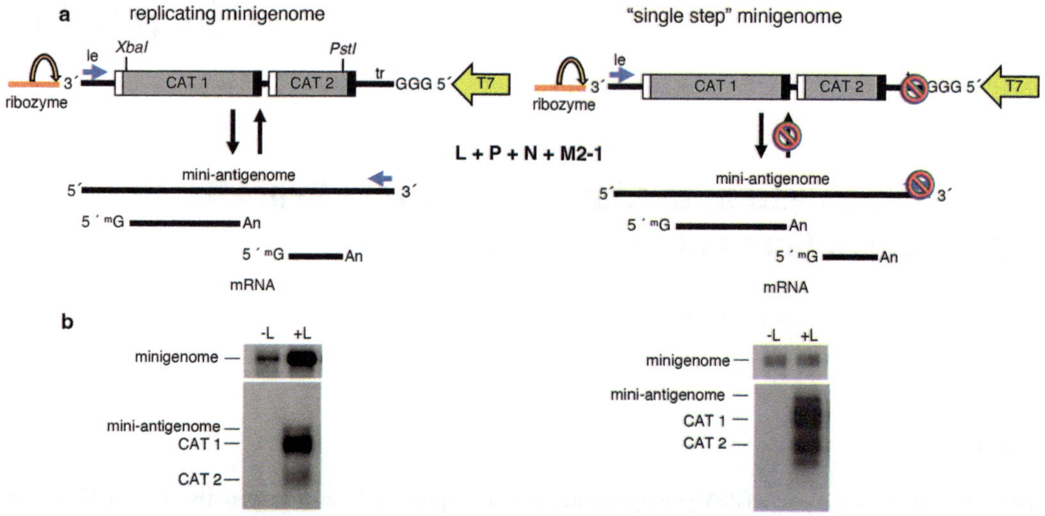

Fig. 1 (**a**) A schematic depiction of the RSV dicistronic CAT minigenome as described in the text (not to scale; *see* **Notes 1** and **2**). The standard replicating minigenome shown on the left-hand side is flanked by a self-cleaving hepatitis ribozyme at the 3′ terminus (orange line and arrow) and a T7 promoter at the 5′ terminus (yellow arrow). Le and tr promoters are shown as blue arrows. CAT 1 and CAT 2 segments (gray boxes) are each flanked by gs and ge sequences (white and black boxes, respectively). Support plasmids expressing L, P, N, and M2–1 in *trans* enable the expression of mini-antigenome and the two CAT mRNAs. The "single step" minigenome, depicted on the right-hand side, is identical to the replicating minigenome except that it has a mutation or deletion (red symbol) in the tr such that the tr promoter at the 3′ terminus of the anti-minigenome is inactivated. In this case, the mini-antigenome cannot act as a template for more minigenome synthesis, such that genome replication and transcription are now independent events. (**b**) Northern blot analysis of RNA expressed from a standard replicating minigenome (left-hand side) and a "single step" minigenome (right-hand side). The upper panel shows the negative-sense minigenome expressed by T7, while the lower panel shows RNA expressed by the RSV polymerase. *Lane 1* is a control in which the pTM1-L support plasmid was omitted from the transfection, while *lane 2* shows mini-antigenome RNA, CAT 1 and CAT 2 mRNAs that are expressed by the RSV polymerase

~19 kb genome are replaced with a reporter gene (such as bacterial chloramphenicol acetyltransferase (CAT), green fluorescent protein (GFP), or firefly luciferase gene), while maintaining the 3′ and 5′ non-coding termini as well as other *cis*-acting sequences that are essential for RSV transcription and genome replication. The resulting RSV minigenome is smaller (e.g., ~ 1 kb in the case of a CAT containing minigenome described below) and therefore easier to manipulate and study. Viral proteins necessary for efficient transcription and genome replication, N, P, L, and M2–1 are provided in *trans* via expression from support plasmids such that all aspects of RSV transcription and genome replication can be reproduced in a cellular context. Furthermore, the minigenome can also be modified such that mRNA synthesis is independent of the levels of RNA replication. This is achieved by mutating the trailer (tr) region that lies at the 5′ terminus of the minigenome; this inactivates the tr promoter at the 3′ end of the mini-antigenome to prevent further rounds of minigenome RNA synthesis (Fig. 1). Here, this minigenome is referred to as a "single step" minigenome (previously referred to as a nonreplicating or replication-deficient minigenome). Key considerations in designing a minigenome system are that the RSV polymerase requires that the minigenome template has 3′ and 5′ termini with fewer than six additional nucleotides [10] and that in the context of infection, RSV mRNAs are not generated in the nucleus, and so if mRNAs for support proteins consist of native RSV sequence, they might not be well expressed from a pol II promoter due to the presence of cryptic splice sites or termination/ polyadenylation signals. To overcome these issues, the minigenome system described below uses minigenome and support plasmids that are driven by intracellular expression of T7 RNA polymerase. A precise minigenome 3′ end can be created by flanking the leader (le) end of the minigenome cassette with a ribozyme sequence. The T7 support plasmids are in a pTM1 plasmid background in which the (uncapped) T7 polymerase-generated mRNA transcript has an internal ribosome entry site to allow for efficient translation of the support protein.

Here, we describe the minigenome assay and one method for analyzing viral RNAs produced in the system, namely northern blotting, which enables the detection of viral transcripts and genomic RNA. Other methods for analyzing minigenome transcription and genome replication such as measuring firefly luciferase expression, performing real-time quantitative PCR or primer extension analyses can also be used and are described elsewhere [8, 12, 13, 18, 19].

2 Materials

Always wear a suitable lab coat, disposable gloves, and protective goggles when working with hazardous chemicals and biological reagents. Consult appropriate material safety data sheets available from product suppliers, and follow all waste disposal regulations. Prepare all solutions using ultrapure water prepared by purifying deionized water to a resistance greater than 18 MΩ-cm at 25 °C and analytical grade reagents.

2.1 Plasmids

1. A standard dicistronic CAT-replicating minigenome plasmid (Fig. 1, *see* **Note 1**).
2. "Singe step" minigenome plasmid (*see* **Note 2**).
3. pTM1 plasmids expressing support viral proteins (*see* **Note 3**).
4. Plasmid templates for making ^{32}P-labeled CAT-specific riboprobes (*see* **Note 4**).

2.2 Cells

Here, we describe an RSV minigenome assay that utilizes BSR-T7 cells, an engineered cell line that constitutively expresses T7 RNA polymerase [20] (*see* **Note 5**). Sterile culture techniques and procedures are essential when preparing cell culture media and handling cells. Prepare reagents and cell culture media at room temperature unless otherwise specified.

1. BSR-T7 cells (*see* **Note 6**).
2. BSR-T7 culture media: G-MEM media containing 10% heat-inactivated fetal bovine serum (FBS; *see* **Note 7**), 1× Gluta-MAX, and 1× of Nonessential Amino Acids. Store at 4 °C, protected from light.
3. BSR-T7 culture media containing geneticin (G148 sulfate; 1 mg/mL). Store geneticin at 4 °C and add to culture media during passaging and cell seeding.
4. Serum-free Opti-MEM. Store at 4 °C, protected from light.
5. Opti-MEM containing 2% heat-inactivated FBS, 1× Gluta-MAX. Store at 4 °C, protected from light.
6. Trypsin–EDTA. Store at 4 °C.
7. 1× phosphate-buffered saline (PBS). Store at room temperature.
8. Liposome-based transfection reagent, e.g., Lipofectamine 3000 (Thermofisher/Invitrogen). Store at 4 °C.
9. Tissue culture grade T75 sterile flasks.
10. Tissue culture grade six-well plate dishes (*see* **Note 8**).
11. Biological safety cabinet for cell culture.
12. Cell culture CO_2 incubator with water reservoir.
13. Light microscope.
14. Hemocytometer or cell counter.
15. Cell scrapers.
16. Centrifuge and rotor capable of reaching 14, 000 × *g* that can be chilled to 4 °C.

2.3 RNA Isolation

To avoid RNase contamination, wear gloves and use RNase-free filter tips, sterile plugged serological pipettes, and nuclease-free water where indicated. RNA can be extracted from transfected cells using a commercially available kit or by phenol-based RNA extraction (e.g., using TRIzol) as described below (*see* **Note 9**).

1. RNA extraction kit (optional).
2. Chemical hood.
3. TRIzol.
4. Chloroform (chloroform:isoamyl alcohol 24:1).

5. Isopropanol.
6. NTE buffer: 2 M NaCl, 40 mM Tris pH 7.4, 1 mM EDTA.
7. Acid-phenol (acid-phenol:chloroform 5:1).
8. Ethanol, 200 proof, molecular biology grade.
9. Glycogen (5 mg/mL).

2.4 CAT-Specific Riboprobes

1. Standard dicistronic CAT minigenome plasmid and pGEM3Zf (−)CAT plasmid (*see* **Notes 1** and **4**).
2. 10× RNA polymerase buffer (NEB).
3. 100 mM DTT.
4. Murine RNase inhibitor (NEB).
5. DNase I.
6. Ribonucleotide mix: 2.5 mM ATP, 2.5 mM GTP, 2.5 mM UTP, 0.5 mM CTP.
7. [α-^{32}P]CTP (3000 Ci/mmol, 10 mCi/mL).
8. T7 RNA polymerase (NEB).
9. 500 mM EDTA.
10. Nuclease-free water.
11. Acid-phenol (acid-phenol:chloroform 5:1).
12. G-50 Sephadex column.
13. TE buffer: 10 mM Tris pH 8, 1 mM EDTA.
14. Heat block.
15. Centrifuge.

2.5 Northern Blotting Analysis

1. Agarose.
2. 10× MOPS ((3-N-morpholino) propanesulfonic acid).
3. Formaldehyde (36.5–38%).
4. Microwave.
5. Chemical hood.
6. Deionized formamide.
7. 6X DNA Loading Dye containing bromophenol blue and xylene cyanol for loading RNA on agarose gels.
8. RNA ladder (DynaMarker or similar).
9. Heat block.
10. Power supply and electrical cables.
11. Agarose gel electrophoresis apparatus suitable for a 11 × 14 cm gel and ten-well combs.
12. Whatman TurboBlotter apparatus (suitable for a 11 × 14 cm gel).

13. Nylon membrane or nitrocellulose membrane. Pore size 0.2 μM.
14. 3MM filter paper cut to same size as gel (11 × 14 cm).
15. 3MM filter paper cut to 22 × 14 cm (wick).
16. GB003 blotting paper (11 × 14 cm).
17. *Transfer buffer*: To make 1 L of transfer buffer, weigh 175.32 g of NaCl in a glass beaker. Add 8 mL of 1 N NaOH and make up to 950 mL with water. Stir with a magnetic stir bar until NaCl has dissolved. Transfer to a graduated cylinder and add water to 1 L.
18. Disposable razor blade.
19. UV cross-linker.
20. Plastic container.
21. <10% Methylene blue staining solution.
22. Benchtop platform rocker or similar.
23. 20× saline-sodium citrate (SSC).
24. Rotisserie hybridization oven.
25. Hybridization glass bottles.
26. *Hybridization buffer*: Prepare fresh 20 mL of 6× SSC, 2× Denhardt's solution, 0.1% SDS per northern blot.
27. Sheared salmon sperm DNA (10 mg/mL).
28. *Low-stringency wash buffer*: Prepare 1 L of 2× SSC, 0.1% SDS.
29. *High-stringency wash buffer*: Prepare 1 L of 0.1× SSC, 0.1% SDS.
30. Plastic wrap.
31. X-ray cassette.
32. X-ray film or phosphor imager.

3 Methods

All cell culture steps should be performed in a biological safety cabinet using aseptic techniques. The amount of DNA and reagents described here is optimized for one well of a six-well plate and can be scaled up or down as needed. A control in which the pTM1-L plasmid is omitted from the transfection mixture, or in which the pTM1-L plasmid has been mutated to generate a D811A substitution in the L protein (to inactivate polymerase activity), should be included in the experimental design such that RNA products expressed by the RSV polymerase can be distinguished from background signal (Fig. 1).

3.1 Minigenome Transfection

1. Seed BSR-T7 cells in a six-well plate such that they reach ~70% confluence either 6 h later for a same-day transfection or by the next day (*see* **Note 10**).
2. Check that cells are evenly distributed and ~ 70% confluent at the time of transfection by viewing under a light microscope (*see* **Note 11**).
3. Prepare a DNA and Opti-MEM mix according to Table 1. The amounts listed are for one well of a six-well plate (*see* **Notes 12** and **13**).
4. Next, prepare a lipofectamine 3000 and Opti-MEM mix according to Table 2. The amount listed is for one well of six-well plate (*see* **Note 14**). Vortex briefly and incubate at room temperature for 5 min.
5. Combine the DNA and lipofectamine mixes in a 1:1 ratio such that the final volume will be 250 μL per well according to Table 3. Briefly vortex and gently shake or tap liquids to the bottom of the tube. Incubate 10–15 min at room temperature.
6. Add 750 μL of Opti-MEM to each tube and mix well.
7. Aspirate media from the cell culture plate using a sterile tip, taking care not to disturb cells, and add 1 mL of transfection mix dropwise into each well, dispensing against the well wall. Rock the plate gently to ensure that cells are covered (*see* **Notes 15** and **16**).
8. Incubate cells at 37 °C overnight in cell culture CO_2 incubator (*see* **Note 17**).
9. After the overnight incubation, aspirate media and replace with 1–1.5 mL of pre-warmed cell growth medium (Opti-MEM + 2% FBS; *see* **Note 18**).
10. Incubate 40–48 h as desired (*see* **Note 19**).

Table 1
DNA mix composition per well of a six-well plate

Plasmid/reagent	Concentration (ng/μL)	μL/well	ng/well
pTM1-N	100	4	400
pTM1-P	100	2	200
pTM1-M2-1	100	1	100
pTM1-L	100	1	100
Minigenome	100	2	200
Opti-MEM	–	115	–
Total	–	125	–

Table 2
Lipofectamine mix composition per well of a six-well plate

Reagent	μL/well
Opti-MEM	121
Lipofectamine 3000	4
Total	125

Table 3
Volume of DNA mix and lipofectamine mix to be combined per well of a six-well plate

Transfection mix	Volume to add (μL)
Lipofectamine mix	125
DNA mix	125
Total	250

3.2 Cell Harvesting

1. Label microcentrifuge tubes and place them on ice.
2. Using a sterile cell scraper, scrape cells into media. Transfer the cell suspension into the labeled tubes and immediately place on ice (*see* **Note 20**).
3. Centrifuge samples at ~8000 × g for 5 min at 4 °C.
4. Aspirate supernatant avoiding cell pellet (*see* **Note 21**). Cell pellets can be stored at −80 °C or processed immediately (*see* **Notes 22** and **23**).

3.3 TRIzol RNA Extraction

All steps should be performed in a chemical hood with appropriate PPE. Replace gloves immediately if contact is made with TRIzol, chloroform, or phenol reagents.

1. Resuspend cell pellet in 1 mL of TRIzol and incubate 5 min at room temperature (*see* **Note 24**).
2. Add 200 μL of chloroform (chloroform:isoamyl alcohol 24:1), cap tube securely, and shake vigorously by hand for ~15 s. Incubate for 2–3 min at room temperature.
3. Centrifuge the sample 12,000 ×g for 15 min at 4 °C. Collect the colorless upper aqueous phase containing the RNA, taking care not to touch the interphase, and transfer to a fresh tube containing 50 μL 2 M NTE and 1 μL glycogen (5 mg/mL). Invert or vortex the tube.
4. Add 500 μL of isopropanol. Mix by inverting or vortexing.

5. Incubate at 4 °C for 10 min, then centrifuge 12,000 × *g* for 10 min at 4 °C. Precipitated RNA forms a transparent to white pellet on the side and bottom of the tube that may be hard to see. Carefully remove supernatant without disturbing the RNA pellet, using a fine pipette tip.
6. Allow RNA pellet to dry briefly at room temperature and add 300 μL nuclease-free water. Ensure the RNA pellet has dissolved by allowing it to stand at room temperature for several minutes. Vortex every now and then to aid dissolving.
7. Add 33 μL 2 M NTE and vortex.
8. Add 330 μL acid-phenol (acid-phenol:chloroform 5:1) and mix by shaking tubes for ~15 s.
9. Centrifuge at 4500 × *g* for 3 min at 4 °C.
10. Collect the upper aqueous phase and transfer to tubes containing 330 μL chloroform. Mix by shaking tubes for ~15 s.
11. Centrifuge at 4500 × *g* for 3 min at 4 °C.
12. Collect the upper aqueous phase and transfer to tubes containing 700 μL 100% ethanol. Mix by inverting several times.
13. Precipitate RNA by storing overnight at −80 °C (*see* **Note 25**).
14. Centrifuge samples at 12,000–14,000 × *g* for 20 min at 4 °C to precipitate RNA.
15. Wash RNA pellet with 1 mL of 70% ethanol and centrifuge samples at 12,000–14,000 × *g* for 20 min at 4 °C.
16. Resuspend RNA pellet in 50 μL of nuclease-free water. To maintain RNA integrity, snap freeze on dry ice before storage at −80 °C.

3.4 CAT-Specific Riboprobes

1. To prepare a DNA template for the synthesis of a negative-sense riboprobe linearize ~5 μg of the standard minigenome DNA plasmid with *XbaI*. To prepare a DNA template for the synthesis of a positive-sense riboprobe, digest ~5 μg of the pGEM3Zf(−)CAT plasmid with *PstI*. Gel extract linearized plasmids and elute the DNA into 50 μL of water or 10 mM Tris–HCl pH 8.
2. To make the RNA riboprobes, combine the following reagents listed in Table 4 (*see* **Notes 26** and **27**).
3. Incubate for 1 h at 37 °C.
4. Add a further 1 μL of Murine RNase inhibitor and 2 μL of DNase I to the reaction and incubate 10 min at 37 °C.
5. Add 1 μL of 500 mM EDTA and 25 μL of water and mix by pipetting.

Table 4
Riboprobe reaction mix

Reagent	Volume (μL)	Final concentration/amount
10× RNA polymerase buffer (NEB)	2.5	1×
100 mM DTT	2.5	10 mM
Murine RNase inhibitor (NEB)	1	40 U
Ribonucleotide mix 2.5 mM ATP, 2.5 mM GTP, 2.5 mM UTP, 0.5 mM CTP	2.5	250 μM ATP 250 μM GTP 250 μM UTP 50 μM CTP
[α-^{32}P]CTP (3000 Ci/mmol, 10 mCi/ml)	10	100 μCi
Linearized and gel-extracted plasmid template	4.5	~ 500 ng
T7 RNA polymerase (NEB)	2	100 U
Total	25	–

6. Next, add 54 μL of acid-phenol, securely cap the tube, and vortex briefly. The upper aqueous phase contains the RNA riboprobe.
7. To remove unincorporated radiolabeled tracer, wash a G-50 Sephadex column with TE buffer, centrifuge 735 × *g* for 1 min at room temperature, discard the flow through, and repeat the wash two more times. Next, apply the upper aqueous containing the RNA riboprobe to the washed G-50 Sephadex column, and centrifuge 735 × *g* for 2 min at room temperature.
8. Collect the flow through and combine with 350 μL of TE buffer to a final volume of ~400 μL.
9. Store radioactive riboprobes at −20 °C until ready for use.

3.5 Northern Blotting Analysis

The minigenome, anti-minigenome, and CAT mRNAs expressed in the minigenome system can be analyzed by northern blotting. This involves the separation of RNA on a denaturing gel, followed by the downward transfer of the RNA onto a nylon or nitrocellulose membrane using a TurboBlotter system (Whatman). Transferred RNAs are UV cross-linked to the membrane and then hybridized to a positive- or negative-sense riboprobe containing CAT sequence. Finally, hybridized RNAs are visualized by autoradiography or phosphor imaging.

1. Prepare denaturing 1.5% agarose MOPS/formaldehyde gel just before. Set up gel rig according to manufacturer's instructions, and tape ends of casting tray to avoid leakage if needed (*see* **Note 28**).

Table 5
RNA sample and ladder preparation for gel electrophoresis

Reagent	Volume (μL)
RNA (~ 2 μg or ~ 1/10 of a well) or RNA ladder	5
10× MOPS	2
Formaldehyde 36.5–38%	7
Deionized formamide	20
Nuclease-free water	6
Total	40

2. Combine 1.5 g agarose with 86.4 mL water in a 250-mL glass bottle, and microwave until the agarose has completely dissolved (*see* **Note 29**).
3. Without cooling, in a chemical hood add 10 mL of 10× MOPS and 3.6 mL of formaldehyde. Mix by swirling and cool for 1–2 min before pouring gel (*see* **Note 30**). Immediately add ten-well comb (*see* **Note 31**) and remove any bubbles in the gel using a pipette before the gel begins to solidify.
4. Prepare each RNA sample according to Table 5 in a chemical hood (*see* **Notes 32** and **33**).
5. Mix well and heat samples to 65 °C for 10 min. Place samples on ice.
6. Add 6.6 μL of 6× loading buffer into each sample. Briefly vortex and pop spin the samples to bring down condensate. Place samples on ice.
7. Once the 1.5% agarose MOPS/formaldehyde gel has solidified (~ 1 h), remove tape if used and pour 1× MOPS running buffer into the rig. Gently remove comb.
8. Load 45 μL of each RNA sample or ladder per well.
9. Run the gel at 90–100 V until the bromophenol blue dye front has migrated about ¾ of the gel.
10. While the gel is running, cut the nylon or nitrocellulose membrane to the same size of the gel (11 × 14 cm). If nylon is being used, pre-wet by floating in transfer buffer at room temperature with gentle rocking. If nitrocellulose is being used, pre-soak the membrane in water heated to 65 °C (*see* **Note 34**). Prepare a wick by cutting 3MM filter paper to 22 × 14 cm.
11. Assemble TurboBlotter according to Fig. 2. Stack 25 sheets of GB003 followed by three dry 3MM papers in the center of the stacking tray.

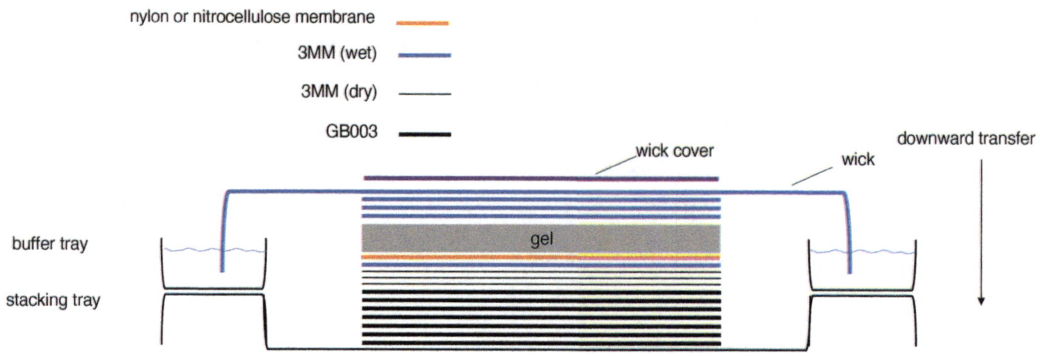

Fig. 2 A side view schematic depiction of a northern blot transfer sandwich. From bottom to top: 25 × dry GB003 sheets (thick black lines), 3 × dry 3MM sheets (thin black lines), 1 × 3MM sheet presoaked in transfer buffer (blue lines), nylon or nitrocellulose membrane (orange line), 1.5% agarose/MOPs gel, 3 × 3MM sheets presoaked in transfer buffer, 1 × 3MM wick presoaked in transfer buffer and a wick cover (purple). The wick is inserted into the transfer buffer on either side of the stack to initiate transfer

12. Briefly wet one 3MM sheet in transfer buffer before placing it on the top of the stack.
13. Next, add the pre-wet nylon or nitrocellulose membrane.
14. Carefully lay the agarose gel on top of the membrane, avoiding the introduction of any bubbles between the gel and the membrane.
15. Trim the meniscus created at the edges of the gel and around the wells using a fresh razor blade, so that the top of the gel is flat with no raised edges.
16. Next, briefly wet three more 3MM papers in transfer buffer and add to the top of the gel, smoothing out any bubbles that may occur (*see* **Note 35**).
17. Attach the buffer tray to the stacking tray and fill with transfer buffer.
18. To initiate the transfer, briefly wet the wick in transfer buffer and place the wick on top of the gel, immersing both sides of the wick in the transfer buffer tray.
19. Finally, place the provided wick cover on top of the stack.
20. Allow transfer to continue for at least 4 h, preferably overnight (*see* **Note 36**).
21. Following transfer, remove the upper wick and 3MM sheets. Mark the wells on the nylon or nitrocellulose membrane with a pencil.
22. Neutralize the membrane blot in 6× SSC buffer for 10–15 min on a rocker at room temperature.

23. Dry the membrane on 3MM filter paper to remove dampness, then immediately UV cross-link at 120,000 μJ/*cm*^{2} (*see* **Notes 37** and **38**).
24. To control for RNA loading, place membrane in a plastic container and apply 50 mL of methylene blue. Gently rock on a benchtop rotator at room temperature for 10 min until bands are visible. Wash with deionized water for 5–20 min and repeat wash two more times. Dry membrane on 3MM filter paper as before and collect an image of the gel.
25. To proceed with probe hybridization, pre-warm a hybridization bottle to 65 °C. This helps to prevent leakage from bottle once buffer is added.
26. Add the UV cross-linked membrane to the hybridization bottle along with 20 mL of hybridization buffer. Place in hybridization oven with rotation at 65 °C.
27. Meanwhile, heat 200 μL of sheared salmon sperm DNA to 95 °C in a heat block for 5 min. Add directly to the hybridization bottle containing the filter and buffer to give a final concentration of 0.1 mg/mL of DNA.
28. Pre-hybridize for 1 h at 65 °C with rotation.
29. Add 20–30 μL of appropriate radiolabeled RNA probe and hybridize overnight at 65 °C (*see* **Note 39**).
30. Carefully remove the RNA probe mix, briefly rinse the membrane with low-stringency buffer, and add a further ~20 mL of low-stringency wash buffer to the hybridization bottle. Wash at 65 °C for at least 1 h with rotation.
31. Decant wash buffer and repeat the low-stringency wash at 65 °C for 1 h.
32. Perform an additional wash using ~20 mL of high-stringency wash buffer, washing at 65 °C for 15 min with rotation.
33. Using tweezers, remove the membrane from the glass bottle and place it on 3MM paper. Allow it to dry to dampness and cover with plastic wrap.
34. Place membrane in an X-ray cassette and either expose to film with an intensifying screen at −80 °C or a phosphor screen at room temperature as desired.

4 Notes

1. *A standard dicistronic CAT-replicating minigenome* (Fig. 1): A functional RSV minigenome consists in 3′ to 5′ order, the 44-nt le region, the 10-nt NS1 gene start (gs) signal, 32-nt of the NS1 gene non-translated region (NTR), a dicistronic

negative-sense copy of the CAT open reading frame (ORF) in which the viral N–P gene junction, consisting of the N gene end (ge) signal, the 1-nt N–P intergenic region, and P gs signal, is inserted into a CAT ORF to create two genes CAT 1 (580-nt, including the gs and ge sequences) and CAT 2 (190-nt, including the gs and ge sequences) that are expressed as two mRNAs, the last 12-nt of the NTR of the L gene, the 12-nt L ge signal, followed by 155-nt tr region. A T7 RNA polymerase promoter abuts three nonviral G residues (to increase T7 expression of the minigenome), followed by the 5′ end of the tr sequence; while a hepatitis delta ribozyme and T7 terminator sequence abut the 3′ terminus of the le. Ribozyme-mediated self-cleavage of the T7-expressed RNA generates a minigenome RNA with a correct 3′ terminus [6, 21].

2. *"Singe step" minigenome plasmid*: To analyze transcription and genome replication as independent events, a "single step" minigenome can be created by either introducing a C-to-G substitution at position 2 relative to the 5′ end of the tr or deleting the 22-nt from the 5′ end of the tr [5, 6, 13].
3. *pTM1 plasmids expressing support viral proteins*: The viral proteins necessary for transcription and replication are N, P, L, and M2–1 [1]. The ORFs of N, P, M2–1, and L protein are inserted into pTM1 plasmid, such that they are under transcriptional control of a T7 RNA polymerase promoter and are under translation regulation by the internal ribosome entry site of encephalomyocarditis virus. Each ORF is inserted such that the ATG that codes for initiation is inserted at the *NcoI* site (CCATGG) [22]. It is recommended that codon-optimized ORFs are used for the support plasmids to increase the overall efficiency of the minigenome system.
4. *Plasmid templates for making ^{32}P-labeled CAT-specific riboprobes*: CAT-specific riboprobes are synthesized in vitro with a T7 RNA polymerase and digested plasmid templates. To generate a negative-sense probe that will bind positive-sense antigenome and mRNAs, the standard dicistronic CAT minigenome plasmid is linearized by restriction digest with *XbaI* (Fig. 1). For a positive-sense probe capable of binding negative-sense minigenome RNA, CAT sequences from the standard dicistronic CAT minigenome are inserted between *XbaI* and *PstI* sites of pGEM3Zf(−) (Promega) to create pGEM3Zf(−)CAT plasmid which is then linearized with *PstI* [11].
5. The RSV minigenome assay also can be performed with HEp-2 cells infected with modified vaccinia virus Ankara-T7 [23] to express the T7 RNA polymerase or co-transfected with a

plasmid expressing T7 such as pCAGGs-T7. However, it should be noted that using plasmid-expressed T7, the minigenome system is unlikely to be sufficiently robust for northern blot RNA analysis.

6. Cells should be cultured at a constant 37 °C, CO_2 at 5–10%, and 100% relative humidity.
7. To heat inactivate complement proteins present in FBS, heat the FBS to 56 °C for 30 min in a sterile setting. Wipe bottle with 70% ethanol before preparing aliquots in sterile tubes that can be stored at 4 °C or at −20 °C for longer periods (allow extra room in the tubes for FBS expansion if freezing).
8. Transfections can be performed in smaller or larger cell culture plates if scaled appropriately. For RNA analyses, it is recommended that a six-well culture plate is used.
9. RNA can be extracted from transfected cells using a commercially available kit. However, the size of the RNA that is to be analyzed needs to be considered, for example, some column-based kits do not capture small RNAs, such as the 190-nt CAT 2 mRNA. An alternative method is to perform a phenol-based RNA extraction (e.g., using TRIzol), which will capture a wider range of RNA sizes.
10. For optimal T7 expression, include geneticin at a final concentration of 1 mg/mL in the cell seeding media.
11. Ensure cells are not overconfluent before transfecting, as this will reduce the transfection efficiency.
12. Scale up as needed if multiple wells are to be transfected, allowing extra for pipetting errors.
13. The same amounts of DNA are used to transfect HEp-2 cells.
14. Scale up as needed, allowing extra for pipetting errors.
15. If antiviral compounds are to be investigated, compounds (usually diluted in Dimethyl sulfoxide) can be added immediately following the addition of the transfection media to the well.
16. Ensure that the plates lie flat in the incubator if they are stacked, to avoid uneven transfection.
17. Avoid prolonged application of the DNA/lipofectamine transfection mix. Media replacement is recommended on day 2 of the transfection to help minimize the toxic effects of the lipofectamine 3000 reagent.
18. If antiviral compounds have been added to the transfection, they should be replaced once the fresh media have been added.
19. Harvesting times may require optimization. The length of the assay will be dependent on the particular research goals, the time it takes for the reporter gene to reach steady-state

expression and may be influenced by the type of cell transfected. In the case of a minigenome expressing GFP, GFP expression can be detected as early as 24 h post-transfection by fluorescence microscopy.

20. Use a fresh cell scraper to harvest each well, taking care to check that there are no cells left behind.
21. If a vacuum line is not available for aspiration, remove culture media from cells by pipetting.
22. If the product of the transfected gene is sensitive to degradation, immediately freeze the harvested cell pellet on dry ice before storing at −80 °C.
23. Micrococcal nuclease (MCN) digestion of RNA: To enrich for N-encapsidated minigenome and mini-antigenomes, cells can be treated with MCN to remove unencapsidated RNA prior to RNA extraction [2]. For MCN analysis of RNA, each transfection should be set up in duplicate wells. Following harvesting, one cell pellet is RNA extracted directly (using either TRIzol or an RNA extraction kit) to generate a "total" or untreated RNA sample while the duplicate cell pellet is digested with MCN. To MCN treat, resuspend the cell pellet in 100 μL of lysis buffer: 10 mM NaCl, 10 mM Tris, pH 7.5, 1.5 mM $MgCl_2$, 1% Triton X-100, 0.5% sodium deoxycholate, 10 mM $CaCl_2$, 0.9 K.I.U aprotinin, and 30 U of MCN S7. Incubate lysate for 1 h at 30 ° C and use either TRIzol or an RNA extraction kit to isolate RNA.
24. Samples in TRIzol can be stored at −80 °C.
25. This is a stopping point in the TRIzol RNA extraction protocol.
26. Follow all appropriate radiation safety guidelines when handling radioactive material.
27. Alternative radiolabel tracers can be used when preparing the riboprobe. If this is done, the ribonucleotide mix must be adjusted such that the final concentration of the unlabeled NTP corresponding to the tracer is now 50 μM, while the other NTPs are all 250 μM each. This is to ensure efficient incorporation of the label.
28. Always handle rigs with gloves. If RNase contamination is possible, wash with a concentrated solution of SDS (10%). Rinse with tap water, followed by deionized water.
29. The amount is adjusted to 11 × 14 horizontal gel rig. The total volume should be calibrated depending on the size of the rig used.
30. Try to avoid introducing bubbles by gently swirling the agarose mixture to mix.

31. If desired, a double-tiered gel can be prepared by inserting a second comb halfway down the gel.
32. When more than one RNA sample is to be analyzed, prepare a master mix containing 10× MOPS, formaldehyde, deionized formamide, and water for all samples, then combine 35 μL of the master mix with each RNA sample.
33. It is recommended that an RNA marker is also run out. Prepare the RNA ladder using the reagents listed in Table 5 and proceed with **step 5**.
34. Ensure the membrane is thoroughly soaked in transfer buffer or water as appropriate. When handling the membrane, grab it at the edge using a tweezers.
35. Bubbles between the 3MM, gel, and membrane will inhibit RNA transfer. If bubbles are present, do not attempt to lift the agarose gel from the membrane, since the RNA transfer starts immediately. Bubbles can be very gently smoothed out using a roller.
36. Transfer time will differ according to the thickness of the gel and the size of nucleic acids.
37. Dry until no more puddles are left. Do not overdry as the membrane will begin to curl.
38. Following UV cross-linking, the membrane can be stored between 3MM filter papers in a plastic bag.
39. Be sure to follow all radiation safety protocols appropriately when handling the riboprobes and hybridized blots.

Acknowledgments

The authors thank Dr. Barbara Ludeke for generously providing an example of a northern blot analysis of RNA expressed in the "single cycle" minigenome assay; Dr. Rachel Fearns and other members of her laboratory for helpful discussions and critical review of the manuscript.

References

1. Grosfeld H, Hill MG, Collins PL (1995) RNA replication by respiratory syncytial virus (RSV) is directed by the N, P, and L proteins; transcription also occurs under these conditions but requires RSV superinfection for efficient synthesis of full-length mRNA. J Virol 69(9): 5677–5686. https://doi.org/10.1128/jvi.69.9.5677-5686.1995
2. Fearns R, Peeples ME, Collins PL (1997) Increased expression of the N protein of respiratory syncytial virus stimulates Minigenome replication but does not Alter the balance between the synthesis of mRNA and Antigenome. Virology 236(1):188–201. https://doi.org/10.1006/viro.1997.8734
3. Fearns R, Collins PL (1999) Role of the M2-1 transcription Antitermination protein of respiratory syncytial virus in sequential transcription. J Virol 73(7):5852–5864. https://doi.org/10.1128/JVI.73.7.5852-5864.1999

4. Fearns R, Collins PL, Peeples ME (2000) Functional analysis of the genomic and Antigenomic promoters of human respiratory syncytial virus. J Virol 74(13):6006–6014. https://doi.org/10.1128/JVI.74.13.6006-6014.2000
5. Peeples ME, Collins PL (2000) Mutations in the 5′ trailer region of a respiratory syncytial virus Minigenome which limit RNA replication to one step. J Virol 74:146
6. Fearns R, Peeples ME, Collins PL (2002) Mapping the transcription and replication promoters of respiratory syncytial virus. J Virol 76(4):1663–1672. https://doi.org/10.1128/JVI.76.4.1663-1672.2002
7. Beavis AC, Tran KC, Barrozo ER et al (2021) Respiratory syncytial virus phosphoprotein residue S156 plays a role in regulating genome transcription and replication. J Virol 95(24): e01206–e01221. https://doi.org/10.1128/JVI.01206-21
8. Braun MR, Deflubé LR, Noton SL et al (2017) RNA elongation by respiratory syncytial virus polymerase is calibrated by conserved region V. PLoS Pathog 13(12):e1006803. https://doi.org/10.1371/journal.ppat.1006803
9. Cressey TN, Shareef AM, Kleiner VA et al (2022) Distinctive features of the respiratory syncytial virus priming loop compared to other non-segmented negative strand RNA viruses. PLoS Pathog 18(6):e1010451. https://doi.org/10.1371/journal.ppat.1010451
10. Cowton VM, Fearns R (2005) Evidence that the respiratory syncytial virus polymerase is recruited to nucleotides 1 to 11 at the 3′ end of the nucleocapsid and can scan to access internal signals. J Virol 79:11311
11. McGivern DR, Collins PL, Fearns R (2005) Identification of internal sequences in the 3′ leader region of human respiratory syncytial virus that enhance transcription and confer replication processivity. J Virol 79:2449
12. Noton SL, Cowton VM, Zack CR et al (2010) Evidence that the polymerase of respiratory syncytial virus initiates RNA replication in a nontemplated fashion. Proc Natl Acad Sci USA 107(22):10226–10231. https://doi.org/10.1073/pnas.0913065107
13. Noton SL, Fearns R (2011) The first two nucleotides of the respiratory syncytial virus antigenome RNA replication product can be selected independently of the promoter terminus. RNA 17(10):1895–1906. https://doi.org/10.1261/rna.2813411
14. Tremaglio CZ, Noton SL, Deflubé LR et al (2013) Respiratory syncytial virus polymerase can initiate transcription from position 3 of the leader promoter. J Virol 87(6):3196–3207. https://doi.org/10.1128/JVI.02862-12
15. Noton SL, Nagendra K, Dunn EF et al (2015) Respiratory syncytial virus inhibitor AZ-27 differentially inhibits different polymerase activities at the promoter. J Virol 89(15): 7786–7798. https://doi.org/10.1128/JVI.00530-15
16. Wang L, Zhu Q, Xiang K et al (2021) Discovery of a novel respiratory syncytial virus replication inhibitor. Antimicrob Agents Chemother 65(6):e02576–e02520. https://doi.org/10.1128/AAC.02576-20
17. Sourimant J, Lieber CM, Aggarwal M et al (2022) 4′-Fluorouridine is an oral antiviral that blocks respiratory syncytial virus and SARS-CoV-2 replication. Science 375(6577): 161–167. https://doi.org/10.1126/science.abj5508
18. Kleiner VA, O. Fischmann T, Howe JA et al (2023) Conserved allosteric inhibitory site on the respiratory syncytial virus and human metapneumovirus RNA-dependent RNA polymerases. Commun Biol 6(1):649. https://doi.org/10.1038/s42003-023-04990-0
19. Teng MN, Tran KC (2016) Use of Minigenome systems to study RSV transcription. Methods Mol Biol 1442:155–164
20. Buchholz UJ, Finke S, Conzelmann K-K (1999) Generation of bovine respiratory syncytial virus (BRSV) from cDNA: BRSV NS2 is not essential for virus replication in tissue culture, and the human RSV leader region acts as a functional BRSV genome promoter. J Virol 73(1):251–259. https://doi.org/10.1128/JVI.73.1.251-259.1999
21. Perrotta AT, Been MD (1991) A pseudoknot-like structure required for efficient self-cleavage of hepatitis delta virus RNA. Nature 350(6317):434–436. https://doi.org/10.1038/350434a0
22. Elroy-Stein O, Fuerst TR, Moss B (1989) Cap-independent translation of mRNA conferred by encephalomyocarditis virus 5′ sequence improves the performance of the vaccinia virus/bacteriophage T7 hybrid expression system. Proc Natl Acad Sci USA 86(16): 6126–6130. https://doi.org/10.1073/pnas.86.16.6126
23. Fuerst TR, Niles EG, Studier FW et al (1986) Eukaryotic transient-expression system based on recombinant vaccinia virus that synthesizes bacteriophage T7 RNA polymerase. Proc Natl Acad Sci USA 83(21):8122–8126. https://doi.org/10.1073/pnas.83.21.8122

Chapter 9

Secretory Expression and Purification of Respiratory Syncytial Virus G and F Ectodomain in Human Cells

Ranjini Sankaranarayanan, Binh Ha, Samadhan J. Jadhao, and Larry J. Anderson

Abstract

Respiratory syncytial virus (RSV) is one of the leading causes of respiratory viral infections in infants, immunocompromised individuals, and the elderly. The important role RSV F and G proteins play in disease progression and their ability to induce neutralizing antibodies make them desirable targets for vaccine production. Access to these expressed and purified proteins is also important to RSV research and diagnostics. Here, we describe methods to produce recombinant, secretory RSV F and G ectodomain in human cells and purify these proteins using Ni-Sepharose affinity chromatography.

Key words Respiratory syncytial virus, Expression, F protein, G protein, Western blot, Affinity chromatography

1 Introduction

The respiratory syncytial virus (RSV) genome encodes 11 distinct proteins, and three of these proteins, the attachment glycoprotein G, the fusion glycoprotein F, and the small hydrophobic protein SH, are expressed at the surface of the virus [1]. The F and G proteins control the initial stages of the disease and are capable of inducing RSV neutralizing antibodies [2]. In addition to inducing antibodies that neutralize virus in vitro and in vivo by preventing virus from infecting cells, F and G proteins may also induce antibodies that neutralize virus in vivo through Fc-mediated mechanisms such as antibody-dependent cellular cytotoxicity (ADCC) and complement-mediated neutralization [3, 4]. Their role in protective immunity makes measuring anti-G and anti-F protein antibodies important to understanding RSV immune responses.

The full-length G protein contains 289–319 amino acid residues, depending on the strain, and is a heavily glycosylated type II

Bo Liang and Rachel Fearns (eds.), *Human Respiratory Syncytial Virus: Methods and Protocols*, Methods in Molecular Biology, vol. 2948, https://doi.org/10.1007/978-1-0716-4666-3_9,

transmembrane protein. The G protein mediates attachment to the cell by binding to heparin-like motifs [5] and the chemokine receptor, CX3CR1 [6]. The predicted molecular weight of G protein is about 32–33 kDa, but heavy glycosylation with N- and O-linked sugars increases its actual molecular weight to about 90 kDa [7–9]. The F protein, involved in fusion and binding to the host cell, is initially synthesized as a precursor protein (F_0) that is 574 amino acids long. The F_0 monomers then form a homotrimer that is activated in the Golgi complex through cleavage by a host furin-like protease, which cleaves the monomer into three peptide subunits—F2 (N-terminal peptide) and F1 (C-terminal peptide) connected by disulfide bonds, and a 27 amino acid peptide, p27, which dissociates after protease cleavage, generating the prefusion form of the F protein [2]. The predicted molecular weight of the F protein under reducing conditions is about 70 kDa [8]. The F protein in contrast to G protein has only N-linked glycosylation [10, 11].

Two major subgroups of RSV, A and B, have been identified with monoclonal antibodies [12–14] and from sequence studies [15, 16] based on the variations in the G protein sequence. The F protein is less variable than the G protein and induces cross-reacting and cross-protective antibodies, while most of the antibodies the G protein induces are subgroup- or genotype-specific. Genotype being strains with distinct G protein sequences within the subgroups. Thus, a sensitive anti-G protein antibody EIA should include the G protein from both subgroups A and B, while anti-F protein antibodies can reliably be detected with the F protein from the subgroup A or B strain [12–14].

A variety of methods have been used to obtain purified RSV F and G proteins, including immune-affinity, immune precipitation, and SDS-PAGE [6, 8, 10, 17, 18]. Some of these techniques have been associated with the loss of some antigenicity of the RSV G protein [18]. A milder, one buffer system used for simultaneous chromatographic purification of RSV F and G proteins from RSV-infected HEp-2 cells that does not denature the protein and may maintain antigenicity has also been suggested for antigen purification [19]. Here, we describe the expression and column affinity purification of the secretory form of RSV G and F ectodomain from human cells.

2 Materials

2.1 Transfection of 293-F Human Cells for Secretion of RSV F and G Ectodomain

1. Suspension FreeStyle™ 293-F cells.
2. Endotoxin-free purified pcDNA/5TO-RSV G-His and pcDNA/5TO-RSV F-His plasmids.
3. Lipofectamine™ 3000.

4. Opti-MEM® I Reduced Serum Medium.
5. FreeStyle™ 293 Expression Medium.
6. 125 mL polycarbonate, disposable, sterile Erlenmeyer flasks with vented caps.
7. Orbital shaker in 37 °C incubator with a humidified atmosphere of 8% CO_2.
8. Blasticidin.
9. Puromycin.

2.2 Purification of RSV G or F Ectodomain

1. Nickel Sepharose Excel (Millipore Sigma; Cat no: GE17371201).
2. High-purity imidazole.
3. Equilibration buffer: 20 mM sodium phosphate, 0.5 M NaCl, pH 7.4.
4. Wash buffer: 20 mM sodium phosphate, 0.5 M NaCl, 10 mM imidazole, pH 7.4.
5. Elution buffer: 20 mM sodium phosphate, 0.5 M NaCl, 500 mM imidazole, pH 7.4.

2.3 Detection of RSV G and F Ectodomain by Western Blotting

1. 2× Laemmli sample buffer (Bio-Rad; Cat no: 161-0737).
2. Running buffer (Tris/glycine/SDS): 50 mM Tris, 190 mM glycine, 0.1% SDS.
3. Transfer buffer: 25 mM Tris, 190 mM glycine, 20% methanol.
4. Tris-buffered saline with Tween-20 (TBST) buffer: 20 mM Tris pH 7.5, 150 mM NaCl, 0.1% Tween-20.
5. Blocking buffer: 5% skim milk powder in TBST.
6. 3D3 anti-RSV-G monoclonal antibody (obtained from Trellis Bioscience).
7. Motavizumab anti-RSV-F monoclonal antibody (MedImmune).
8. Goat-anti-human IRDye800CW-conjugated secondary antibody (LicorBio; Cat no: 926-32232).

3 Methods

3.1 Transfection of FreeStyle 293-F Cells with pcDNA-G-His and pcDNA-F-His

1. Seed FreeStyle™ 293-F 6/TR cells at a density of 7×10^5 viable cells/ml in FreeStyle Expression Medium supplemented with Blasticidin (5 μg/mL) and incubate at 37 °C on an orbital shaker. Change media every 3–4 days until the concentration is $3–4 \times 10^6$ cells/mL (*see* **Note 1**).

Fig. 1 Schematic diagram of a cytomegalovirus promoter-driven plasmid construct encompassing the ectodomain of RSV G (or F) gene open reading frame flanked upstream by a mammalian secretory signal sequence and downstream by 6× -Histidine tag with transcription termination signal. The transfection of plasmid construct in 293-F human cells results in the secretion of RSV G/F protein in cell culture supernatant, as shown in Fig. 2

2. For each 15 mL transfection, prepare lipid–DNA complexes by diluting 15 μg of plasmid DNA (Fig. 1) in Opti-MEM® I to a total volume of 0.5 mL and mix by vortexing.
3. Dilute 30 μL of Lipofectamine™ in Opti-MEM® I to a total volume of 0.5 mL and incubate for 10 min.
4. Mix the DNA–Opti-MEM® mixture with the Lipofectamine™–Opti-MEM® mixture by vortexing. Incubate for 30 min at RT.
5. Transfer a total of 15×10^6 FreeStyle™ 293-F 6/TR cells into a sterile disposable 125-mL Erlenmeyer flask and add 15 mL pre-warmed complete media. Add DNA–Lipofectamine mixture and incubate the cells at 37 °C and 8% CO_2 under shaking condition for 48–72 h. To the transfection control flask (serving as a negative control), add 1 mL of Opti-MEM® I to the cells to get a total volume of 15 mL (*see* **Note 2**).
6. Change the media and select the transfected cells by adding appropriate antibiotics (in this case, Puromycin at a concentration of 1 μg/mL). Change the selection media every 3–4 days until cells reach a stable growth phase in the selection media (usually 2–3 weeks; *see* **Note 3**).
7. Following selection, after the transfected cells have reached a density of $3–4 \times 10^6$ cells/mL, induce protein expression by adding doxycycline at a concentration of 2 μg/mL, incubate for 72 h. Harvest cell supernatant by centrifuging the cell culture at 850 × g (*see* **Note 4**).
8. Filter cell supernatant through a 0.45-μm filter to remove cell debris. Chill the medium containing secreted RSV G or RSV F proteins on ice. This preparation can now be used in EIAs to detect serum antibodies against the RSV F or G proteins (as explained in the Chapter 2).

3.2 Purification of RSV G or F Ectodomain

1. Transfer 1 mL of Nickel Sepharose Excel medium slurry sediment into the purification column (*see* **Note 5**) and rinse the column with distilled water.

2. Equilibrate the column by adding 5 column volumes (CVs) of equilibration buffer 3–4 times. Following equilibration, remove the beads (in equilibration buffer) from the column, add them to the filtered media containing the secreted protein, and rotate the tube overnight at 4 °C.
3. Add the bead–media mixture to the purification column, allowing the beads to settle and collect the flow through in a fresh tube (*see* **Note 6**).
4. Once the beads have settled and are packed well, wash the column at least 4–5 times with 5× CVs of wash buffer. To prevent nonspecific binding of the unwanted cellular proteins, the concentration of imidazole in the wash buffer could be increased to 45 mM imidazole, pH 7.4 (*see* **Notes 7, 8,** and **9**).
5. The protein can then be eluted by adding 1 mL of elution buffer and allowing it to sit for 2–3 min before collecting the eluate.
6. Collect the 1 mL elution fractions for a total of five fractions.
7. Determine the protein concentration for each eluate either by measuring the absorbance at 280 nm and/or SDS-PAGE and Coomassie staining.

3.3 Detection of RSV G and F Ectodomain by Western Blotting

1. Take 1–2 μg of each sample and add an equal volume of 2× Laemmli sample buffer and boil each sample at 95 °C for 5 min.
2. Load purified recombinant RSV F or RSV G protein onto an SDS-PAGE gel, along with molecular weight markers.
3. Run the gel at 100 V till the proteins reach the separating gel, and then increase the voltage to 150 V till the dye runs off.
4. Assemble the transfer sandwich, using a nitrocellulose membrane of appropriate size, and make sure no air bubbles are trapped in the sandwich.
5. Transfer at 100 V for 1.5–2 h at 4 °C.
6. Block in blocking buffer (5% skimmed milk in TBST) for 1 h at RT.
7. Incubate the membrane with 3D3 mAb (for detecting G protein) and motavizumab mAb (for detecting F protein) diluted in blocking buffer (1:2000 dilution) for 2 h at RT or 4 °C overnight.
8. Wash the membrane three times for 15 min with TBST.
9. Incubate with goat-anti-human IRDye800CW-conjugated secondary antibody diluted in blocking buffer (1:10,000) for 30–45 min at RT.
10. Wash the membrane three times for 15 min with TBST.

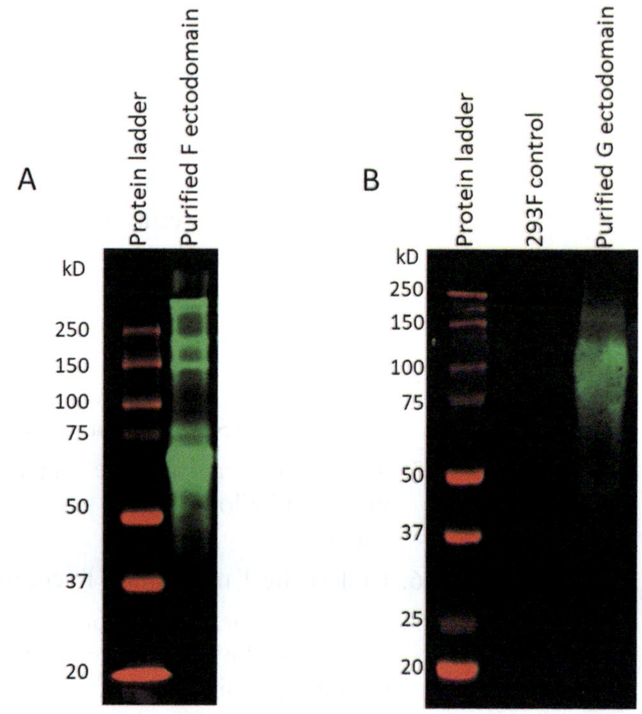

Fig. 2 Western blot reactivity of 293-F human cell culture supernatant, 72 h post-induction for (**a**) recombinant secreted RSV A2 virus F protein labeled with anti-RSV F mAb, motavizumab, and (**b**) recombinant secreted RSV A2 virus G protein with anti-RSV G mAb 3D3

11. Scan the membrane using LI-COR Imaging system. Figure 2 represents a western blot, detecting the secreted RSV-F and -G proteins, from the 293-F human cell culture supernatant, 72 h post-induction.

4 Notes

1. FreeStyle 293-F 6/TR cells are suspension cells that need to be continuously shaken in culture to be healthy. Make sure the cells are not stagnant for more than 1 h.
2. Be sure to have healthy cells for transfection, as this determines how efficient the transfection will be. Indications of healthy cells include cells being rounded and cells not aggregating.
3. Growing the transfected cells in complete media for 24–48 h will ensure cell recovery and multiplication. This is essential before selecting the transfected cells. Note that this flask contains both transfected and non-transfected cells.
4. The cell density will increase for the first several days and decrease from the effects of the antibiotic in non-transfected

cells (about 1 week) following selection. This indicates that the non-transfected cells have been eliminated and what remains is the transfected cells. Growing these cells further until cells reach log phase to ensure complete selection of transfected cells at appropriate density.

5. Centrifugation of the cell culture media can be done twice to ensure maximal cell pelleting before filtration of the cell media.
6. The tip of the purification column can be cut to facilitate larger and/or faster flow of buffers.
7. It is good practice to collect and check for any residual unbound His-tagged protein in the flow through. If found, the purification steps can be repeated using the flow through.
8. Washing can be done by disturbing the beads to ensure contact with the wash buffer and then allowing the beads to resettle.
9. The concentration of imidazole can be increased up to 45 mM in the washing buffer to prevent nonspecific binding of proteins to the column. One should also keep in mind the affinity of the protein of interest to the column when deciding the imidazole concentration of the wash buffer. It is better to batch test the concentration of imidazole compatible with the protein of interest before scaling up purification.

Acknowledgments

This work was supported in part by funding from Emory University to Dr. Larry Anderson and from NIH through the Vaccine and Treatment Evaluation Units (VTEU) awarded to Emory University (HHSN272201300018I). The content is solely the responsibility of the authors and does not necessarily reflect the official views of the National Institutes of Health. The authors thank Dr. Paul Spearman and Dr. Xuemin Chen for providing 293-F 6/TR human cell line.

Conflict of Interest Statement In the last 3 years, Dr. Larry Anderson has done paid consultancies on RSV vaccines for GSK, Janssen, and AstraZeneca, on RSV antiviral drugs for Enanta, and on influenza virus vaccines for Pfizer, and Dr. Anderson's laboratory is currently receiving funding through Emory University from Pfizer for laboratory studies for RSV surveillance studies in adults and maternal antibody transfer to infants, and from Sciogen for animal studies of RSV vaccines. Previously, Dr. Anderson received funding from Vernagen, LLC for animal studies of RSV and other vaccines. Dr. Anderson is a co-inventor on several CDC patents on the RSV G protein and its CX3C chemokine motif relative to immune therapy and vaccine development and on a device to capture coughed droplets from the lung, as well as a co-inventor on Emory patent filing for use of RSV platform VLPs with the F and G proteins for vaccines and G protein constructs for RSV vaccines.

References

1. Collins PL, McIntosh K, Chanock RM (1996) Respiratory syncytial virus. Fields virology, 3 edn. Lippincott Raven Press, Philadelphia
2. McLellan JS, Ray WC, Peeples ME (2013) Structure and function of respiratory syncytial virus surface glycoproteins. Curr Top Microbiol Immunol 372:83–104. https://doi.org/10.1007/978-3-642-38919-1_4
3. Hiatt A, Bohorova N, Bohorov O, Goodman C, Kim D, Pauly MH, Velasco J, Whaley KJ, Piedra PA, Gilbert BE, Zeitlin L (2014) Glycan variants of a respiratory syncytial virus antibody with enhanced effector function and in vivo efficacy. Proc Natl Acad Sci USA 111(16):5992–5997. https://doi.org/10.1073/pnas.1402458111
4. Miao C, Radu GU, Caidi H, Tripp RA, Anderson LJ, Haynes LM (2009) Treatment with respiratory syncytial virus G glycoprotein monoclonal antibody or F(ab')2 components mediates reduced pulmonary inflammation in mice. J Gen Virol 90(Pt 5):1119–1123. https://doi.org/10.1099/vir.0.009308-0
5. Krusat T, Streckert HJ (1997) Heparin-dependent attachment of respiratory syncytial virus (RSV) to host cells. Arch Virol 142(6):1247–1254. https://doi.org/10.1007/s007050050156
6. Tripp RA, Jones LP, Haynes LM, Zheng H, Murphy PM, Anderson LJ (2001) CX3C chemokine mimicry by respiratory syncytial virus G glycoprotein. Nat Immunol 2(8):732–738. https://doi.org/10.1038/90675
7. Collins PL, Mottet G (1992) Oligomerization and post-translational processing of glycoprotein G of human respiratory syncytial virus: altered O-glycosylation in the presence of brefeldin A. J Gen Virol 73(Pt 4):849–863. https://doi.org/10.1099/0022-1317-73-4-849
8. Walsh EE, Brandriss MW, Schlesinger JJ (1985) Purification and characterization of the respiratory syncytial virus fusion protein. J Gen Virol 66(Pt 3):409–415. https://doi.org/10.1099/0022-1317-66-3-409
9. Wertz GW, Krieger M, Ball LA (1989) Structure and cell surface maturation of the attachment glycoprotein of human respiratory syncytial virus in a cell line deficient in O glycosylation. J Virol 63(11):4767–4776. https://doi.org/10.1128/JVI.63.11.4767-4776.1989
10. Lambert DM (1988) Role of oligosaccharides in the structure and function of respiratory syncytial virus glycoproteins. Virology 164(2):458–466. https://doi.org/10.1016/0042-6822(88)90560-0
11. Pastey MK, Samal SK (1997) Analysis of the bovine respiratory syncytial virus fusion protein (F) using monoclonal antibodies. Vet Microbiol 58(2–4):175–185. https://doi.org/10.1016/s0378-1135(97)00171-5
12. Anderson LJ, Hierholzer JC, Tsou C, Hendry RM, Fernie BF, Stone Y, McIntosh K (1985) Antigenic characterization of respiratory syncytial virus strains with monoclonal antibodies. J Infect Dis 151(4):626–633. https://doi.org/10.1093/infdis/151.4.626
13. Gimenez HB, Hardman N, Keir HM, Cash P (1986) Antigenic variation between human respiratory syncytial virus isolates. J Gen Virol 67(Pt 5):863–870. https://doi.org/10.1099/0022-1317-67-5-863
14. Mufson MA, Orvell C, Rafnar B, Norrby E (1985) Two distinct subtypes of human respiratory syncytial virus. J Gen Virol 66(Pt 10):2111–2124. https://doi.org/10.1099/0022-1317-66-10-2111
15. Collins PL, Melero JA (2011) Progress in understanding and controlling respiratory syncytial virus: still crazy after all these years. Virus Res 162(1–2):80–99. https://doi.org/10.1016/j.virusres.2011.09.020
16. Peret TC, Hall CB, Schnabel KC, Golub JA, Anderson LJ (1998) Circulation patterns of genetically distinct group A and B strains of human respiratory syncytial virus in a community. J Gen Virol 79(Pt 9):2221–2229. https://doi.org/10.1099/0022-1317-79-9-2221

17. Roberts SR, Lichtenstein D, Ball LA, Wertz GW (1994) The membrane-associated and secreted forms of the respiratory syncytial virus attachment glycoprotein G are synthesized from alternative initiation codons. J Virol 68(7):4538–4546. https://doi.org/10.1128/JVI.68.7.4538-4546.1994
18. Routledge EG, Willcocks MM, Samson AC, Morgan L, Scott R, Anderson JJ, Toms GL (1988) The purification of four respiratory syncytial virus proteins and their evaluation as protective agents against experimental infection in BALB/c mice. J Gen Virol 69(Pt 2):293–303. https://doi.org/10.1099/0022-1317-69-2-293
19. Roder C, Krusat T, Reimers K, Werchau H (2000) Purification of respiratory syncytial virus F and G proteins. J Chromatogr B Biomed Sci Appl 737(1–2):97–106. https://doi.org/10.1016/s0378-4347(99)00442-9

Chapter 10

Expression and Purification of Recombinant Respiratory Syncytial Virus Proteins

Eliseo Salas, Nicholas Keung, Zhenhang Chen, Yunrong Gao, Dongdong Cao, and Bo Liang

Abstract

Recombinant protein expression and purification are crucial for studying pathogen–host interactions, especially in therapeutic contexts. This chapter comprehensively overviews various expression systems and chromatography techniques to express and purify respiratory syncytial virus (RSV) proteins. These proteins include nonstructural protein 1 (NS1), nonstructural protein 2 (NS2), nucleoprotein (N), phosphoprotein (P), matrix protein (M), M2–1 protein, M2–2 protein, and the large RNA-dependent RNA polymerase protein (L). Choosing an appropriate expression system, such as bacterial *E coli* cells or insect cells, is critical for protein expression and will depend on the specific size and properties of the particular protein. Chromatography is a common technique in protein purification that leverages the protein's binding affinity, charge, hydrophobicity, or size, but it often requires significant optimization. We outline the protocol for expressing RSV proteins and detail the purification procedures using affinity chromatography, ion exchange chromatography, and size exclusion chromatography to achieve optimal purification of these RSV proteins.

Key words Respiratory syncytial virus (RSV), Recombinant protein expression, *E. coli* cells, Insect cells, Protein purification, Affinity chromatography, Ion exchange chromatography, Size exclusion chromatography, NS1, NS2, N, P, M, M2–1, M2–2, L

1 Introduction

Pathogens evolve rapidly due to many environmental and genetic factors, such as changing modes of infection, environmental stress, and rapid growth rates. One example of a recurring virus that alters its genetic sequence yearly is the respiratory syncytial virus (RSV). RSV hospitalizations have increased almost twofold in 2023, according to the CDC RSV-Associated Hospitalization Surveillance Network [1, 2]. This virus targets nearly 100% of children under the age of two. Infection is also predominant in elderly patients who are immunocompromised, and it can, in severe cases, result in premature mortality [2]. The FDA has approved only three treatments against RSV. The first (Arexvy) and second

Bo Liang and Rachel Fearns (eds.), *Human Respiratory Syncytial Virus: Methods and Protocols*, Methods in Molecular Biology, vol. 2948, https://doi.org/10.1007/978-1-0716-4666-3_10,

(Abrysvo) protein-based vaccines are only administered to older individuals and pregnant women. In contrast, the third available long-acting monoclonal antibody treatment (Nirsevimab) is given to children older than 8 months (GlaxoSmithKline Biologicals GSK, Pfizer, Sanofi, and AstraZeneca) [3]. Despite years of research, it is still poorly understood how the RSV proteome overcomes the adaptive and innate immune system [4].

The RSV genome is an encapsidated negative sense RNA that encodes 11 proteins. RSV proteins include nonstructural protein 1 (NS1) [5, 6], nonstructural protein 2 (NS2) [6, 7], nucleoprotein (N) [8–10], phosphoprotein (P) [11–13], matrix protein (M) [14–16], transcriptional processivity and antitermination factor M2–1 [17–21], M2–2 protein [22–24], and RNA-dependent RNA polymerase (L) [25, 26] as non-membrane proteins. The genome also encodes for small hydrophobic (SH) protein [27, 28], glycoprotein (G) [29], and fusion protein (F) [30] as membrane proteins (Fig. 1a). The initial step in deciphering viral pathogen–host interactions involves mapping the genetic material of the viral genome. Once the open reading frames of genes are identified using DNA or RNA sequencing, these genes can then be chemically synthesized. This method bypasses the traditional need to cultivate the virus until sufficient genetic material accumulates for extraction. Through this reductive analysis, we gain insights into the essential components of a virus, understanding its structure and the elements crucial for its infectivity. An advantage of this synthetic approach is that it significantly lowers the risk of viral infection and spread, making it possible for biosafety level 1 and 2 laboratories to handle otherwise hazardous viruses safely.

The second step in the process is protein expression and purification. Molecular biologists have developed methods that allow for the expression and purification of viral proteins in a relatively safe and controlled environment using model organisms. Recombinant protein expression is the term used to describe the expression of proteins in a model organism different from the original organism from which the gene was taken. We use *E. coli* BL21 strains or BL21 derivatives for recombinant protein expression for our bacterial cell culture. *Escherichia coli* works exceptionally well to mass produce viral proteins that would otherwise take a long time to harvest from RSV-infected cells and reduces manufacturing time and cost.

Most RSV genes are codon-optimized for desired cell systems (e.g., *E. coli*, or insect cells) and cloned individually into commercially acquired vector DNA (MacroLab) except for the P and L proteins co-expressed in insect cells [31]. These MacroLab vectors contain either an ampicillin or kanamycin resistance gene that allows us to select transformed *E. coli* cells. In addition, in these vectors, the expression of our RSV proteins is regulated by a T7 promoter that is inducible by IPTG (Fig. 1b). The ability to induce protein expression is a crucial aspect of any protein overexpression

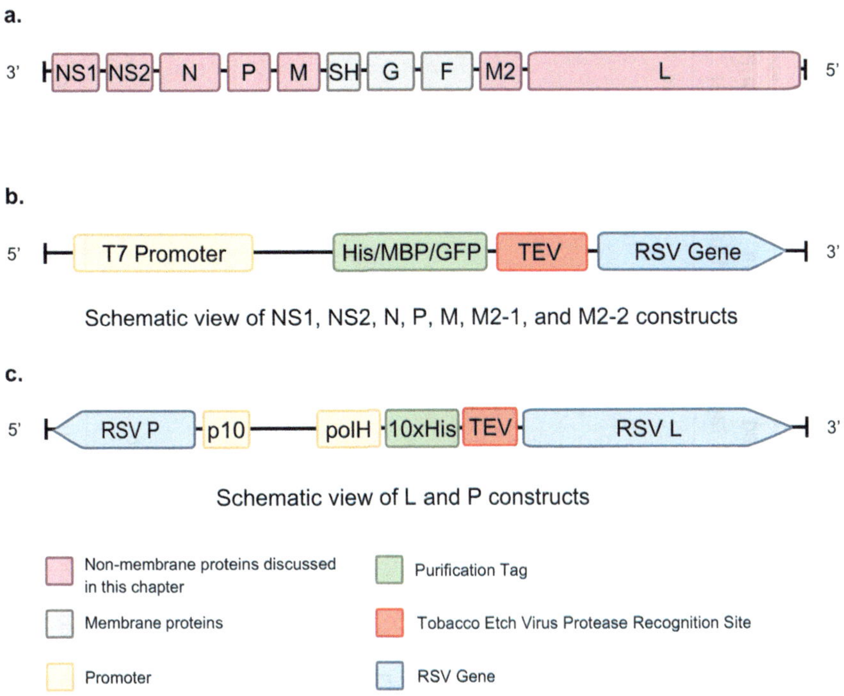

Fig. 1 Representation of the negative sense RNA genome of RSV and schematic representations of coding regions within plasmids used for protein expression. **a** The RSV genome is a non-segmented negative strand RNA. The genome encodes nonstructural proteins 1 and 2 (NS1 and NS2), nucleoprotein (N), phosphoprotein (P), matrix protein (M), small hydrophobic protein (SH), glycoprotein (G), fusion protein (F), and the large polymerase subunit (L). The M2 gene contains two open reading frames that encode for M2–1 and M2–2. Non-membrane proteins, which are discussed in this paper, are highlighted in Pink. Genes highlighted in gray are beyond the scope of this paper. **b** Plasmids transformed and expressed in *E. coli* were positioned downstream from a T7 promoter and had either a 10× histidine, maltose-binding protein (MBP), and/or green fluorescent protein (GFP) tag at the N-terminus. **c** A dual expression vector encoding the RSV L RNA-dependent RNA polymerase and the RSV P phosphoprotein were transformed into Sf9 or Sf21 insect cells. An additional tobacco etch virus (TEV) protease cleavage site was introduced between the N-terminal tags and the gene of interest

plasmid, since foreign-expressed proteins might interact with the endogenous *E. coli* proteins, which could result in apoptosis [32]. In addition, a peptide or protein tag for affinity chromatography is usually added at the beginning or end of the gene of interest to help simplify the purification process.

The RSV proteins expressed in bacterial culture are NS1, NS2, N, P, M, M2–1, and M2–2 which contain either a 10× or 6× histidine tag (Table 1) The histidine tag has a high affinity for Ni-NTA beads which can only be disrupted by a chemically similar compound as histidine, imidazole. Immobilized Ni-NTA beads in complex with histidine tagged proteins can be washed to remove most of the endogenous *E. coli* proteins [33]. In some cases where separation from native *E. coli* protein was challenging, we used a

Table 1
Relevant protein information for each construct

Protein	Tag	Buffer	Calculated pI (without tag)	MW, kDa (without tag)	Net charge in buffer
NS1	10× His (N-terminal)	50 mM Na3PO4 pH 7.4, 150 mM NaCl, 10% glycerol, 2 mM βME	5.71	15.6	Negative
NS2	10× His, MBP (N-terminal)	25 mM HEPES pH 7.4, 150 mM NaCl, 10% glycerol, 2 mM βME	8.73	14.7	Positive
N	10× His (N-terminal)	20 mM HEPES pH 7.4, 200 mM NaCl	6.87	43.5	Negative
P	10× His (N-terminal)	50 mM Tris-HCl pH 7.4, 300 mM NaCl, 0.5 mM tris(2-carboxyethyl) phosphine [TCEP], 10% glycerol	4.52	27.1	Negative
M	6× His, MBP (N-terminal)	50 mM HEPES, pH 7.5, 200 mM NaCl, 5% glycerol	8.66	28.7	Positive
M2–1	6× His, GFP (N-terminal)	50 mM HEPES, pH 7.5, 200 mM NaCl, 5% glycerol	9.13	22.2	Positive
M2–2	6× His, MBP (N-terminal)	30 mM Tris-HCl pH 8.0, 300 mM NaCl, 10% glycerol	6.23	10.7	Negative
LP	6× His (L protein N-terminal)	25 mM HEPES pH 7.4, 300 mM NaCl, 6 mM MgSO4, 0.5 mM tris(2-carboxyethyl) phosphine [TCEP]	L: 8.94 P: 4.48	L: 250.4 P: 27.1	L: Positive P: Negative

The RSV constructs are listed in the table with information about the specific purification tags used. The information about the pI (isoelectric point), MW (molecular weight), and buffer used are also described

10× histidine tag, which allows for a higher affinity to the Ni-NTA resin. The stronger affinity of the tag to the resin helps remove nonspecific resin-binding proteins from eluting together with our protein of interest. In the case of M, M2–2, and NS2, an additional maltose-binding protein (MBP) tag was added after the histidine sequence at the N-terminal to increase protein solubility [34]. Recently, it has been shown that in some cases, a poly-histidine tag may affect the oligomeric states of the proteins and their functions [35]. For this reason, a short peptide sequence recognized by the tobacco etch virus (TEV) protease is included between the poly-histidine tag and the N-terminus of the protein. Following purification via the first Ni-NTA column, the eluted protein is incubated with TEV protease to cleave the histidine tag from the target protein chemically. The resulting mixture, containing the cleaved histidine tag and the overexpressed protein, is then subjected to dialysis before being passed through another Ni-NTA column to remove the tag remnants. Subsequent separation of the tag-free proteins is achieved through ion exchange chromatography and size exclusion chromatography, based on their charge and size, respectively (Table 2). Obtaining highly pure protein samples is critical for studying protein–protein interactions between hosts and pathogens and for therapeutic evaluations. Therefore, a comprehensive array of purification tools is essential to achieve protein purity levels exceeding 95%.

While an *E. coli* expression system is sufficient for most proteins, insect cell expression systems, such as Bac-to-Bac Baculovirus Expression Systems, can be used for complex proteins requiring eukaryotic posttranslational modifications [36]. This is the case for the L protein of RSV that can only be co-expressed with the RSV P

Table 2
Purification steps for each RSV construct

Protein constructs	Ni-NTA	Co-NTA	TEV digestion	Amylose	Heparin	SP	Q	SEC
NS1	●		●	●			●	●
NS2	●		●			●		●
N	●						●	●
P	●						●	●
M	●		●			●		●
M2–1	●		●		●			●
M2–2	●		●				●	●
L-P (insect cells)		●	●		●			●

Each construct is listed in the table. Various purification steps are listed from left to right, and a ● indicates that the purification step was used for the construct protein (Fig. 1c). The RSV polymerase comprises the L and P proteins, which are co-expressed in insect cells [25]. The L protein is around 250 kDa in size and cannot be successfully expressed in *E. coli* or insect cells alone. In contrast, the P protein can be independently expressed in *E. coli* and insect cells, highlighting the importance of using the correct model organism (Tables 1 and 2).

In this chapter, we will focus on the second step of understanding viral pathogen–host interactions: protein purification. We will explore the purification of soluble proteins encoded by the RSV through recombinant protein expression in bacterial and insect cell cultures. Our goal is to provide a broad understanding of the chromatography methods applicable to the mass production of RSV proteins while offering insights and solutions to challenges encountered during their purification.

2 Materials

2.1 Equipment

1. P10, P200, and P1000 pipettes.
2. Serological pipette controller.
3. Shaking incubator.
   (a) New Brunswick Innova 44 (Eppendorf).
   (b) MaxQ 6000 Incubated Shakers (Thermo Scientific).
4. Leica DMil inverted microscope (Leica Microsystems).
5. Fisherbrand Model 505 Sonicator (Fisher Scientific).
6. Dounce homogenizer (DWK Life Sciences).
7. Centrifuges (1000–40,000 × g).
   (a) Allegra X-14R centrifuge (Beckman Coulter).
   (b) J6-MI centrifuge (Beckman Coulter).
   (c) Sorvall RC 6+ centrifuge (Thermo Scientific).
8. Nalgene 1-L bottle compatible with J6-MI centrifuge.
9. Nalgene 50-mL tubes for high-speed centrifugation compatible with Sorvall RC 6+ centrifuge.
10. NanoDrop One-C Spectrophotometer (Thermo Scientific).
11. Electrophoresis chamber with power supply (Bio-Rad).
12. Gravity column (Bio-Rad).
13. Dynamax RP-1 peristaltic pump (Rainin).
14. Prepacked affinity, ion exchange, and size columns (Cytiva Life Sciences).
15. AKTA Pure fast protein liquid chromatography (FPLC) system (Cytiva Life Sciences).

2.2 Supplies

1. P10, P200, and P1000 pipette tips.
2. Serological pipettes.
3. 1.7-mL microcentrifuge tubes
4. 15-mL conical centrifuge tubes
5. 50-mL conical centrifuge tubes
6. Petri dishes.
7. Ni-NTA resin.
8. Co-NTA resin.
9. Dialysis tubing with various molecular weight cutoffs (MWCO).
10. Centrifugal concentrators (MWCO: 10, 30, 50, 100 kDa).
11. SDS–PAGE protein standards.

2.3 Chemicals

1. Agar.
2. Isopropyl β-D-1-thiogalactopyranoside [IPTG] (1 M).
3. Bluo-gal (100 mg/mL).
4. 2-Mercaptoethanol [βME]
5. Trizma.
6. Glycine.
7. Coomassie blue G250.
8. Acrylamide/Bis-acrylamide solution (37.5:1) (40%).
9. Sodium dodecyl sulfate [SDS] (20%).
10. Ammonium persulfate [APS] (10%).
11. Ammonium sulfate.
12. Tetramethylethylenediamine (TEMED).
13. Sodium chloride.
14. Glycerol.
15. 4-(2-Hydroxyethyl)-1 piperazine ethanesulfonic acid (HEPES)
16. Tris–HCl.
17. Sodium monophosphate.
18. Sodium diphosphate.
19. Phosphoric acid.
20. Methanol.
21. Dithiothreitol (DTT).
22. 2-(N-morpholino) ethanesulfonic acid (MES)
23. Tris (2-carboxyethyl) phosphine (TCEP).
24. Imidazole.

2.4 Media

1. Luria broth (LB).
2. Cellfectin II reagent (Thermo Scientific).
3. Sf-900 II Serum-Free Media (Thermo Scientific).
4. Powder Ex-Cell 405 media.
5. Fetal bovine serum (FBS).

2.5 Cells

1. TOP10 competent cells (Thermo Scientific).
2. BL21(DE3) competent cells (Thermo Scientific).
3. DH10Bac competent cells (Thermo Scientific).
4. Sf9 or Sf21 insect cells (Thermo Scientific).

2.6 Plasmids

1. LIC vector.
2. pFastBac Dual vector.

2.7 Antibiotics

1. Ampicillin (100 mg/mL, 1000×).
2. Carbenicillin (100 mg/mL, 1000×).
3. Kanamycin (50 mg/mL, 1000×).
4. Gentamicin (7 mg/mL, 1000×).
5. Tetracycline (10 mg/mL).
6. Penicillin.
7. Streptomycin.

2.8 Protease Inhibitors

1. Aprotinin (2 mg/mL, 1000×).
2. Leupeptin (1 mg/mL, 1000×).
3. Pepstatin in ethanol (1 mg/mL, 1000×).
4. Phenylmethylsulfanoxide (PMSF) in 2-propanol (100 mM, 100×).

2.9 Buffers for Affinity Chromatography

1. 30 mM Tris–HCl pH 8.0, 300 mM NaCl, 10% glycerol
2. 30 mM Tris–HCl pH 8.0, 1 M NaCl, 10% glycerol
3. 30 mM Tris–HCl pH 8.0, 750 mM NaCl, 10% glycerol
4. 30 mM Tris–HCl pH 8.0, 300 mM NaCl, 10% glycerol
5. 30 mM Tris–HCl pH 8.0, 300 mM NaCl, 10% glycerol, 25 mM imidazole
6. 30 mM Tris–HCl pH 8.0, 300 mM NaCl, 10% glycerol, 50 mM imidazole
7. 30 mM Tris–HCl pH 8.0, 300 mM NaCl, 10% glycerol, 70 mM imidazole
8. 30 mM Tris–HCl pH 8.0, 300 mM NaCl, 10% glycerol, 90 mM imidazole

9. 30 mM Tris–HCl pH 8.0, 300 mM NaCl, 10% glycerol, 110 mM imidazole
10. 30 mM Tris–HCl pH 8.0, 300 mM NaCl, 10% glycerol, 150 mM imidazole
11. 30 mM Tris–HCl pH 8.0, 300 mM NaCl, 10% glycerol, 450 mM imidazole
12. 30 mM Tris–HCl pH 8.0, 300 mM NaCl, 10% glycerol, 1 M imidazole.

2.10 TEV Protease Digestion

1. TEV dialysis buffer: 30 mM Tris–HCl pH 8.0, 300 mM NaCl, 10% glycerol, 2 mM 2-mercaptoethanol (βME).

2.11 Buffers for Ion Exchange

1. Q buffer.
   (a) 30 mM Tris–HCl pH 8.0, 10% glycerol, 2 mM βME, 0 mM NaCl
   (b) 30 mM Tris–HCl pH 8.0, 10% glycerol, 2 mM βME, 1 M NaCl
2. SP buffer.
   (a) 30 mM MES pH 6.0, 10% glycerol, 2 mM βME, 0 mM NaCl
   (b) 30 mM MES pH 6.0, 10% glycerol, 2 mM βME, 1 M NaCl.

2.12 Buffers for Size Exclusion Chromatography

1. L-P: 25 mM HEPES pH 7.4, 300 mM NaCl, 6 mM $MgSO_4$, 0.5 mM TCEP.
2. M2–1: 50 mM HEPES pH 7.5, 200 mM NaCl, 5% glycerol.
3. M2–2: 30 mM Tris–HCl pH 8.0, 150 mM NaCl, 0.5 mM TCEP.
4. NS1: 50 mM Na_3PO_4 pH 7.4, 150 mM NaCl, 10% glycerol, 2 mM βME.
5. NS2: 25 mM HEPES pH 7.4, 150 mM NaCl, 10% glycerol, 2 mM βME.

2.13 SDS–Polyacrylamide Gel Electrophoresis (PAGE)

1. SDS loading dye (6×): 375 mM Tris–HCl pH 6.8, 12% SDS, 60% glycerol, 0.06% bromophenol blue, 0.6 M DTT.
2. SDS–PAGE running buffer (10×): 248 mM Trizma, 1.92 M glycine, 1% SDS.
3. Coomassie blue stain solution: 0.1% (w/v) Coomassie Brilliant Blue R-250, 40% (v/v) Methanol, 10% (v/v) Acetic acid.
4. Coomassie (R250) destain solution: 10% acetic acid, 20% ethanol.

3 Methods

3.1 Bacterial Cell Culture

3.1.1 Plasmids and Transformation

1. Mix 30 μL of chemically competent BL21(DE3) *E. coli* cells with approximately 100 ng of plasmid (*see* **Notes 4.1.1** and **4.1.2**),
2. Leave the cells on ice for 30 min.
3. Heat shock cells in a 42 °C water bath for 45 s.
4. Leave the cells on ice for 3 min to recover.
5. Add 400 μL of LB media and incubate the cells for 1 h at 37 °C and 220 rotations per minute (rpm).
6. Warm two LB agar plates containing the antibiotic needed for selection by incubating them at 37 °C.
7. Plate 75 μL of cells on one plate and 250 μL on the other (*see* **Note 4.1.3**).
8. Incubate the plates upside down at 37 °C overnight.
9. Store the plates upside down at 4 °C.

3.1.2 Inoculation of Cell Culture

1. Scrape a single colony into a 500-mL Erlenmeyer flask containing 250 mL of LB with antibiotic (Fig. 2).
2. Incubate the culture in the bacterial shaker at 220 rpm and 37 ° C overnight (12–16 h).
3. Add 10–20 mL of overnight culture into a 2-L Erlenmeyer flask containing 1 L of LB and antibiotic.
4. Incubate the flask in a 37 °C shaker at 220 rpm.
5. After 2–3 h, monitor the OD_{600} every 15–30 min until it reaches approximately 0.6 (3–4 h) (*see* **Note 4.1.4**).

3.1.3 Induction and IPTG Optimization

1. After reaching OD_{600} of 0.6, place all 2-L flasks in ice water to cool.
2. Add IPTG to each flask to reach a final 0.1–1 mM concentration (*see* **Notes 4.1.5** and **4.1.6**).
3. Incubate the flasks at 16 °C, shaking at 220 rpm overnight (*see* **Note 4.1.7**).

3.1.4 Harvesting and Sonication

1. Using 1-L bottle, spin down 1-L cell culture at 2670 × g using a J6-MI centrifuge for 30 min at 4 °C.
2. If storing the cell pellets for later purification, transfer them to 50-mL conical vials, flash freeze them using liquid nitrogen, and store them at −80 °C (*see* **Notes 4.1.8** and **4.1.9**).

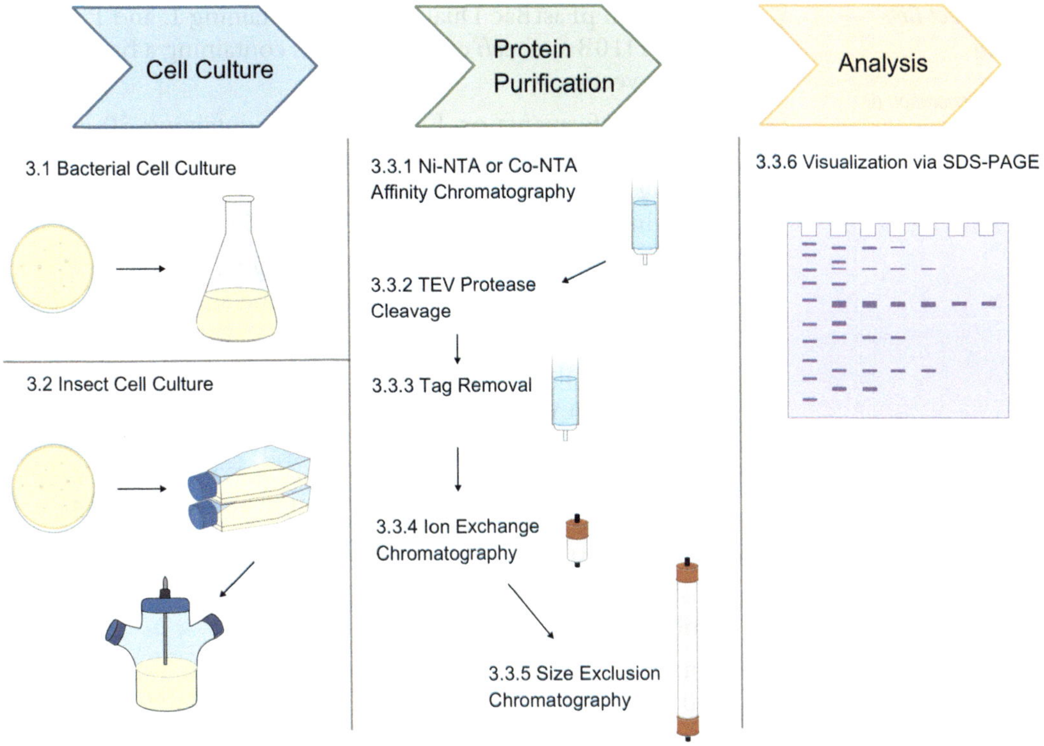

Fig. 2 RSV protein purification workflow. The proteins of interest are expressed using an *E. coli* or insect cell expression system. First, each gene is cloned into a vector that is transformed into *E. coli* competent cells (Subheading 3.1) or transfected into insect cells (Subheading 3.2). After the protein has been expressed, the cells are then lysed, and the protein is purified using Ni-NTA or Co-NTA affinity chromatography (Subheading 3.3.1). Then, the purification tag on the protein is cleaved using TEV protease (Subheading 3.3.2). Next, the sample is further purified with additional chromatography steps, including ion exchange chromatography and size exclusion chromatography (Subheadings 3.3.3, 3.3.4, and 3.3.5). Finally, the purified protein is analyzed via SDS–PAGE (Subheading 3.3.6). Each process step is labeled with the corresponding section number in the methods

3. Resuspend the cell pellet using 50 mL of lysis buffer per liter of cells with leupeptin, aprotinin, pepstatin, and PMSF protease inhibitors (*see* **Note 4.1.10**).
4. Prepare an ice water bath and place a metallic beaker inside of it. Allow the beaker to cool for 5 min.
5. Add the resuspended cells to the beaker and sonicate at 50% amplitude, 3-s on/3-s off, for a total sonication time of 12–15 min.
6. Transfer the lysed cells to 50-mL Nalgene centrifuge tubes and spin at 38,759 × g for 30 min at 4 °C using a Sorvall centrifuge (*see* **Note 4.1.11**).
7. Transfer the supernatant to a clean glass beaker and proceed with purification (*see* **Note 4.1.12**).

3.2 Insect Cell Culture

3.2.1 Generation of Recombinant Bacmid

1. Transform pFastBac Dual plasmid (containing L and P genes) into DH10Bac *E. coli* competent cells containing a baculovirus shuttle vector.
2. Plate transformants on LB agar plates containing 50 μg/mL kanamycin, 7 μg/mL gentamicin, 10 μg/mL tetracycline, 100 μg/mL Bluo-gal, and 40 μg/mL IPTG. Incubate the plates in 37 °C incubator for 2–3 days (Fig. 2).
3. Select recombinant bacmids by choosing white colonies. Blue colonies indicate an unaltered bacmid and should not be used (*see* **Note 4.2.1**).
4. Isolate recombinant bacmids containing L and P genes and confirm them with PCR analysis (*see* **Note 4.2.2**).

3.2.2 Transfection of Insect Cells (Fig. 2)

1. Seed 9×10^5 Sf9 or Sf21 insect cells per 35-mm well (of a six-well plate) in 2 mL of Sf-900 II SFM (*see* **Note 4.2.3**).
2. Allow cells to attach at 27 °C for 1 h.
3. Wash the cells once with 2 mL of Sf-900 II SFM without antibiotics.
4. Solution A: Mix ~5 μL recombinant bacmid (~5 μg) with 100 μL Sf-900 II SFM media.
5. Solution B: Mix 8 μL Cellfectin II reagent with 100 μL Sf-900 II SFM media (*see* **Note 4.2.4**).
6. Mix Solutions A and B gently and incubate for 15–45 min at room temperature, then add to insect cell culture (*see* **Note 4.2.5**).
7. Incubate insect cells for 4–5 h in a 27 °C incubator, then replace media with 2 mL fresh Sf-900 II SFM containing penicillin/streptomycin and 3% FBS.
8. After incubating for several days, harvest the supernatant to obtain recombinant baculovirus P1 (*see* **Note 4.2.6**).
9. The P1 virus is used to generate P2 and P3 viruses by infecting more insect cells and harvesting the supernatant.
10. Store the virus at 4 °C while protected from light. Add 2% fetal bovine serum (FBS) for long-term storage and store at −70 °C.

3.2.3 Insect Cell Protein Expression (Fig. 2)

1. Prepare 8 L of insect cells cultured in Sf-900 II SFM media (*see* **Note 4.2.7**).
2. Infect insect cells with baculovirus using a multiplicity of infection (MOI) of 5 (*see* **Note 4.2.8**).
3. Harvest cells 72 h postinfection by centrifugation for 15 min at $1000 \times g$ (*see* **Note 4.2.9**).
4. Resuspend the cells with 100 mL of lysis buffer per liter of cells.
5. Lyse cells using a homogenizer (*see* **Note 4.2.10**).

6. Spin down cell lysate for 60 min at 16,000 × g using 50-mL Nalgene tubes.
7. Collect the supernatant and advance to the next step in purification (*see* **Note 4.2.11**).

3.3 Protein Purification Using Chromatography Methods

3.3.1 Ni-NTA or Co-NTA Affinity Chromatography (Fig. 2)

1. Use approximately 1 mL of Ni-NTA or Co-NTA resin for every 2 L of cells (*see* **Note 4.3.1**).
2. For every 1 mL of resin needed, transfer 2 mL of slurry into a 50-mL gravity column.
3. Allow ethanol to drain through the column.
4. Wash the column with water using five times the volume of resin or five column volumes (CV).
5. Equilibrate the Ni-NTA or Co-NTA resin with 5 CV of lysis buffer and allow it to sit for at least 5 min. Drain the column.
6. Transfer the supernatant from the cell lysate into the column (*see* **Notes 4.3.2** and **4.3.3**).
7. Allow the lysate to flow through the beads at a rate of three drops every 10 s. Collect the flowthrough for SDS–PAGE analysis.
8. Wash the resin with 20 CVs of wash buffer as described in Subheading 2.9, **items 1–4** and collect the fractions. Repeat the wash steps with decreasing NaCl concentrations (750 mM and 300 mM NaCl) (*see* **Notes 4.3.4–4.3.7**).
9. Elute the column with an increasing imidazole gradient and collect the fractions. Load 3 CV of buffers **items 5–11** listed in Subheading 2.9 (*see* **Notes 4.3.8–4.3.10**).
10. Wash the column with the buffer listed in Subheading 2.9, **item 12**. Then, wash the column with 40 CVs of water and store the resin at 4 °C for future use.

3.3.2 TEV Protease Cleavage Optimization (Fig. 2)

1. Pool elution fractions and use a spectrophotometer to measure the concentration of your desired protein fractions.
2. Prepare 1 L of dialysis buffer and add fresh βME to a concentration of 2 mM in a large beaker. If the NaCl concentration is low, consider increasing it to 300 mM NaCl (*see* **Note 4.3.11**).
3. Add TEV protease to a 1:100 or 1:50 mass ratio with the protein of interest (*see* **Notes 4.3.12** and **4.3.13**).
4. Transfer the sample to dialysis tubing and place the sample in 1 L of buffer.
5. Stir the sample at room temperature for 1 h.
6. Transfer the dialysis tubing to a fresh 1 L buffer and stir at 4 °C overnight (*see* **Note 4.3.14**).

3.3.3 Tag Removal: Second Ni-NTA or Amylose Column Purification (Fig. 2)

1. Clean the resin thoroughly with water and equilibrate the column with the buffer.
2. Load the dialyzed sample into the gravity column with Ni-NTA or Amylose resin. If possible, use the same resin as used in the first column to maintain similar chemical environments.
3. Collect the flowthrough and save it to collect untagged protein of interest (*see* **Note 4.3.15**).
4. Wash the Ni-NTA column with lysis buffer containing 10 mM, 20 mM, and 30 mM of imidazole, or wash the Amylose column with lysis buffer only.
5. Elute histidine tagged TEV protease and histidine tag alone with 500 mM of imidazole, or elute MBP-tagged proteins with 20 mM maltose.
6. Run all the follow through fractions on SDS–PAGE gel to identify the target protein.

3.3.4 Ion Exchange Chromatography: Heparin, Capto Q and Capto SP (Fig. 2)

1. Prepare a low salt (0 mM NaCl) and high salt (1 M NaCl) buffer with the desired pH.
2. Load column on AKTA Pure system.
3. Wash FPLC pumps with DI water, then wash the column with at least 10 CV of DI water.
4. Equilibrate the column with 5 CV of low salt buffer.
5. Remove the column and load the sample onto the column using a peristaltic pump.
6. Load the column on the FPLC system.
7. Elute the column using a gradient or step increasing NaCl and monitor the UV absorbance at 280 and 260 nm (*see* **Note 4.3.16**).
8. Maintain the same NaCl concentration when the UV absorbance indicates the protein is eluting. Hold the concentration until the peak falls completely before resuming the NaCl gradient.

3.3.5 Size Exclusion Chromatography (Fig. 2)

1. Concentrate fractions picked from previous steps using appropriate concentrators (*see* **Note 4.3.17**).
2. Equilibrate the SEC column with 1.2 CVs of filtered DI water followed by 1.3 CVs of SEC buffer.
3. Stop the AKTA run once the column is equilibrated and place the AKTA pump into the LOAD position.

4. Using a 5-mL syringe, clean out the sample loop in the AKTA system by injecting 3 mL of sample buffer into the loop until no bubbles are seen in the waste loop.
5. Disconnect the waste loop from the clear disposal line and push a few drops of buffer using the syringe.
6. Place your concentrated sample in a 1.7-mL tube and place the waste loop into your sample.
7. Pull on the syringe until all the samples are entered in the AKTA, and reconnect the waste loop.
8. Using the presettings program of the AKTA system for desired column and start the run.
9. Collect desired fractions and check them with the SDS–PAGE gel. Collect the fractions containing pure target protein and concentrate them to the desired concentration.
10. Clean the column with 1.2 CVs of filtered DI water.

3.3.6 Visualization Via SDS–PAGE (Fig. 2)

1. Mix the sample with 6× SDS loading dye. Mix 10 μL of sample with 2 μL of 6× loading dye.
2. Place the sample on a heat block at 98 °C for 5 min.
3. Fill the gel box with approximately 500 mL of 1× SDS–PAGE running buffer.
4. Open the arms of the electrode unit and align the SDS–PAGE gel with the gasket, ensuring that the shorter plate faces inward.
5. Align the buffer dam with the gasket and close the arms of the electrode unit.
6. Place the electrode unit in the gel box, ensuring the negative and positive electrodes align with the box.
7. Fill the electrode unit with 1× SDS–PAGE running buffer. Check for any leaks in the electrode unit.
8. Remove the comb from the gel and load protein standards and the other samples into the wells. Secure the lid on the gel box and connect the electrodes to the power supply. Run the gel at 220 V for 30 min.
9. Remove the SDS–PAGE gel and place it in a container with 100 mL of Coomassie blue stain solution. Microwave the gel for 30 s, then transfer it to a shaker and incubate for 5 min.
10. Discard the stain solution and destain the gel using Coomassie (R250) destaining solution. Protein bands should become visible after destaining (Fig. 3).

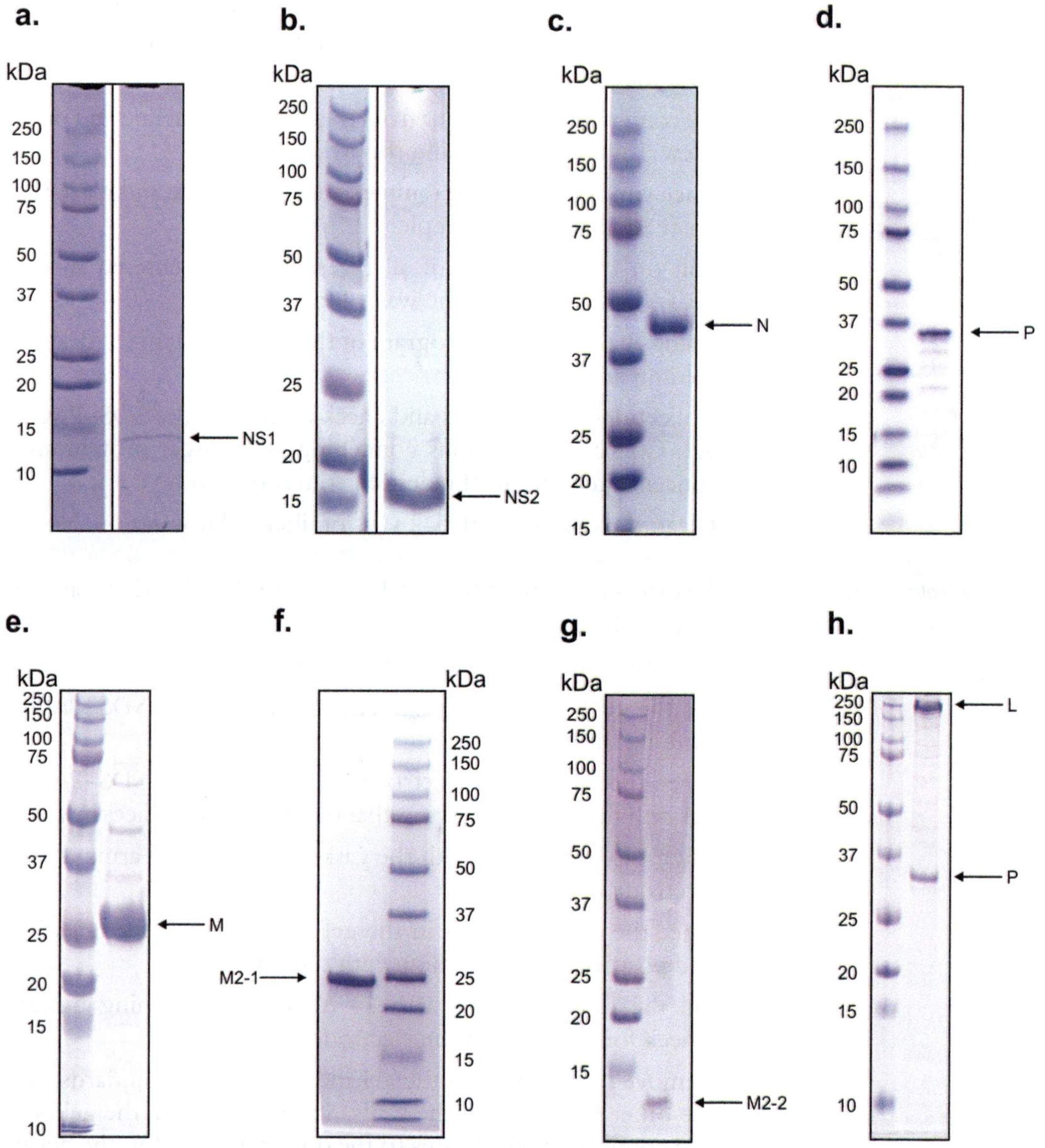

Fig. 3 SDS–PAGE analysis of RSV proteins in post-size exclusion chromatography (SEC) fractions. RSV proteins were analyzed via 4–20% gradient SDS–PAGE gels post-SEC before concentration. SDS–PAGE gel stained with Coomassie blue and destained are shown in the figure. Each panel depicts (**a**) NS1, (**b**) NS2, (**c**) N, (**d**) P, (**e**) M, (**f**) M2–1, (**g**) M2–2, and (**h**) L-P at the last step of purification

4 Notes

4.1.1. Always sequence the plasmids you are using for protein expression. This has been one of the most prevalent recurring issues we have seen with errors in protein purification.

4.1.2. Competent cells over a year old have reduced protein expression, so be conscious of when the competent cells were bought.

4.1.3. Use a spreader or glass beads to spread your cells completely over the agar plates during transformation. This will allow single colonies to form when plating bacterial cells. Do not leave the culture on the plate stagnant for over 3 min.

4.1.4. Healthy cells grow in 3–4 h, while unhealthy cells usually take 5–6 h.

4.1.5. Adjust IPTG concentration to optimize protein expression and maintain protein solubility. For optimization, look for increased soluble protein expression and low protein presence in the cell pellet after lysis.

4.1.6. If induction is performed with a final IPTG concentration of 1 mM for the smaller RSV proteins NS1 and NS2, a high degree of protein aggregation is observed.

4.1.7. A sign that your cells are dying after induction is excessive foam accumulation at the top of your flask before harvest.

4.1.8. We observed around 7 grams of cells per liter of LB after harvest for healthy bacterial cells.

4.1.9. A decrease in cell pellet mass at the time of harvest indicates toxicity of your overexpressed protein. Analyzing your cell pellet fraction after sonication will help identify the degree to which your protein is denatured.

4.1.10. Add leupeptin, pepstatin, and aprotinin to the lysis buffer immediately before usage. PMSF should be added just before sonication due to its high instability in water.

4.1.11. For the pellet fraction after sonication, take a small scoop of the cell debris that is found at the bottom of your Nalgene tubes and transfer it to a 1.7-mL microcentrifuge tube. Add 1 mL of lysis buffer to your pellet fraction and vortex until most of the membrane is resuspended and analyzed via SDS–polyacrylamide gel.

4.1.12. If you allow your supernatant to come to room temperature or take too long to load it into your column, you will see the solution turn from a transparent brown to a milky opaque brown. This is protein precipitation and will cause your flow rate to be slow.

4.2.1. Always select the large round white colonies to avoid the selection of false positives. A dark background would benefit the selection of white colonies.

4.2.2. It is not recommended to store the purified bacmid DNA by freezing at −20 °C, as it decreases the transfection efficiency. You can store the purified bacmid DNA for up to 2 weeks at 4 °C in TE buffer, pH 8.0.

4.2.3. Use only cells in the mid-log phase with a viability of >97%.

4.2.4. Cellfectin II reagent is a lipid suspension and should be mixed thoroughly by inverting the tube 5–10 times to ensure a homogeneous sample.

4.2.5. Mix gently. Aspirate wash media from cells and overlay the diluted lipid–DNA complexes onto the cells.

4.2.6. Small-scale expression tests can be done with the P1 or P2 virus, while large-scale expression should be done with the P3 virus.

4.2.7. The volume of the cells is dependent on the expression level of the target protein.

4.2.8. MOI refers to the ratio of viruses to cells. MOI can be optimized to get a better protein expression.

4.2.9. A time course expression test should be performed to select the best harvest time point.

4.2.10. Sonicator can also be used to break down insect cells.

4.2.11. Optimizing infection conditions by varying the MOI is recommended, if protein expression yield is low.

4.3.1. HisPur Ni-NTA beads were purchased with a capacity of up to 60 mg of 6× histidine tag protein per mL of settled resin beads. For a 12-L purification, use 4–5 mL of Ni-NTA resin to ensure saturation and reduce nonspecific binding.

4.3.2. As an alternative, allow the Ni-NTA beads to interact with your lysate for at least 30 min for optimal binding. This step can be carried out in a 50-mL falcon tube with gentle rocking at 4 °C, if lysate is under 40 mL. If you use a bigger bottle for incubation of beads with protein, make sure rocking does not cause excess foam (foam is protein aggregating).

4.3.3. The cell lysate should be incubated overnight with Co-NTA beads to maximize binding for the L protein.

4.3.4. If nonspecific protein bands are still observed after Ni-NTA washing steps, your protein of interest did not overexpress and is found in a very low abundance. Optimize IPTG concentration or lower the amount of Ni-NTA resin used.

4.3.5. After your lysate has been passed through the column, the resin may change from blue-green to brown. This brown color should be mainly gone by the end of your washes.

4.3.6. When purifying L and P in the complex, the NaCl concentration should not fall below 50 mM to avoid precipitation.

4.3.7. If the NaCl concentration falls below 100 mM, N protein precipitation is highly likely.

4.3.8. We have observed 10× his tag protein elution as early as 50 mM imidazole; however, these fractions are often saturated with other *E. coli* proteins, so we discard them most of the time and only keep the 450 mM imidazole fractions.

4.3.9. A protein at around 75 kDa is frequently overexpressed and should be expected. We suspect from other publications that this is heat shock protein 70 that is overexpressed when we rapidly cool down the *E. coli* cells to 16 °C.

4.3.10. For 10× His NS1, precipitation is observed within 15 min of elution from the Ni-NTA column using 450 mM imidazole. To increase stability, reduce the concentration to 350 mM imidazole and increase elution volume. Dialyze fractions before adding TEV.

4.3.11. Imidazole in high concentrations has been shown to reduce TEV cleavage and the ability of the proteins to bind to other affinity columns. Consider using 2 L of room temperature dialysis buffer to dialyze out the imidazole for a minimum of 1 h before adding TEV.

4.3.12. It is important to maximize the TEV cleavage efficiency when purifying M2–2. Because the protein can form oligomers, cleaved M2–2 can interact with tagged M2–2, making further isolation difficult.

4.3.13. His-MBP-NS2 is stable before TEV cleavage. After cleavage, keep cleaved NS2 in HEPES or sodium phosphate pH 7.5 buffer with 10% glycerol, 300 mM NaCl, 4 mM βME to prevent N-terminal degradation (~2–3 kDa degradation).

4.3.14. After TEV cleavage, the M protein undergoes partial precipitation, necessitating removal via high-speed centrifugation before proceeding to the next step.

4.3.15. The M2–1 construct was designed with a His-GFP tag. After TEV cleavage, it is important to thoroughly remove GFP, as residual GFP remnants may impact downstream experimental assays involving fluorescent labeling.

4.3.16. When using the AKTA system for ion exchange chromatography, use the hold function to maintain the ionic strength constant. This will help in separating proteins of different ionic strengths better. Resume when A280 peak has finished straightening out.

4.3.17. When using the concentrators, gently invert tube to avoid protein aggregation at the bottom of the concentrator. If you are concentrating your protein of interest to less than 1 mL, do not spin for more than 5 min at a time at 3500 RPM. If the white spaces was observed at the bottom, stop concentration. These specs are protein aggregates, and they are an indication that your protein has reached the limit of concentration.

Acknowledgments

The authors thank the members of the Liang laboratory for critical discussions and helpful support. The research programs in the Liang laboratory at Emory are supported by the US National Institute of General Medical Sciences (NIGMS), the National Institutes of Health (NIH) under award number R01GM130950, and the American Lung Association Innovation Award, grant number IA-831472. The content is solely the responsibility of the authors and does not necessarily represent the official views of the National Institutes of Health.

Conflict of Interest Statement The authors declare no competing interests.

References

1. (NCIRD), C (2022) RSV-NET database. CDC. https://www.cdc.gov/rsv/research/rsv-net/overview-methods.html
2. Statistics, N.C.f.H (2024) Trends in deaths for viiral respiratory illness. CDC. https://www.cdc.gov/respiratory-viruses/data-research/dashboard/most-impacted-deaths.html
3. Simoes EAF et al (2022) Prefusion F protein-based respiratory syncytial virus immunization in pregnancy. N Engl J Med 386(17):1615–1626
4. Battles MB, McLellan JS (2019) Respiratory syncytial virus entry and how to block it. Nat Rev Microbiol 17(4):233–245
5. Chatterjee S et al (2017) Structural basis for human respiratory syncytial virus NS1-mediated modulation of host responses. Nat Microbiol 2:17101
6. Pei J et al (2021) Structural basis for IFN antagonism by human respiratory syncytial virus nonstructural protein 2. Proc Natl Acad Sci USA 118(10)
7. Ling Z, Tran KC, Teng MN (2009) Human respiratory syncytial virus nonstructural protein NS2 antagonizes the activation of beta interferon transcription by interacting with RIG-I. J Virol 83(8):3734–3742
8. Tawar RG et al (2009) Crystal structure of a nucleocapsid-like nucleoprotein-RNA complex of respiratory syncytial virus. Science 326(5957):1279–1283
9. Gao Y et al (2020) In vitro trackable assembly of RNA-specific nucleocapsids of the respiratory syncytial virus. J Biol Chem 295(3):883–895

10. Wang Y et al (2023) Cryo-EM structure of the nucleocapsid-like assembly of respiratory syncytial virus. Signal Transduct Target Ther 8(1): 323
11. Lu B et al (2002) The major phosphorylation sites of the respiratory syncytial virus phosphoprotein are dispensable for virus replication in vitro. J Virol 76(21):10776–10784
12. Galloux M et al (2012) Characterization of a viral phosphoprotein binding site on the surface of the respiratory syncytial nucleoprotein. J Virol 86(16):8375–8387
13. Galloux M et al (2015) Identification and characterization of the binding site of the respiratory syncytial virus phosphoprotein to RNA-free nucleoprotein. J Virol 89(7): 3484–3496
14. Cao J et al (2023) The matrix protein of respiratory syncytial virus suppresses interferon signaling via RACK1 association. J Virol 97(10): e0074723
15. Forster A et al (2015) Dimerization of matrix protein is required for budding of respiratory syncytial virus. J Virol 89(8):4624–4635
16. Bajorek M et al (2014) The Thr205 phosphorylation site within respiratory syncytial virus matrix (M) protein modulates M oligomerization and virus production. J Virol 88(11): 6380–6393
17. Tanner SJ et al (2014) Crystal structure of the essential transcription antiterminator M2-1 protein of human respiratory syncytial virus and implications of its phosphorylation. Proc Natl Acad Sci USA 111(4):1580–1585
18. Gao Y et al (2020) Structure of the human respiratory syncytial virus M2-1 protein in complex with a short positive-sense gene-end RNA. Structure 28(9):979–990 e4
19. Leyrat C et al (2014) Drastic changes in conformational dynamics of the antiterminator M2-1 regulate transcription efficiency in Pneumovirinae. elife 3:e02674
20. Richard CA et al (2018) RSV hijacks cellular protein phosphatase 1 to regulate M2-1 phosphorylation and viral transcription. PLoS Pathog 14(3):e1006920
21. Tran TL et al (2009) The respiratory syncytial virus M2-1 protein forms tetramers and interacts with RNA and P in a competitive manner. J Virol 83(13):6363–6374
22. Blanchard EL et al (2020) Polymerase-tagged respiratory syncytial virus reveals a dynamic rearrangement of the ribonucleocapsid complex during infection. PLoS Pathog 16(10): e1008987
23. Bermingham A, Collins PL (1999) The M2-2 protein of human respiratory syncytial virus is a regulatory factor involved in the balance between RNA replication and transcription. Proc Natl Acad Sci USA 96(20):11259–11264
24. Cheng X et al (2005) Overexpression of the M2-2 protein of respiratory syncytial virus inhibits viral replication. J Virol 79(22): 13943–13952
25. Cao D et al (2024) Structures of the promoter-bound respiratory syncytial virus polymerase. Nature 625(7995):611–617
26. Cao D et al (2020) Cryo-EM structure of the respiratory syncytial virus RNA polymerase. Nat Commun 11(1):368
27. Huong TN et al (2023) The respiratory syncytial virus SH protein is incorporated into infectious virus particles that form on virus-infected cells. Virology 580:28–40
28. Fuentes S et al (2007) Function of the respiratory syncytial virus small hydrophobic protein. J Virol 81(15):8361–8366
29. Tripp RA et al (2018) Respiratory syncytial virus: targeting the G protein provides a new approach for an old problem. J Virol 92(3)
30. McLellan JS, Ray WC, Peeples ME (2013) Structure and function of respiratory syncytial virus surface glycoproteins. Curr Top Microbiol Immunol 372:83–104
31. QB3 MacroLab Plasmids. 2024, UC Berkeley: https://qb3.berkeley.edu/facility/qb3-macrolab/
32. Erental A et al (2014) Apoptosis-like death, an extreme SOS response in Escherichia coli. MBio 5(4):e01426–e01414
33. Gao Y et al (2023) Optimal conditions for in vitro assembly of respiratory syncytial virus nucleocapsid-like particles. Viruses 15(2)
34. Jin T et al (2017) Design of an expression system to enhance MBP-mediated crystallization. Sci Rep 7:40991
35. Booth WT et al (2018) Impact of an N-terminal polyhistidine tag on protein thermal stability. ACS Omega 3(1):760–768
36. Bac-to-Bac Baculovirus Expression System. 2002

Chapter 11

Single-Particle Electron Microscopy Analysis of Respiratory Syncytial Virus Protein Complexes

Dongdong Cao, Yunrong Gao, Zhenhang Chen, Natalie Horgan, and Bo Liang

Abstract

Cryogenic electron microscopy (cryo-EM) has been an important technique for determining the structures of macromolecules at near-atomic resolution in their native states. In recent years, the structures of some respiratory syncytial virus (RSV) protein complexes, such as the RSV polymerase (L:P complex) and nucleoprotein-RNA complex, have been determined by cryo-EM, providing insights into the mechanisms of viral RNA synthesis and facilitating antiviral development. Here, we describe the step-by-step protocols for the negative-stain and cryo-EM analysis of the RSV protein complexes.

Key words Respiratory syncytial virus, Single-particle analysis, Negative-stain EM, Cyro-EM, Negative staining, Cryo-freezing, EM data processing, Model building

1 Introduction

Cryogenic electron microscopy (Cryo-EM) is a version of electron microscopy that helps people determine the structures of macromolecules at near-atomic resolution in their native states. In the mid-1970s, scientists invented the idea of freezing samples to preserve the natural structures of biological specimens and reduce damage from the electron beam, and cryo-EM was born [1]. In 2017, the Nobel Prize in Chemistry was awarded to three scientists, Jacques Dubochet, Joachim Frank, and Richard Henderson, "for developing cryo-electron microscopy for the high-resolution structure determination of biomolecules in solution" [2]. Recent advances in detector technology and software algorithms allow images of unprecedented quality to be recorded and structures to be determined at near-atomic resolution [1].

Depending on how samples are prepared and imaged under cryo conditions, cryo-electron tomography (cryo-ET) is another technique used for studying the three-dimensional structures of

Bo Liang and Rachel Fearns (eds.), *Human Respiratory Syncytial Virus: Methods and Protocols*, Methods in Molecular Biology, vol. 2948, https://doi.org/10.1007/978-1-0716-4666-3_11,

biological macromolecules and cellular structures at the nanoscale. Single-particle cryo-EM is ideal for determining the high-resolution structures of individual macromolecules or complexes, while cryo-ET specializes in capturing 3D structures of complex cellular or subcellular structures within their native environments. Cryo-EM involves freezing and imaging isolated particles to create 3D reconstructions, while cryo-ET captures multiple 2D images from different angles to generate 3D tomograms of entire cells or cellular components. These techniques have greatly advanced structural and cell biology, providing detailed insights into the structures of biological molecules and cellular content.

We use complementary methods to study the high-resolution details of RSV proteins. Although crystallography is a good method to determine the structures of some respiratory syncytial virus (RSV) proteins, such as M2–1 [3–6], and N [7, 8], single-particle analysis (SPA) cryo-EM is a proven technique for the structure determination of RSV protein complexes, such as RSV polymerase [9–11], and N-RNA complexes [12]. Cryo-ET is used for the structure analysis of lipid-encapsulated viruses, such as RSV [13] and SARS-CoV2 [14]. The structures of the viral polymerases from some other nonsegmented negative-sense (NNS) viruses, such as vesicular stomatitis virus (VSV), rabies virus (RABV), human metapneumovirus (HMPV), parainfluenza virus (PIV), Newcastle disease virus (NDV), and Ebola virus (EBOV), were also determined by single-particle cryo-EM recently [15–23]. These structures have provided critical insights into the enzymatic domains responsible for RNA synthesis.

Here, we describe the step-by-step protocols of sample preparation, grid preparation, negative-stain EM analysis, single-particle cryo-EM grids making, data collection, data processing, and model building of the RSV protein complexes (Fig. 1). We used Uranyl Formate to prepare negative-stain EM grids and Vitrobot Mark IV to prepare cryo-EM grids. CryoSPARC and Relion were used for the cryo-EM data processing [24–26]. Model building was carried out by UCSF Chimera, PHENIX, and COOT [27–29], and the model geometries were validated using MolProbity [30].

2 Materials

The reagents used in this protocol are analytical grade. The solutions are prepared with ddH_2O and filtered through 0.22 micron filters.

2.1 Chemical Reagents

1. Sodium chloride.
2. Magnesium sulfate.
3. HEPES (4-(2-hydroxyethyl)-1-piperazineethanesulfonic acid).

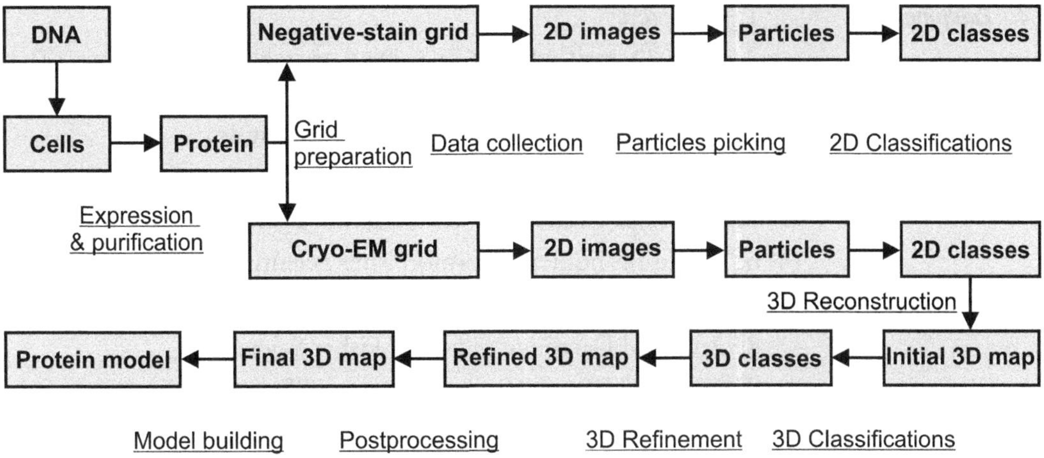

Fig. 1 Overall process of cryo-EM data analysis. Negative-stain EM is used to estimate the quality of the sample. Once the sample quality is adequate, the sample will be applied for further structure determination using single-particle cryo-EM. Significant steps in structure determination include 2D classification, 3D classification, 3D refinement, Post processing, and Model building

4. TCEP (Tris(2-carboxyethyl)phosphine).
5. Uranyl formate.

2.2 Sample and Solutions

1. 0.025 mg/mL of diluted RSV proteins on ice for Negative Stain.
2. 0.3–1.5 mg/mL of RSV proteins for cryo-EM.
3. Buffer A: 25 mM HEPES pH 7.4, 300 mM NaCl, 6 mM $MgSO_4$, 0.5 mM TCEP.
4. Buffer B: 25 mM HEPES pH 7.4, 300 mM NaCl, 0.5 mM TCEP.
5. Negative-stain solution: 0.75% Uranyl Formate.
6. Purified RSV proteins in a −80 °C freezer (0.6–5 mg/mL), including RSV polymerase (L:P), N:RNA.

2.3 Supplies and Consumables

1. UltrAuFoil 300 mesh R1.2/1.3 grids.
2. Quantifoil 300 mesh R1.2/1.3 grids.
3. Empty grid boxes.
4. Glass petri dish.
5. Microscope glass slide.
6. Liquid nitrogen.
7. Ethane gas.

2.4 Equipment

1. Tweezers.
2. PELCO easiGLOW Glow Discharge Cleaning System (Ted Pella).
3. Vitrobot Mark IV (Thermo Fisher Scientific).
4. FEI Talos 120 KV (Thermo Fisher Scientific).
5. FEI Talos Arctica 200 KV (Thermo Fisher Scientific).
6. Single-tilt holder (Thermo Fisher Scientific).
7. 626 cryo-transfer Holder (Gatan).
8. Autogrid Cartridge (Thermo Fisher Scientific).
9. Autoloader cassette transfer station (Thermo Fisher Scientific).
10. Autoloader cassette (Thermo Fisher Scientific).

2.5 Software

1. SBGrid.
2. EMAN2.
3. REgularised LIkelihood OptimisatioN (RELION).
4. Cryo-EM Single Particle Ab-Initio Reconstruction and Classification (CryoSPARC).
5. Python-based Hierarchical ENvironment for Integrated Xtallography (PHENIX).
6. UCSF ChimeraX.
7. Crystallographic Object-Oriented Toolkit (COOT).
8. Alphafold.
9. Phyre^2.
10. MolProbity.

3 Methods

3.1 Sample Preparation

1. Thaw the protein sample on ice for 30 min before sample preparation (*see* **Note 4.1.1**).
2. Dilute the RSV proteins with buffer A (RSV polymerase L-P complex) or B (RSV N:RNA, M2–1:RNA, SH) to a proper concentration (0.025 mg/mL for negative-stain EM samples and 0.3–1.5 mg/mL for cryo-EM samples). The sample was incubated at room temperature for 1 h before grid preparation (*see* **Note 4.1.2**).

3.2 Negative-Stain EM Grid Preparation

1. Glow-discharge EM grids coated with continuous carbon film using the PELCO easiGLOW (*see* **Note 4.2.1**).
2. Prepare the wash and staining buffer for use (*see* **Note 4.2.2**).
3. Apply 3–4 μL of protein sample to the carbon-coated side of the EM grid and wait for 1 min.

4. Blot the edge of the grid on a piece of filter paper, then wash the grid by tapping the top of the grid with the water or buffer drop and then blotting the edge of the grid on the edge of the filter paper to dry (*see* **Note 4.2.3**). Repeat with another water or buffer drop.
5. Repeat **step 3.2.4** with two drops of staining solution with the second drop of stain solution incubated for 30–60 s.
6. Let the grid air dry for 5 min.
7. Discard the parafilm and filter paper in the uranyl waste basket.
8. Collect grids and take the grid to the microscope.
9. Clean the workbench.

3.3 Negative-Stain EM Imaging Using 120 kV Electron Microscope (TFS Talos L120)

1. Prepare the microscope (*see* **Note 4.3.1**).
2. Load the grid into the single-tilt holder (*see* **Note 4.3.2**).
3. Start the turbo pump and wait for it to reach 100%. Align and insert the holder straight into the stage load lock area (holder pin toward the closed position) and wait for the load lock to pump down. Select the single-tilt holder when prompted (*see* **Note 4.3.3**).
4. Rotate the holder toward the open position and fully insert it into the microscope column area.
5. Open the column valves and find your beam.
6. Set the microscope to 3400× magnification.
7. Spread the beam so that the size of it is slightly larger than the screen, so the edges of the beam are not visible (*see* **Note 4.3.4**).
8. Bring the specimen to Eucentric height: find a reference object and activate the "wobbler," and minimize the movement with the Z up and down buttons on the right-hand panel (*see* **Note 4.3.5**).
9. Click the "Int Zoom" button in the "Beam Settings" OCX (*see* **Note 4.3.6**).
10. Go to high magnification (120kx to start).
11. Press the "Eucentric focus" button on the right-hand panel.
12. Condense the beam to a spot (*see* **Note 4.3.7**).
13. Fix the beam if it is not centered (*see* **Note 4.3.8**).
14. Expand the beam again just until the edges are not visible.
15. Lift the fluorescent screen.
16. Insert the Ceta camera in the CCD/TV OCX and hit the "Search" button. Click on the "Live, FFT" button. You should now see the Fourier transform (FFT) of your live image.

17. Adjust the focus so that you see the first few rings of the FFT (under focus, not overfocus).
18. Find the stained area on the grid and start screening the grid by taking images with the Acquire button in the CCD/TV OCX (*see* **Note 4.3.9**).
19. Saving single images on TIA: Right-click on the image and select Export. Select the format of the image to be saved and proceed.
20. When finished, insert the fluorescence screen, go to 3400× magnification, spread out the beam until you cannot see the edges, remove used objective apertures, retract the camera, and reset the holder.
21. Close the column valve, start the turbo pump, then remove the holder.
22. If you are the last user, turn off the filament, remove the dewar, and start the Cryo-cycle.
23. Examine the micrographs using the example micrographs provided (Fig. 2).

3.4 Negative-Stain EM Data Processing with EMAN2

1. Create a project folder for your sample (*see* **Note 4.4.1**).
2. Open EMAN2. When in a Linux-based system, use the code e2projectmanager.py &.
3. Select Project, and Edit Project in the upper left toolbar
4. Enter the proper project name, estimated kDa, microscope CS (mm), Microscope Voltage, and Microscope apix (*see* **Note 4.4.2**).
5. Under Raw Data in the SPR list, select Evaluate and Import Micrographs.
6. Click on Browse and select all the micrographs.
7. Select Import Micrographs and establish defocus (*see* **Note 4.4.3**).
8. Under the particles tab, select interactive particle picking (*see* **Note 4.4.4**).
9. Select Good Ref when the control panel launches.
10. Click through the micrographs to find good references.
11. Autobox based on the good references you selected. Select autobox all.
12. Generate the output (*see* **Note 4.4.5**) and autoprocess the CTF. Select midres and use 80 threads.
13. Complete reference-free class averaging. This will be cycled with the next step. Chose the input from the output of the previous job. Set the ncls to 100. Parallelize thread:80.

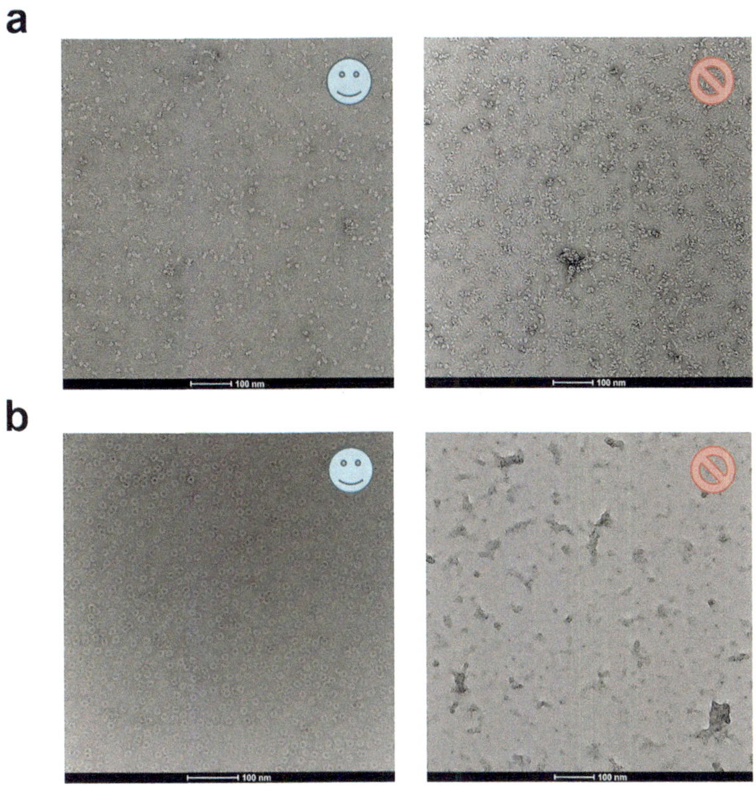

Fig. 2 Negative-stain EM screening. **a** Good and bad examples of negative staining micrographs of RSV-LP. The left has little aggregation, and the particles are easily distinguishable. The right micrograph has aggregated particles that make it more challenging to use. **b** Good and bad examples of negative staining micrographs of N:RNA complex. The left image shows a good distribution of the particles. There is also little aggregation and few broken particles. The right micrograph has staining problems, making it impossible to see the particles with the eye

14. Open the class selection panel, and in Linux, the code is e2evalparticles.py &.
15. Select all and make a new set from the latest class.
16. Select the good classes. Example provided (Fig. 3).
17. Complete steps 15–18 again with the new particle set as the new input.

3.5 Cryo-EM Grid Preparation with a Cryo-Plunger (Vitrobot Mark IV)

1. Document the process using the example provided (Fig. 4).
2. Open the chamber, install the blot papers on the pad pair, and then close the chamber (*see* **Note 4.5.1**).
3. Mount the humidifier carefully and fill it with 40 mL of distilled water using a syringe through the plastic tube at the bottom part of the humidifier (*see* **Note 4.5.2**).

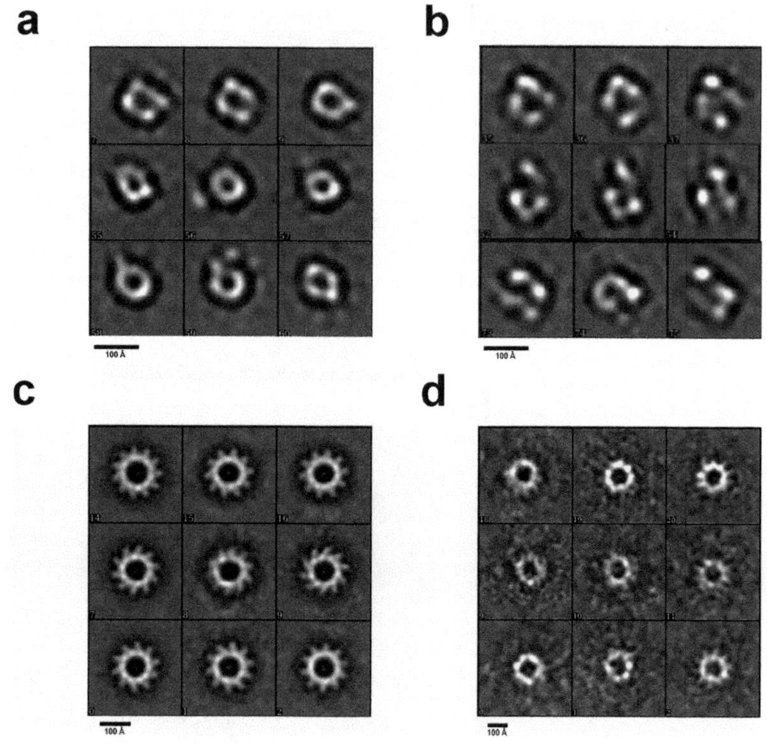

Fig. 3 Negative-stain EM 2D classification. **a** Relatively good quality negative-stain EM 2D class images of RSV-LP. **b** Relatively bad negative-stain EM 2D class images of RSV-LP. A and B are from the same classification job and are split between good and bad classes. **c** Excellent negative-stain EM 2D classes of N:RNA complex. **d** Bad negative-stain EM 2D classes of N:RNA complex

4. Turn on the Vitrobot Mark IV by switching the button at the back of the instrument and wait until the user interface becomes operational (*see* **Note 4.5.3**).
5. Set the chamber temperature to 22 °C. Manually increase the humidity to 100% and maintain it at 100%. Set the blotting time, blotting force, and wait time (*see* **Note 4.5.4**).
6. Assemble and cool the cryogen container (*see* **Note 4.5.5**).
7. Glow-discharge the grids (*see* **Note 4.5.6**).
8. Pick up one glow-discharged grid with the Vitrobot tweezer and properly mount it to the metal rod. Operate the Vitrobot to lift the grid to the chamber (*see* **Note 4.5.7**).
9. Operate the Vitrobot to lift the ethane container and move the tweezers to a lower position in the chamber (*see* **Note 4.5.8**).
10. Open the side window and carefully apply ~3 μL of sample solution onto the sample side of the grids. Quickly start the automatic blotting and plunge freezing (*see* **Note 4.5.9**).

a

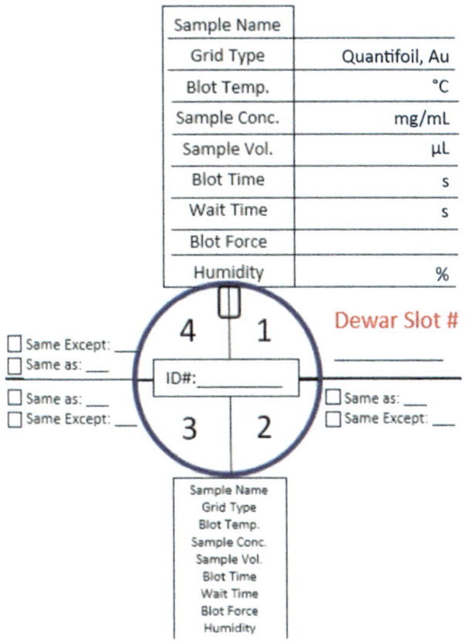

b

Example Values:

Grid Type:	Quantifoil, Cflat, Au, graphene
Blot Temp.:	4 °C, RT (22°C)
Sample Conc.:	Typically 0.3-2 mg/mL
Sample Vol.:	3μL , 3.5μL , 4μL
Blot Time:	2s , 3s , 3.5s , 4s , 4.5s , 5s
Wait Time:	0s , 5s , 10s , 30s
Blot Force:	0
Humidity:	100%

c

Glow Discharge			By:			Date:	
Grid ID # ______			Dewar Slot # ______				
Grid Position	Grid Type	Sample Name	Sample Conc. (mg/mL)	Sample Vol. (μL)	Blot Time (s)	Blot Temp. (°C)	
1							
2							
3							
4							

Fig. 4 Cryo-EM screening documentation. **a** Example sheet to fill out when making cryo grids. Each quadrant is for a separate grid, and each grid position is labeled in the center. **b** Example values that are commonly used when making cryo grids. We do not vary the humidity when making cryo grids. **c** Spreadsheet version of **a**. Either can be used when making cryo grids. It is useful to bring the sheet when screening on the Talos or Arctica, so screening notes can be added

11. Place the frozen grid into the grid box in the cryogen reservoir (*see* **Note 4.5.10**).
12. Repeat steps 9–12 to freeze more grids if needed. When a grid box is finished, make sure to screw well the cap and transfer it to a 50 mL tube filled with liquid nitrogen (*see* **Note 4.5.11**).
13. Transfer the 50 mL tube to a storage Dewar filled with liquid nitrogen (*see* **Note 4.5.12**).

3.6 Cryo-EM Grids Screening with Talos

1. Insert the cryo-holder into the cryo-transfer workstation and connect the Smartset cable to the cryo-holder to monitor the temperature.
2. Simultaneously, start filling the cryo-holder dewar and the cryo-transfer workstation with liquid nitrogen until the temperature is below −190 °C.
3. Load the grid into the cryo-holder (*see* **Note 4.6.1**).
4. Prepare the Microscope (*see* **Note 4.6.2**).
5. Start the Turbo pump (Vacuum OCX) and wait for it to reach full speed (100%).
6. Tilt the stage to −55 degrees by typing this value and pressing "Set Alpha."
7. Remove the cryo-holder from the cryo-transfer workstation and swiftly and carefully insert it into the microscope with the holder pin toward the closed position (*see* **Note 4.6.3**).
8. Slowly rotate the holder using the right hand (from right to left) until the microscope automatically inserts the holder into the airlock.
9. Press the "Set Alpha" button again to tilt the stage back to 0 degrees.
10. Fill the dewar with liquid nitrogen and close it with the lid.
11. Open the grid lid to expose the grid.
12. Open the column valves and find your beam.
13. Set the microscope to 3400× magnification.
14. Bring specimen to eucentric height (*see* **Note 4.6.4**).
15. Click the "Int Zoom" button in the "Beam Settings" OCX.
16. Go to high magnification (120kx to start).
17. Press the "Eucentric focus" button on the right-hand panel.
18. Lift the fluorescent screen.
19. Insert the Ceta camera in CCD/TV OCX and hit the "Search" button. Click on the "Live, FFT" button (*see* **Note 4.6.5**).
20. Adjust the focus so that you see the first few rings of the FFT (under focus not overfocus).
21. Click on the "Low does" tab.
22. Screen for searching and Focus and Exposure button for getting the image.
23. Saving single images on TIA: Right-click on the image and select Export. Select the format of the image to be saved and proceed.
24. When finished, insert the fluorescence screen, go to 3400× magnification, spread out the beam until you cannot see the

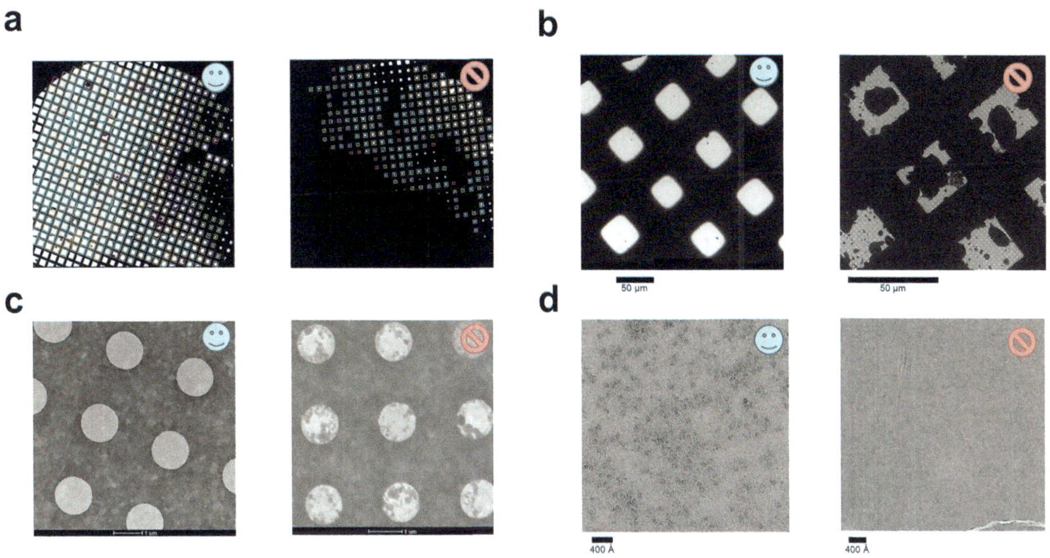

Fig. 5 Cryo-EM screening. **a** The Atlas on the left shows a good grid, with many squares outlined in blue. The one on the right shows much more ice, smaller, and fewer large blue squares. **b** The left low-magnification micrograph shows very little contamination in the center of the squares, whereas the ice contamination on the right micrograph is apparent. **c** There is good ice in the center of the holes in the left square, whereas the one on the right has contamination in the center of the holes. **d** The left image displays a high-quality micrograph with good contrast and well-defined particles, while the right image shows bad ice and low-quality particles

edges, remove used objective apertures, retract the camera, and reset the holder.

25. Close the column valve, start the turbo pump, then remove the holder by reversing **steps 3.6.6** to **3.6.9**.
26. If you are the last user, turn off the filament, remove the dewar, and start the Cryo-cycle.
27. Examine the micrographs using the screening examples provided (Fig. 5).

3.7 Cryo-EM Imaging Using Electron Microscope (Arctica)

1. Cool the autoloader cassette in the transfer station with liquid nitrogen.
2. Carefully clip each grid with a clip ring and C-clip, the clipped grids are referred to as AutoGrids (*see* **Note 4.7.1**).
3. Properly insert well-clipped grids into the cassette and load them into the autoloader of the microscope (*see* **Note 4.7.2**).
4. Open the column valves and find your beam.
5. Set the microscope to 3400× magnification.
6. Bring the specimen to the eucentric height (*see* **Note 4.7.3**).
7. Open EPU, under "Preparation", choose the option of "Grid square." Click the "preview" to get the image (*see* **Note 4.7.4**).

8. Move the mouse to the center of four holes, right-click to choose "move stage here" (*see* **Note 4.7.5**).
9. In the "Preset," choose "Hole/Eucentric Height," click the button of "Preview."
10. Under the "Auto Functions," do the "Auto-eucentric by beam tilt" and then the "Autofocus."
11. Move the stage to the hole area and click the "Preview" to get the screening images.
12. Find another good area to collect more images (*see* **Note 4.7.6**).
13. After finishing the screening, acquire the Atlas for all interested grids.
14. Create a new session for the data collection.
15. Select squares from the Atlas in the square selection task. Examples of good and bad squares are provided (Fig. 5).
16. Go to the automated acquisition task and start data collection.

3.8 Cryo-EM Data Analysis with CryoSPARC and RELION

1. Import the dataset using the Import Movies job in CryoSPARC (*see* **Note 4.8.1**). A comparison of high-quality magnification images is provided (Fig. 6).
2. Determine the correct gain flip (*see* **Note 4.8.2**). Example provided (Fig. 7).
3. Complete the Patch Motion Correction job by inputting your imported movies. Parallelize as many GPUs as possible.
4. Use the Patch CTF estimation job. Input the result of the Patch Motion Correction job and parallelize as many GPUs as possible.

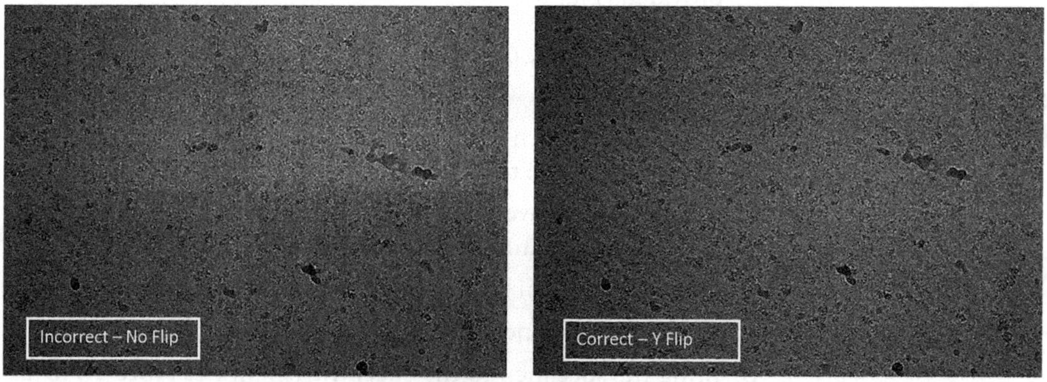

Fig. 6 Gain reference. The Micrograph on the left shows a micrograph where the correct flip of the gain reference was not determined. It appears to be split into four different quadrants. The micrograph on the right did not show these artificial lines and had the collect gain reference flip chosen (in this case, a Y Flip)

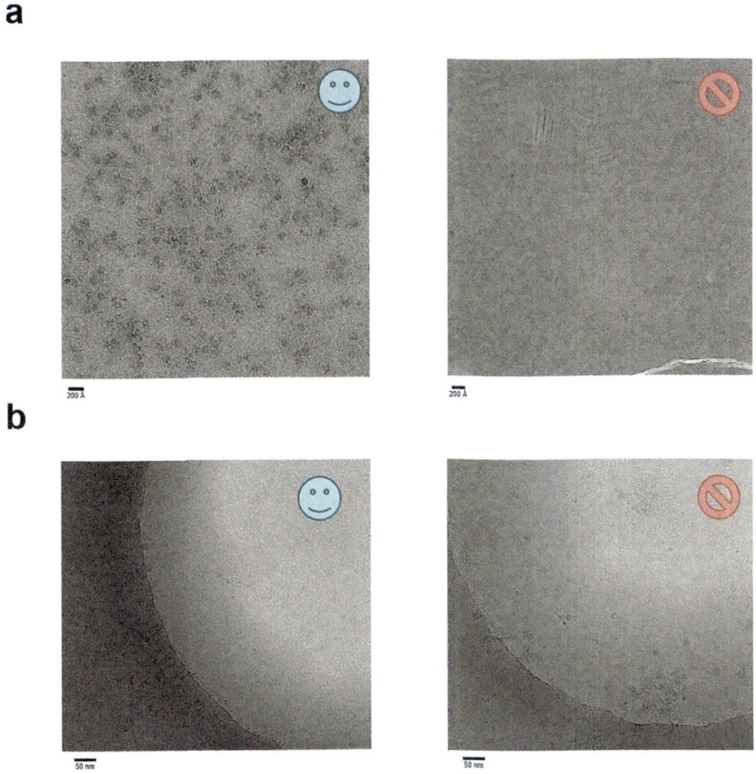

Fig. 7 Particle quality. **a** Good Particle distribution of RSV-LP on the left and bad particle distribution on the right. The dark dots on the left micrograph are individual RSV-LP particles, while the right micrograph is missing these and has some streak-like ice on the top half of the micrograph. **b** Good particle distribution of NP is on the left. There are many unbroken "loops" that are apparent. There is not much aggregation. The right micrograph has bad particle distribution, with a couple of instances of aggregation and many broken particles. The distribution of the particles is also not consistent across the micrograph when compared to the left micrograph

5. Use the Curate Exposures job. Remove bad micrographs. Limit the removed micrographs to ~5% of the dataset (*see* **Note 4.8.3**).
6. Autopick particles from your dataset using the Blob Picker job. A minimum and maximum particle diameter is essential. Select ~10% of the dataset as the micrographs to process. 150–400 micrographs.
7. Use the Inspect Particle Picks job to tailor the autopicking (*see* **Note 4.8.4**).
8. Extract the Auto-picked particles using the Extract from Micrographs (GPU) job in cryoSPARC. Bin heavily for this particle training section (to 128 or 64 pixels). Determine how many micrographs to use and what to bin using Fig. 8.

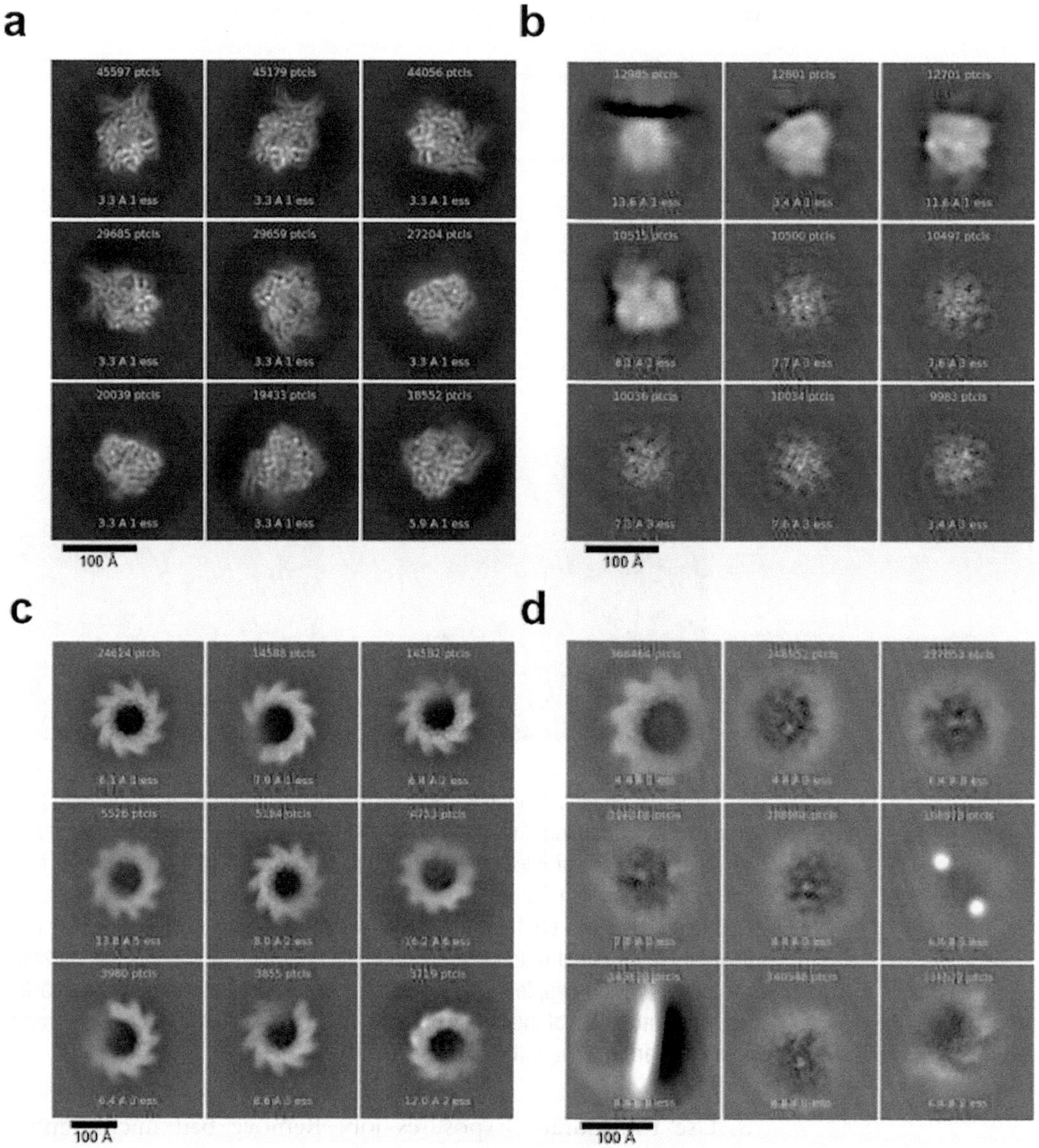

Fig. 8 Cryo-EM 2D classification. **a** Good quality 2D classes of the RSV-LP protein. The protein is about 100 Angstroms in diameter. **b** Bad quality 2D classes of the RSV-LP protein. There are no features visible in the classes. **c** Good quality cryo-EM 2D class images of the N:RNA complex. The protein is between 150 and 200 Angstroms in diameter. **d** Bad quality cryo-EM 2D class images of the N:RNA complex. The box size here could be increased to improve the overall quality of the classes

9. Run the 2D classification job and select the best classes with Select 2D. Repeat once. Examples of good and bad 2D classes are provided (Fig. 9).
10. Use the Topaz Train job to tell the Topaz system what kinds of particles to pick (*see* **Note 4.8.5**).

a

Step	Pre-Topaz Train	During Topaz Cycles 1&2	Topaz Train 3	Topaz Extraction 3	Extract From Mics. 3	Re-Extraction of Final Particles
Bin to X pix	Bin to 128 or 64	Bin to 128 or 64	Bin to 128 or 64	N/A	Bin to 128 or 64	$1.1\ Å/pixel$
Dataset	200-500 Mics.	200-500 Mics.	200-500 Mics.	Entire Dataset	Entire Dataset	Entire Dataset

b

	Example			
Pixel Size (Å/pixel)	0.85	0.94	1.03	1.1
Extraction Box Size of 200 pix				
Pixel Size for Bin to 128 pix	1.33 Å/pix	1.47 Å/pixel	1.63 Å/pix	1.72 Å/pix
Box Size for Bin to 1.1 Å/pix	155 pix	171 pix	187 pix	200 pix
Extraction Box Size 150 pix				
Pixel Size for Bin to 128 pix	1.00Å/pix	1.10 Å/pix	1.21 Å/pix	1.29 Å/pix
Box Size for Bin to 1.1 Å/pixel	116 pix	128 pix	140 pix	150 pix

$$\frac{200\ pix(1.03\ Å/pixel)}{x} = 1.1\ Å/pixel$$

$$x = \frac{150\ pix(1.1\ Å/pixel)}{128\ pix}$$

Fig. 9 The Nyquist limit at different steps. The pixel size times two determines the best resolution possible at that specific pixel size (Nyquist). This is important for the final re-extraction. This final bin should be to a final pixel size of 1.1 Å/pixel. The pixel size does not need to be optimized until this final step is reached. Combining this with the small subset dataset saves time and space during the data processing cycle

11. Use the Topaz Extract job to autopick based on the training model (Topaz Train Output).
12. Use the Extract from Micrographs (GPU) job to extract the Topaz-picked particles. Make sure to keep the heavy bin (128 or 64 pixels).
13. Do 1–2 rounds of 2D classification and select 2D to pick the best particles to put back into the Topaz Train cycle.
14. Complete Topaz Train, Topaz Extract, and Extract from micrographs (GPU) with the new starting particles (3.8.13).
15. Do 1–2 rounds of 2D classification and select 2D to pick the best particles.
16. Complete a Topaz Train job with the new particles.
17. Complete a Topaz Extract job with the 3.8.16 Topaz Train Model and the 3.8.5 micrographs.
18. Do an Extract from micrographs (GPU) job, but make sure to keep the data binned (128 or 64 pix).
19. Do two rounds of 2D classification and select 2D and select the best particles.
20. Do an Extract from micrographs (GPU) job to re-extract the particles at a barely binned box size (Pixel Size = 1.1).

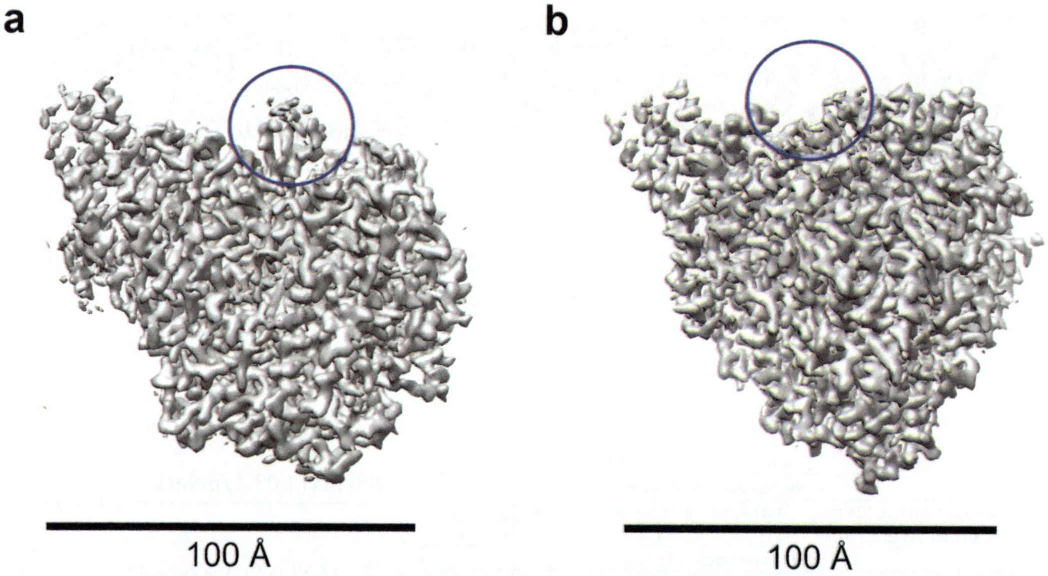

Fig. 10 Demonstration of Z-Flip. **a** The original map. **b** The same map after Z-flip. Z-Flipped changed its handedness. The artifact highlighted with a circle in **a** is flipped to the backside and no longer visible on the front side

21. Complete one final round of 2D classification and select 2D.
22. Move the final particles into the RELION directory (*see* **Note 4.8.6**).
23. Import the particle star file in RELION.
24. Complete 1× 2D classification job and select the best classes with subset selection.
25. Complete an Initial model job using the particles selected in 3.7.24.
26. Complete multiple rounds of 3D classification and Refine 3D. Make sure to check for Z-flip at the Refine 3D step (Fig. 10).
27. Open the Refine 3D map in ChimeraX to determine the proper initial binarization threshold (*see* **Note 4.8.7**).
28. Use the Refine 3D map and Initial Binarization to create a mask.
29. Complete a Post-Process job.

3.9 Model Building

1. Generate an initial model (*see* **Note 4.9.1**). If no homology structure is available, you may need to build the model ab initio with PHENIX and COOT [26, 27].
2. Fit the initial model into the cryo-EM map with ChimeraX [27].
3. Complete a Real-space Refinement with PHENIX [29].

4. Use COOT to refine and fit the model into the map (*see* **Note 4.9.2**) and tweak the entire model residue by residue [28].
5. Complete a Real-space Refinement with PHENIX [29]
6. Estimate the final model with Molprobity [30] and do a further refinement with COOT [28].
7. Complete a Real-space Refinement with PHENIX [29].

4 Notes

4.1.1. The protein samples should be stored in small aliquots to avoid multiple freeze–thaw cycles and prepared freshly before grid preparation.

4.1.2. The buffer, concentration, and incubation time for a protein–RNA complex can be optimized depending on the proteins.

4.2.1. Be careful when moving the glass chamber because it is very fragile. Do not let the grids sit for more than 30 min.

4.2.2. Cut a parafilm strip and put two drops of water or buffer (20–30 μL each) and two drops of staining solution (20–30 μL each) on it. Make sure the parafilm strip is flat enough to hold the radioactive UF solution drops.

4.2.3. Hold the edge of the grid with tweezers with one hand steady to keep it flat.

4.3.1. Wait at least 10 min after adding liquid nitrogen. You should wait 30 min after turning on the filament before operating the microscope. Vacuum status should be shown as green.

4.3.2. The tool should not be let go of while opening and closing the holder. The grid should be placed upside down in the holder.

4.3.3. Wear gloves when handling a sample holder to prevent contamination. Never touch the brass part of the holder.

4.3.4. If the beam does not get larger when turning the intensity knob clockwise, condense the beam and pass the crossover point.

4.3.5. If necessary, repeat the previous steps at higher magnification to tune the Z position.

4.3.6. When you increase the magnification, the beam diameter will be the same (not the intensity).

4.3.7. The beam must become brighter (smaller) when turning the intensity knob counterclockwise. If this is not the case, keep condensing the beam (i.e., making it smaller) and pass the crossover point (smaller beam size).

4.3.8. Click “Beam Shift” on the “Direct Alignments” OCX (Tune tab) and center the beam with the multifunction knobs. Press “Done” when the beam is centered.

4.3.9. Scan the whole grid to try to find a gradient of the stain’s thickness. You should be able to get good images in a medium area of the gradient.

4.4.1. Combine negative-stain EM data collection and EMAN2 processing to know the low-resolution conformation of the protein of interest.

4.4.2. Diameter, oligomerization state, and related information should also be prepared before EMAN2 processing, especially the particle picking step.

4.4.3. Make sure to have edgenorm, xraypixel, and ctfest selected under ###filtering options###.

4.4.4. Browse to select good micrographs. Boxsize and ptclsize = −1. Select gui. Use 80 threads and apix = −1.

4.4.5. Input the micrographs from the previous step and the boxsize and ptclsize. Select write_dbbox and write_ptcls.

4.5.1. Air dry the gloves before installing the blot paper to make sure it is completely dried. The blot papers should be replaced before usage and can commonly be used for eight blottings.

4.5.2. Please make sure the humidifier is well sealed. Only refill with max 40 mL of distilled water after the message is displayed in the user interface.

4.5.3. Please remember to power off the Vitrobot after usage.

4.5.4. Let the chamber and Vitrobot equilibrate to the set temperature and humidity for approximately 30 min. The temperature and humidity should be well maintained during the grid freezing. The temperature can be set to range from 4 to 60 °C, depending on the samples. The parameters of the blotting can be optimized depending on the samples.

4.5.5. Fill the reservoir with liquid nitrogen and the brass cryogen cup with ethane liquid. Precool a grid box in the metal stand and put the cryogen on the holder under the chamber. *Ethane is flammable and toxic.* Thus, keep it away from fire and heat and wear proper protection, such as eye protection and PPE, when handling liquid ethane. Constantly cover the grid box with liquid nitrogen.

4.5.6. Parafilm can be used to wrap around the glass slide to support the grids, which helps pick up the grids with a tweezer. Do not use parafilm if you use Solarus for glow discharge. Remember to put back the glass chamber and turn off the equipment after it is finished.

4.5.7. The tip of the Vitrobot tweezer is very sensitive and should be protected from touching anything hard, such as glass, a table, metal, and so on.

4.5.8. The foot pedal can be used to operate the Vitrobot. On the options interface, the "Humidifier off During Process" and "Skip Grid Transfer" buttons can be checked.

4.5.9. Make sure the sample solution is well spread onto the surface of the grids and avoid touching the surface of the grid with a pipette tip.

4.5.10. Do not bend the grids during transferring, and they should always be inside liquid nitrogen afterward to prevent crystallization or melting of vitreous ice.

4.5.11. Make sure the ethane solution is clear during grid freezing.

4.5.12. Create one hole on the cap and two holes on the top of the 50 mL tube to allow the liquid nitrogen to enter. Make sure the liquid nitrogen level in the Dewar is higher than that in the 50 mL tube.

4.6.1. Make sure the grid is securely clipped to the holder. Keep the grid under liquid nitrogen during the grid loading process.

4.6.2. Before using the microscope, check the vacuum and temperature status, fill the Dewar with liquid nitrogen, verify that the proper software is running, and turn on the filament. Vacuum status should show as green. Make sure to wait at least 10 min after adding liquid nitrogen. Wait for 30 min after turning on the filament before operating the microscope.

4.6.3. While turning the holder, be careful about the liquid nitrogen spilling from the dewar.

4.6.4. Find a reference object and activate the "wobbler", then minimize the movement with the Z up and down buttons on the right-hand panel.

4.6.5. You should now see the Fourier transform (FFT) of your live image.

4.7.1. For visual learners, caltech has an in depth video course that includes grid clipping, as well as the entire grid making process [31].

4.7.2. Inspect the grid to ensure the clipping succeeded before inserting it into the cassette. An improperly clipped grid can come loose in the microscope.

4.7.3. Find a reference object, activate the "wobbler," and minimize the movement with the Z up and down buttons on the right-hand panel.

4.7.4. Make sure the specimen is accurately set to eucentric height.

4.7.5. Find the good areas, assessed by a sufficient density of particles and a thin layer of vitreous ice.

4.7.6. Every exposure causes damage to the specimen; try to take more images from different areas.

4.8.1. When importing movies in cryoSPARC, you need to obtain the Raw Pixel Size, Accelerating Voltage (kV), Spherical Aberration, and Total Exposure Dose. You will also need to run it with four parallelized GPUs.

4.8.2. The proper gain reference is checked by importing the images with no flip, x-flip, and y-flipped gain references. Comparison in Fig. 7.

4.8.3. The "curate exposures" job is a manual job. In the interactive job, you can determine the quality of different parameters (like ice thickness) and remove bad micrographs by setting upper and lower limits.

4.8.4. Using the Inspect particle picks is another interactive job. Use the slider bar to the left of the "Power Score" to tailor the autopicker. Check multiple micrographs on the left side to ensure that the Power Score you chose is not causing under- or over-picking.

4.8.5. You will need the path to the Topaz executable for all Topaz training steps. For each topaz train, you also need to come up with an Expected number of particles. For this, you take the current number of particles you got per micrograph on average (from the Select 2D output used for the training) and multiply that number by 2–3. One thing that also changed during Topaz train, as you get to the final two jobs of the Topaz train, is increasing the number of epochs to 30 or 50.

4.8.6. A script is needed to move data from cryoSPARC into RELION. The process, in general, is changing the selected 2D or 2D classification output into a star file that can be read by RELION. You will need the location in the computer cluster for both the extract job and the 2D classification/Select 2D jobs. The conversion can be done with csparc2star.py in Linux.

4.8.7. The 3D refine job can be downloaded onto your personal computer and then opened in ChimeraX. The binarization threshold is determined by the value in the level space in the volume viewer. This value changes as you move the bar on the volume viewer.

4.9.1. If the target protein already has a structure available on the RCSB PDB website, you can download it as an initial model. If the target protein has a homologous structure, you can generate an initial model with the protein sequence using 3D structure prediction tools, such as Alphafold [32] or Phyre^2 [33]. The programs can use one target protein domain as an initial model. Sequence alignment can provide more clues for the model building.

4.9.2. Some validation tools in COOT [28] can be used during model building, such as "Rotamer analysis," "Ramachandran Plot," and "Density fit analysis."

Acknowledgments

We thank the members of the Liang laboratory for critical discussions and helpful support. The research programs in the Liang laboratory at Emory are supported by the US National Institute of General Medical Sciences (NIGMS), the National Institutes of Health (NIH) under award number R01GM130950, and the American Lung Association Innovation Award, grant number IA-831472. The content is solely the responsibility of the authors and does not necessarily represent the official views of the National Institutes of Health.

Conflict of Interest The authors declare no competing interests.

References

1. Renaud JP et al (2018) Cryo-EM in drug discovery: achievements, limitations and prospects. Nat Rev Drug Discov 17:471–492
2. Shen PS (2018) The 2017 Nobel prize in chemistry: cryo-EM comes of age. Anal Bioanal Chem 410:2053–2057
3. Tanner SJ et al (2014) Crystal structure of the essential transcription antiterminator M2-1 protein of human respiratory syncytial virus and implications of its phosphorylation. Proc Natl Acad Sci USA 111:1580–1585
4. Selvaraj M et al (2018) The structure of the human respiratory syncytial virus M2-1 protein bound to the interaction domain of the phosphoprotein P defines the orientation of the complex. MBio 9
5. Muniyandi S, Pangratiou G, Edwards TA, Barr JN (2018) Structure and function of the human respiratory syncytial virus M2-1 protein. Subcell Biochem 88:245–260
6. Gao Y et al (2020) Structure of the human respiratory syncytial virus M2-1 protein in

complex with a short positive-sense gene-end RNA. Structure 28:979–990 e974

7. Tawar RG et al (2009) Crystal structure of a nucleocapsid-like nucleoprotein-RNA complex of respiratory syncytial virus. Science 326: 1279–1283
8. Renner M et al (2016) Nucleocapsid assembly in pneumoviruses is regulated by conformational switching of the N protein. elife 5: e12627
9. Cao D et al (2020) Cryo-EM structure of the respiratory syncytial virus RNA polymerase. Nat Commun 11:368
10. Gilman MSA et al (2019) Structure of the respiratory syncytial virus polymerase complex. Cell 179:193–204 e114
11. Cao D et al (2024) Structures of the promoter-bound respiratory syncytial virus polymerase. Nature 625:611–617
12. Gonnin L et al (2023) Structural landscape of the respiratory syncytial virus nucleocapsids. Nat Commun 14:5732
13. Liljeroos L, Krzyzaniak MA, Helenius A, Butcher SJ (2013) Architecture of respiratory syncytial virus revealed by electron cryotomography. Proc Natl Acad Sci USA 110:11133–11138
14. Yao H et al (2020) Molecular architecture of the SARS-CoV-2 virus. Cell 183:730–738 e713
15. Liang B et al (2015) Structure of the L protein of vesicular stomatitis virus from electron cryomicroscopy. Cell 162:314–327
16. Jenni S et al (2020) Structure of the vesicular stomatitis virus L protein in complex with its phosphoprotein cofactor. Cell Rep 30:53–60 e55
17. Horwitz JA, Jenni S, Harrison SC, Whelan SPJ (2020) Structure of a rabies virus polymerase complex from electron cryo-microscopy. Proc Natl Acad Sci USA 117:2099–2107
18. Pan J et al (2020) Structure of the human metapneumovirus polymerase phosphoprotein complex. Nature 577:275–279
19. Xie J et al (2024) Structural basis for dimerization of a paramyxovirus polymerase complex. Nat Commun 15:3163
20. Abdella R, Aggarwal M, Okura T, Lamb RA, He Y (2020) Structure of a paramyxovirus polymerase complex reveals a unique methyltransferase-CTD conformation. Proc Natl Acad Sci USA 117:4931–4941
21. Cong J et al (2023) Structure of the Newcastle Disease Virus L protein in complex with tetrameric phosphoprotein. Nat Commun 14:1324
22. Peng Q et al (2023) Molecular mechanism of de novo replication by the Ebola virus polymerase. Nature 622:603–610
23. Yuan B et al (2022) Structure of the Ebola virus polymerase complex. Nature 610:394–401
24. Punjani A, Rubinstein JL, Fleet DJ, Brubaker MA (2017) cryoSPARC: algorithms for rapid unsupervised cryo-EM structure determination. Nat Methods 14:290–296
25. Scheres SH (2012) A Bayesian view on cryo-EM structure determination. J Mol Biol 415: 406–418
26. Scheres SH (2012) RELION: implementation of a Bayesian approach to cryo-EM structure determination. J Struct Biol 180:519–530
27. Pettersen EF et al (2021) UCSF ChimeraX: structure visualization for researchers, educators, and developers. Protein Sci 30:70–82
28. Emsley P, Cowtan K (2004) Coot: model-building tools for molecular graphics. Acta Crystallogr D Biol Crystallogr 60:2126–2132
29. Afonine PV et al (2018) Real-space refinement in PHENIX for cryo-EM and crystallography. Acta Crystallogr D Struct Biol 74:531–544
30. Williams CJ et al (2018) MolProbity: more and better reference data for improved all-atom structure validation. Protein Sci 27:293–315
31. Jensen G (2017) Cryo-EM course – unit 2: sample preparation. Caltech, Youtube
32. Jumper J et al (2021) Highly accurate protein structure prediction with AlphaFold. Nature 596:583–589
33. Kelley LA, Mezulis S, Yates CM, Wass MN, Sternberg MJ (2015) The Phyre2 web portal for protein modeling, prediction and analysis. Nat Protoc 10:845–858

Chapter 12

In Vitro Biochemical Assays to Study Respiratory Syncytial Virus Polymerase Activities

Sarah L. Noton and Barbara Ludeke

Abstract

Here, we detail a biochemical assay that assesses respiratory syncytial virus (RSV) RNA-dependent RNA polymerase activities at the viral promoters in vitro. The assay consists of a purified recombinant RSV polymerase complex comprising the L protein and its cofactor P, combined with a short RNA template and radiolabeled NTPs. RNA synthesis by the L-P complex yields radiolabeled RNA products that can be separated at the nucleotide level on denaturing polyacrylamide gels and detected by autoradiography or phosphor image analysis. Each component of the assay can be readily manipulated. As such, this simple assay has aided the understanding of early polymerase transcription and genome synthesis initiation events, revealed previously unknown polymerase functions, and determined how various RSV polymerase inhibitors function.

Key words RSV, Negative-sense RNA virus, RNA-dependent RNA polymerase, In vitro biochemical assay, de novo RNA synthesis, Polymerase initiation, Polymerase inhibitors

1 Introduction

The RSV polymerase is a complex of two proteins, the ~250 kDa large (L) protein that contains the enzymatic domains responsible for RNA synthesis, capping and methylation of mRNA [1–4], and the viral phosphoprotein (P) [3, 4], which is required for efficient expression of L and as an adaptor protein [5, 6]. The natural template for RSV RNA synthesis is the nucleocapsid, in which the genome or antigenome RNA is encapsidated along its length in nucleoprotein (N protein). However, for RNA synthesis to occur, the N protein is transiently removed from the template to allow the RNA to enter the template entrance channel of the polymerase. While reconstituting polymerase activity on a purified nucleocapsid template is challenging, reconstituting polymerase activity on a short naked RNA template is tractable and captures events that occur following local displacement of N protein and RNA entry

Bo Liang and Rachel Fearns (eds.), *Human Respiratory Syncytial Virus: Methods and Protocols*, Methods in Molecular Biology, vol. 2948, https://doi.org/10.1007/978-1-0716-4666-3_12,

into the active site. There are two different types of RNA synthesis assays using naked RNA templates: (1) de novo RNA synthesis initiation assays, in which the polymerase is provided with an RNA template containing a promoter, such that the polymerase has the RNA signals necessary to initiate RNA synthesis, and (2) primer elongation assays in which the polymerase is provided with a template (which can be of irrelevant sequence), and primer to which the polymerase adds nucleotides in a templated manner. Details on how to perform the primer elongation assay can be found in multiple references [3, 7–12]. Here, we describe a de novo RNA synthesis initiation assay.

RSV de novo RNA synthesis initiation and subsequent elongation can be recapitulated in vitro by combining recombinant, purified L-P complexes [3, 4, 13, 14], *see* Chapter 10, an oligonucleotide RNA template, and NTPs, including a radioactive NTP. RNA synthesis by the RSV polymerase leads to the production of radiolabeled RNA products ([4]; Fig. 1). Following the RNA synthesis reaction, the products are separated by denaturing gel electrophoresis and can be visualized by autoradiography or phosphorimage analysis, allowing both qualitative and quantitative assessment of polymerase activity.

This in vitro RNA synthesis assay enables detailed mechanistic analyses of activities at the promoter. Because the assay contains only the minimal polymerase complex consisting of L and P proteins, template, and NTPs, it allows testing of the effect of altering the RNA template sequence, introducing mutations into the polymerase, and varying NTP concentrations [15–20]. Since the assay captures all products present in the unsynchronized reaction at the time it is terminated, it is possible to detect intermediates in RNA synthesis, from the first dinucleotide product that is made to products that extend to the end of the template [13, 19] by separating the radiolabeled products by denaturing gel electrophoresis. This allows the initiation and early elongation steps of polymerase activity to be distinguished. The assay described here, or a variation of it, has identified key features of the RSV polymerase, such as the use of two different initiation sites in the promoter [4, 19, 20], the ability to select the initiating NTP independently of the template sequence [19], and the ability of the polymerase to extend the 3′-terminus of an RNA template containing the tr promoter sequence (the promoter for genome synthesis) via a back-priming mechanism ([4]; Fig. 1). The assay has also been used to confirm the targeting of the RSV polymerase by antiviral compounds and to determine their mechanism of inhibition [12, 13, 17, 21–25].

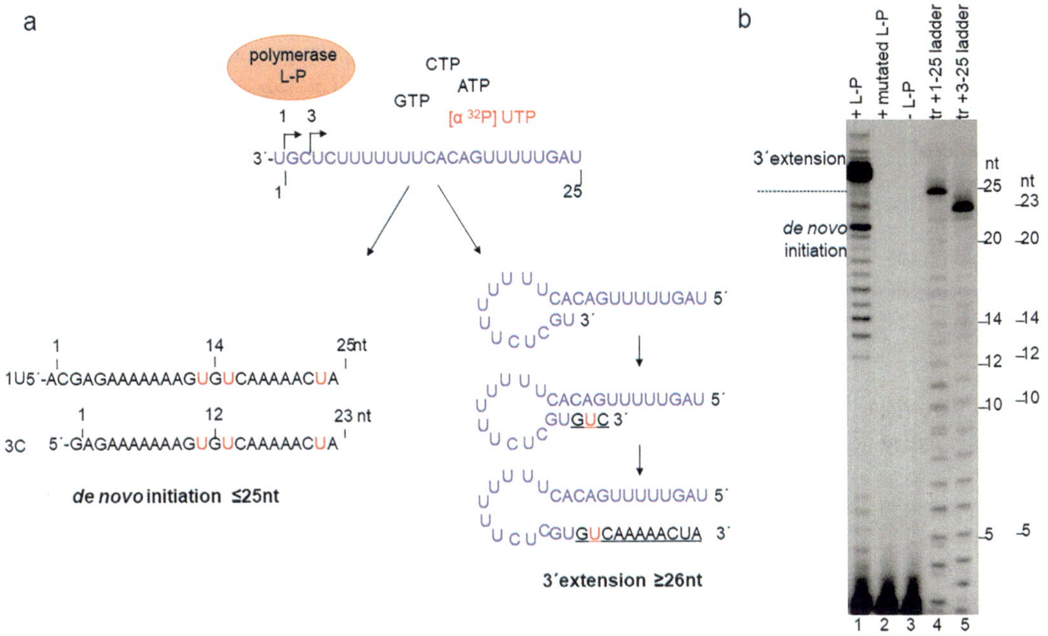

Fig. 1 Example of an in vitro RNA synthesis assay. **a** Schematic diagram of a standard in vitro RNA synthesis assay. Purified recombinant L-P polymerase (oval) is combined with an RNA oligonucleotide (blue sequence) encoding the first 25 nt of the tr promoter for genome synthesis, NTPs (black) and [α^{32}P]-UTP tracer (red). During de novo RNA synthesis initiation, the polymerase initiates either opposite position 1U or internally opposite position 3C of the promoter (reviewed in [29]) resulting in products of up to 25 or 23 nt in size, respectively. Using a radiolabeled UTP tracer, the shortest products that can be detected from the 1U and 3C initiation sites are 14 nt and 12 nt, respectively. In addition, the polymerase can extend the 3′ terminus of the template via a back-priming mechanism in which the 3′ terminal UG bases pair internally with nt 13C and 14A (or nt 15C and 16A) to prime synthesis of RNA products 26 nt and longer, with 1–3 nucleotides predominantly added. **b** An example autoradiogram of a standard in vitro RNA synthesis assay. *Lane 1* is a reaction in which wild-type polymerase was included in the assay. Products up to 25-nt size are a result of de novo RNA synthesis initiation. The "ladder" of products reflects a mix of products obtained by premature abortion of synthesis by the polymerase and products being synthesized at the point the reaction was terminated by heat inactivation of the polymerase. Products longer than 26 nt are due to extension of template 3′ termini via back-priming. *Lane 2* is a negative control in which a catalytic L-P mutant L D811A was included in the reaction. *Lane 3* is a negative control in which L-P complexes were omitted from the reaction. *Lanes 4* and *5* show RNA ladders representing products initiated at position 1U or 3C of the template, respectively

2 Materials

Always wear a lab coat, disposable gloves, and protective goggles when working with hazardous chemicals and biological reagents. Consult appropriate material safety data sheets available from product suppliers and follow all waste disposal regulations. Use appropriate Plexiglass shielding when working with ^{32}P. Note that RNA is highly sensitive to RNase degradation, so whenever working with RNA, clean gloves should be worn, and ultrapure nuclease-free water should be used to prepare all reaction buffers to avoid introducing RNase contamination. To prepare gel mixes and buffers, use ultrapure water prepared by purifying deionized water to a resistance greater than 18 MΩ·cm at 25 °C.

2.1 General Materials

1. RNase-free filter-tipped pipette tips or dedicated pipettes to avoid RNase contamination.
2. Sterile, plugged serological pipettes.
3. Sterile, RNase, and DNase-free 0.5- and 1.5-mL microcentrifuge tubes.
4. Gloves.

2.2 In Vitro RNA Synthesis Assay

1. Nuclease-free water (*see* **Note 1**).
2. Affinity tag purified RSV L-P polymerase preparation (100–400 ng L protein/μL, *see* Chapter 10).
3. Ribonucleotides (NTPs).
4. Radiolabeled NTP tracer (for options, *see* Table 1).
5. RNA oligonucleotide templates (*see* Table 1 and **Note 2**).
6. Ultrapure 1 M Tris–HCl pH 7.4 or 7.5.
7. 50% v/v glycerol in nuclease-free water.
8. 1 M $MgCl_2$.
9. 1 M dithiothreitol (DTT).

Table 1
RNA templates with native sequence for in vitro RNA synthesis assays

Template	RNA sequence 3′-5′	+1 de novo initiation	+3 de novo initiation	3′ extension back priming	Product sizes detected (nt)	References
tr 1–25	*UGCUC* UUUUUUU*CACAG* UUUUUGAU	[γ^{32}P]-ATP	[γ^{32}P]-GTP	[α^{32}P]-CTP, [α ^{32}P]-GTP, or [α ^{32}P]-UTP	>1	[4, 18]
tr 1–14	UGCUC UUUUUUUCA	[α^{32}P]-UTP	[α^{32}P]-UTP		12 and 14	[19]
le 1–25	*UGCGC* UUUUUUA*CGCA* UGUUGUUG		[γ^{32}P]-GTP	[α^{32}P]-ATP or [α ^{32}P]-CTP	>1	[18]
le 1–14	UGCGC UUUUUUACG	[α^{32}P]-UTP	[α^{32}P]-UTP		>12	[30]

Standard RNA templates containing RSV promoter sequence representing the genome (le) and antigenome (tr) promoters are listed. Nucleotides in *italics* can self-anneal to form a loop structure that primes the extension of the 3′-terminus of the RNA template (back-priming). Radiolabeled tracers we have used to detect specific polymerase products/activities are listed for each template

10. Heat block.
11. Bromophenol blue and xylene cyanol mix: 2% w/v bromophenol blue, 2% w/v xylene cyanol solution in water. Store tightly capped at room temperature.
12. *STOP buffer*: Deionized formamide containing 20 mM EDTA and 0.01% each bromophenol blue and xylene cyanol.
13. Refrigerated microcentrifuge.

2.3 Phenol-Chloroform RNA Extraction

1. Acid-phenol (acid-phenol:chloroform with isoamyl alcohol (124:24:1)), pH 4.5.
2. 0.2% sodium dodecyl sulfate (SDS) in nuclease-free water.
3. 4 M sodium chloride.
4. 100% anhydrous ethanol, USP grade.
5. Glycoblue co-precipitant (Invitrogen, 15 mg/mL) or glycogen (5 mg/mL).
6. 70% ethanol, USP grade, in nuclease-free water.

2.4 RNA Size Markers

1. 5′-OH RNA oligonucleotide (Table 2).
2. 10× polynucleotide kinase (PNK) buffer (NEB).
3. T4 Polynucleotide kinase (NEB).
4. [γ^{32}P]-ATP (3000 or 6000 Ci/mmol, 10 mCi/mL).
5. Yeast tRNA (9.4 mg/mL).
6. *Alkaline hydrolysis buffer:* 50 mM sodium carbonate ($NaHCO_3/Na_2CO_3$) pH 9.2, 1 mM EDTA.

2.5 Denaturing Polyacrylamide Gels for RNA Analysis

1. 40% acrylamide/bisacrylamide 19:1.
2. 3% ammonium persulfate (APS) solution in water, freshly prepared for each gel.
3. N,N,N′,N′-tetramethylethane-1,2 diamine (TEMED).
4. 10× Tris-borate-EDTA (TBE) buffer.
5. Magnetic stirring hot plate.

Table 2
Anticipated RNA products if the RSV polymerase initiates at position +1 (tr +1–25) or +3 (tr +3–25) and terminates at position 25 of the tr 1–25 RNA template

Product	RNA sequence 5′-3′
tr +1–25	ACGAGAAAAAAAGUGUCAAAAACUA
tr +3–25	GAGAAAAAAAGUGUCAAAAACUA

These RNA oligonucleotides can be used to generate an RNA ladder to size products generated in the in vitro assays using purified polymerase preparations

6. Glass plates (20 × 40 cm or similar), 0.4 mm spacers, and combs.
7. Large binder clips for gel assembly.
8. Sequencing gel electrophoresis apparatus.
9. High-voltage power supply that supports an output of 3000 V and electrical cables.
10. Heating block.
11. Whatman 3MM blotting paper.
12. Plastic wrap.
13. Vacuum gel dryer.
14. X-ray cassette.
15. X-ray film or phosphor storage screen.

3 Methods

3.1 In Vitro RNA Synthesis Reactions Performed with tr 1–25 Template and [α^{32}P]-UTP Tracer

Here, we describe a standard RNA synthesis assay in which purified recombinant L-P complexes are combined with a tr 1–25 RNA template, NTPs, and [α^{32}P]-UTP to serve as a tracer. This assay captures both RNA synthesis by de novo initiation at the promoter as well as template back priming (Table 1 and Fig. 1). Alternative template and tracer combinations are listed in Table 1 along with references that detail NTP conditions that can be used.

1. Combine 2 μM RNA oligonucleotide with ATP, CTP, and GTP each at 1 mM and UTP at 50 μM with 5 μCi of [α^{32}P]-UTP (3000 Ci/mmol) in freshly prepared transcription buffer (50 mM Tris–HCl at pH 7.4; 8 mM $MgCl_2$; 5 mM DTT; 10% glycerol) in a final volume of 24 μL (*see* **Notes 3**–**7**).
2. Mix well, equilibrate to 30 °C for 5–10 min.
3. Add 1 μL of RSV L/P containing ~100–300 ng of L protein. Mix by slow up and down pipetting or stirring with the pipette tip (*see* **Note 8**).
4. Incubate reactions at 30 °C for 1–3 h as desired (*see* **Note 9**).
5. Heat reaction to 90 °C for 3 min to inactivate the polymerase and cool briefly on ice.
6. Add an equal volume of STOP buffer. Alternatively, purify RNA by extraction with acid-phenol:chloroform (*see* **Notes 10** and **11**; *see* method below) before combining with an equal volume of STOP buffer.
7. Store RNA samples at −20 °C until ready to use.

3.2 Acid-Phenol: Chloroform Extraction and Ethanol Precipitation

1. Bring the sample volume to 50 μL with water or transcription buffer.
2. Add 200 μL of 0.2% SDS.
3. In a chemical fume hood, add 250 μL of acid-phenol: chloroform.
4. Vortex for 30 s and place on ice for 5 min.
5. Separate the aqueous from the organic phase by centrifugation at room temperature for 5 min at 20,000 × *g*.
6. Transfer 200 μL of the aqueous phase to a 1.5 mL microtube containing 10 μL of 4 M NaCl and 1 μL of Glycoblue or 3 μL of glycogen.
7. Add 500 μL of absolute ethanol and precipitate RNA overnight at −20 °C.
8. Pellet the RNA at 4 °C for 30 min at 21,000 × *g*.
9. Wash the pellet once with cold 70% ethanol and spin down for 10 min at 21,000 × *g*.
10. Carefully remove all of the supernatant and air dry the RNA for ~10 min.
11. Resuspend in nuclease-free water.

3.3 Generation of RNA Size Markers by Partial Alkaline Hydrolysis

1. To generate 5′-end ^{32}P-labeled RNA with the sequence of the expected +1 or + 3 initiated full-length product from RNA synthesis reactions (*see* Table 2), combine 5 μL of RNA (100 μM) with 1 μL of 10× polynucleotide kinase (PNK) buffer (NEB) supplied with T4 PNK enzyme (final 1× concentration 70 mM Tris–HCl pH 7.6, 10 mM $MgCl_2$, 5 mM DTT), 30 μCi [γ^{32}P]-ATP (3000 or 6000 Ci/mmol), 1 μL (10 U) T4 PNK (NEB) supplemented with water to a final volume of 10 μL.
2. Incubate the reaction at 37 °C for 30 min.
3. Mix 10 μL of ^{32}P-labeled RNA oligonucleotide from the previous step with 0.6 μL yeast tRNA and 20 μL chilled alkaline hydrolysis buffer.
4. Aliquot 6 μL into 5 additional tubes. Label tubes 1–6 and place on ice.
5. To perform the alkaline hydrolysis, place all tubes on a heat block at 95 °C.
6. After 2 min, remove tube #1 from the heat block, place on ice and add 6 μL of STOP buffer.
7. After 3 min, remove tube #2, place on ice and add 6 μL of STOP buffer.
8. Continue removing tubes at 1-min intervals until there are no tubes left, placing samples on ice and adding 6 μL of STOP buffer.

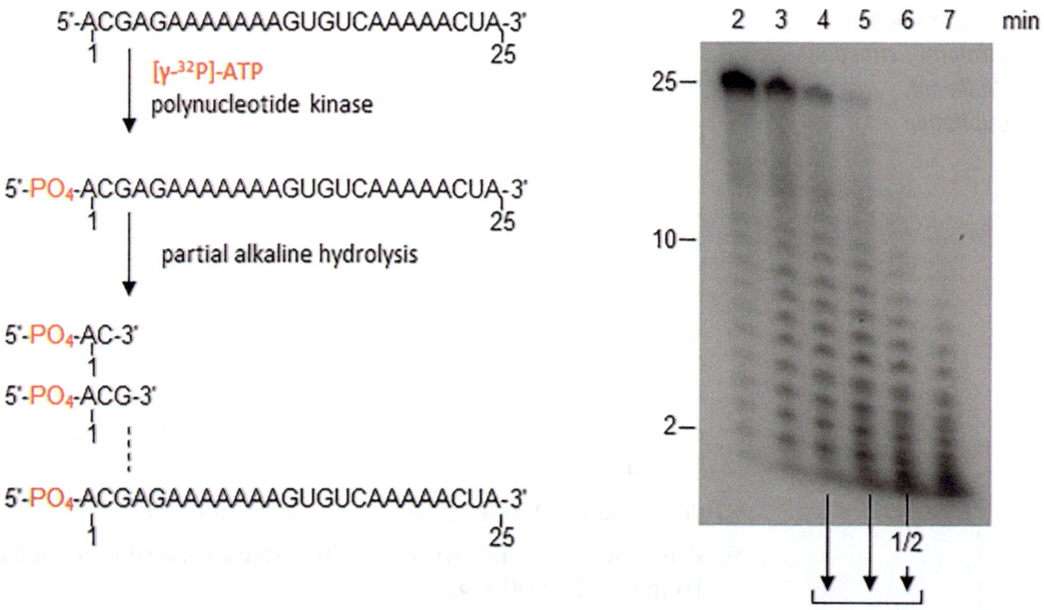

Fig. 2 Preparation of a molecular weight ladder for use in analyzing in vitro RNA synthesis reaction products. An example of a molecular weight marker following partial alkaline hydrolysis over time. **a** Oligonucleotide tr +1–25, corresponding to the full-length product of RNA synthesis originating at 1U of the tr 1–25 template was labeled with ^{32}P at the 5′-end and randomly fragmented by partial alkaline hydrolysis. The schematic indicates hydrolysis products bearing the radioactive label. **b** Hydrolysis products were separated on a denaturing 20% polyacrylamide 7 M urea, 1× TBE gel. Each lane contains 1 μL of a hydrolysis reaction allowed to proceed for the indicated time. Radiolabeled fragments were detected by autoradiography. Arrows indicate which reactions were pooled to create the final RNA ladder, with half of the 6 min sample being included

9. Heat samples to 95 °C for 5 min.
10. Check for hydrolysis of RNA by examining 1–2 μL of each tube on a 20% polyacrylamide 7 M urea TBE gel (*see* Subheading 3.3).
11. Visualize hydrolysis products by exposing the gel to film or to a phosphor image screen for 2–4 h at room temperature.
12. Combine tubes which show RNA fragments representing +1 or + 3 initiated products terminated at each residue along the RNA template (Fig. 2; *see* **Note 12**).
13. Store RNA ladders at −20 °C.

3.4 Denaturing Polyacrylamide Gel Electrophoresis

1. To prepare gel mix, combine 10× TBE and 40% (w/v) 19:1 acrylamide/bisacrylamide according to Table 3 in a glass beaker. Slowly add urea while stirring with a stir bar on a heated magnetic stirrer on the lowest heat setting. Gently heat it while stirring until the urea has dissolved. Add water.

Table 3
Mix for denaturing acrylamide gels in 7 M urea in TBE for resolution of RNA fragments 4 to 35-nt in size

Component	Amount	Final concentration
Urea (g)	48.3	7 M
10× TBE (mL)	10.5	1×
40% (w/v) 19:1 acrylamide/bisacrylamide (mL)	52.5	20%
Water (mL)	8.4	–
Final volume (mL)	~105	–

2. Filter through a 0.22 μm cellulose nitrate membrane and store at room temperature in a glass bottle protected from light for up to 12 months.
3. Thoroughly wash plates with a low-sudsing laboratory detergent, then dry and clean with 100% ethanol.
4. Assemble with spacers according to the manufacturer's instructions. In lieu of a gasket to seal the edges, plates can be sealed with masking tape (Scotch 3 M High Performance Green Masking tape) and clamped with large binder clips.
5. For a 0.4 mm thick gel, combine, on ice, 30 mL gel mix with 300 μL 3% APS and 6 μL TEMED to initiate polymerization (*see* **Note 13**). Swirl to mix, avoiding the introduction of air bubbles.
6. Immediately pour the acrylamide gel mixture between the glass plates and insert the comb. Clamp the comb in place with a large binder clip to prevent separation of the gel from the plates during polymerization.
7. Allow gel to polymerize for at least 1 h (preferably overnight).
8. Following polymerization, remove the comb, any spacers from the bottom of the gel and any tape.
9. Clamp the gel plates to the electrophoresis tank and fill the upper and lower reservoir chambers with 1× TBE buffer such that the wells are covered.
10. Rinse wells with running buffer using a syringe fitted with needle to remove unpolymerized material and urea.
11. Pre-run gel at constant 30 W for at least 1 h (*see* **Note 14**).
12. Dilute RNA ladder in transcription buffer or water as appropriate, such that the ladder and samples are in the same diluent to enable accurate migration during electrophoresis (*see* **Note 15**). Combine the RNA ladder with an equal volume of STOP buffer.

13. Heat denature reaction samples and RNA ladder at 95 °C for 5 min, chill on ice and spin down condensation moisture before loading on pre-run gel.
14. Rinse the wells of the gel with running buffer again just before loading.
15. Load 7 μL of denatured sample or RNA ladder per well.
16. Run the gel at 30 W until the lower bromophenol dye front has migrated to 75% of the length of the gel or as desired (*see* **Note 16**).
17. Drain buffer from upper and lower chambers.
18. Disassemble the gel sandwich by removing the side spacers and one of the glass plates. Cover the gel with plastic wrap.
19. Flip the gel over and remove the other glass plate such that the gel is lying on the plastic wrap.
20. Place a 3 MM paper cut to size on top of the gel and flip again (*see* **Note 17**).
21. Vacuum dry gel at 80 °C for at least 4 h or until dry.
22. Place dried gel in an X-ray cassette.
23. To visualize products, either expose to X-ray film at −80 °C or room temperature or use a phosphor storage screen and expose at room temperature. Following sufficient exposure time, remove film or phosphor screen and process (*see* **Note 18**).

4 Notes

1. DEPC-treated water is not recommended as it can inhibit polymerase and other enzyme activities if the DEPC is not properly inactivated.
2. Deprotect any protected RNA oligonucleotides according to the supplier's instructions.
3. Each component of the transcription buffer can be altered as desired [18].
4. If an alternative tracer is used in the assay, the ribonucleotide mix must be adjusted to ensure efficient incorporation of the tracer. The final concentration of the unlabeled NTP corresponding to the tracer should be adjusted to 50 μM, while the other NTPs are all 1 mM each.
5. If reactions contain the tr 1–14 RNA template and [α^{32}P]-UTP, we recommend 1 mM each ATP, CTP, and GTP and omitting additional UTP to ensure efficient tracer incorporation.

6. For reactions performed with [γ^{32}P]-ATP, we recommend 500 μM ATP, 1 mM CTP, 500 μM UTP, and 10 μM GTP with 5 μCi [γ^{32}P]-ATP (6000 Ci/mmol).
7. [γ^{32}P]-GTP (6000 Ci/mmol) has been discontinued by Perkin Elmer (now Revvity) and we have been unable to find a US supplier. A protocol for the synthesis of [γ^{32}P]-GTP is described by Beutler and Guinto [26].
8. Avoid overmixing as the polymerase is sensitive to mechanical stress.
9. Optimal time for assay will be dependent on the goal of the assay and the specific activity of the polymerase preparation. Examples of time courses can be found in the following references: [18, 19].
10. It may be necessary to purify the RNA samples by extraction with acid-phenol:chloroform before analysis by gel electrophoresis if the product signals are weak. Sample clean up reduces background signals by removing unincorporated radiolabeled tracer and contaminants present in the radionucleotide preparation.
11. Another way to reduce the amount of background signal is to treat the transcription reactions with calf intestinal phosphatase (CIP; NEB), which removes 5′-terminal phosphates from unincorporated NTPs. To do this, briefly cool heat-inactivated reactions on ice, then add 10 U of CIP to each reaction and incubate at 37 °C for 1 h. Samples can be further purified by acid-phenol:chloroform extraction (*see* Subheading 3.1). Note that RNA products following CIP treatment will have a 5′-hydroxyl group, which will impact their migration pattern during gel electrophoresis (*see* **Note 12**).
12. For oligonucleotides less than ~12-nt, migration during gel electrophoresis is impacted by sequence and phosphorylation status of the 5′-terminus (Fig. 1; [27]). Note that the RNA ladder products possess a 5′-monophosphate at their 5′-termini while RNA products generated by the RSV polymerase will have 5′-triphosphate (unless they have been treated with CIP, *see* above).
13. For thicker or smaller-sized gels, scale as appropriate. For a 20 × 40 cm 0.8 mm thick gel, combine 55 mL of gel mix with 550 μL of 3% APS and 55 μL of TEMED. Note that denaturing 20% polyacrylamide gels in 7 M urea and 1× TBE buffer that are 0.8 mm thick cannot be dried down. However, autoradiography can be performed with the gel exposed to film at room temperature or −80 °C.

14. For gels of a different size than described here, a wattage should be chosen that ensures that the glass plates are hot to the touch but do not exceed 60 °C.
15. We suggest diluting the RNA ladder in water or buffer (as appropriate), aiming to load ~1000–3000 counts per minute (determined with a Geiger counter) per well. The amount loaded may require optimization.
16. The bromophenol blue dye co-migrates with 5 to 6-nt RNA while xylene cyanol dye co-migrates with 20-nt RNA. The goal is to run the gel such that unincorporated radiolabeled tracer has migrated off the gel in order to reduce the background signal.
17. Alternatively, the plastic wrap can be replaced with cellophane (Invitrogen) before drying the gel. This ensures a flat surface and reduces static that can appear on film during autoradiography. Thoroughly wet the cellophane sheets in water. Carefully peel the plastic wrap from the gel, adding a few drops of water as necessary to facilitate the release of the plastic from the gel. Place cellophane on the gel and cover with plastic wrap before drying. Remove the plastic wrap before exposing to film.
18. Reactions can also be analyzed using a denaturing 25% polyacrylamide gel containing 6 M urea in Tris–taurine–EDTA (TTE) buffer for resolution of RNA fragments 2 to 20-nt in size (detailed in [13, 19, 28]). The bromophenol blue and xylene cyanol dyes co-migrate with 6-mer and 17-mer RNA fragments, respectively.

Acknowledgments

We thank Dr. Rachel Fearns and Heesu Kim for their critical review of the manuscript.

References

1. Poch O, Sauvaget I, Delarue M, Tordo N (1989) Identification of four conserved motifs among the RNA-dependent polymerase encoding elements. EMBO J 8(12): 3867–3874. https://doi.org/10.1002/j.1460-2075.1989.tb08565.x
2. Fix J, Galloux M, Blondot ML, Eleouet JF (2011) The insertion of fluorescent proteins in a variable region of respiratory syncytial virus L polymerase results in fluorescent and functional enzymes but with reduced activities. Open Virol J 5:103–108. https://doi.org/10.2174/1874357901105010103
3. Gilman MSA, Liu C, Fung A, Behera I, Jordan P, Rigaux P, Ysebaert N, Tcherniuk S, Sourimant J, Eleouet JF, Sutto-Ortiz P, Decroly E, Roymans D, Jin Z, McLellan JS (2019) Structure of the respiratory syncytial virus polymerase complex. Cell 179(1): 193–204 e114. https://doi.org/10.1016/j.cell.2019.08.014
4. Noton SL, Deflube LR, Tremaglio CZ, Fearns R (2012) The respiratory syncytial virus polymerase has multiple RNA synthesis activities at the promoter. PLoS Pathog 8(10):e1002980. https://doi.org/10.1371/journal.ppat.1002980

5. Cao D, Gao Y, Liang B (2021) Structural insights into the respiratory syncytial virus RNA synthesis complexes. Viruses 13(5). https://doi.org/10.3390/v13050834
6. Sutto-Ortiz P, Eleouet JF, Ferron F, Decroly E (2023) Biochemistry of the respiratory syncytial virus L protein embedding RNA polymerase and capping activities. Viruses 15(2). https://doi.org/10.3390/v15020341
7. Deval J, Hong J, Wang G, Taylor J, Smith LK, Fung A, Stevens SK, Liu H, Jin Z, Dyatkina N, Prhavc M, Stoycheva AD, Serebryany V, Liu J, Smith DB, Tam Y, Zhang Q, Moore ML, Fearns R, Chanda SM, Blatt LM, Symons JA, Beigelman L (2015) Molecular basis for the selective inhibition of respiratory syncytial virus RNA polymerase by 2'-fluoro-4'-chloromethyl-cytidine triphosphate. PLoS Pathog 11(6):e1004995. https://doi.org/10.1371/journal.ppat.1004995
8. Ludeke B, Fearns R (2020) The respiratory syncytial virus polymerase can perform RNA synthesis with modified primers and nucleotide analogs. Virology 540:66–74. https://doi.org/10.1016/j.virol.2019.11.002
9. Tchesnokov EP, Raeisimakiani P, Ngure M, Marchant D, Gotte M (2018) Recombinant RNA-dependent RNA polymerase complex of Ebola virus. Sci Rep 8(1):3970. https://doi.org/10.1038/s41598-018-22328-3
10. Tchesnokov EP, Feng JY, Porter DP, Gotte M (2019) Mechanism of inhibition of Ebola virus RNA-dependent RNA polymerase by Remdesivir. Viruses 11(4). https://doi.org/10.3390/v11040326
11. Warren TK, Jordan R, Lo MK, Ray AS, Mackman RL, Soloveva V, Siegel D, Perron M, Bannister R, Hui HC, Larson N, Strickley R, Wells J, Stuthman KS, Van Tongeren SA, Garza NL, Donnelly G, Shurtleff AC, Retterer CJ, Gharaibeh D, Zamani R, Kenny T, Eaton BP, Grimes E, Welch LS, Gomba L, Wilhelmsen CL, Nichols DK, Nuss JE, Nagle ER, Kugelman JR, Palacios G, Doerffler E, Neville S, Carra E, Clarke MO, Zhang L, Lew W, Ross B, Wang Q, Chun K, Wolfe L, Babusis D, Park Y, Stray KM, Trancheva I, Feng JY, Barauskas O, Xu Y, Wong P, Braun MR, Flint M, McMullan LK, Chen SS, Fearns R, Swaminathan S, Mayers DL, Spiropoulou CF, Lee WA, Nichol ST, Cihlar T, Bavari S (2016) Therapeutic efficacy of the small molecule GS-5734 against Ebola virus in rhesus monkeys. Nature 531(7594): 381–385. https://doi.org/10.1038/nature17180
12. Sourimant J, Lieber CM, Yoon JJ, Toots M, Govindarajan M, Udumula V, Sakamoto K, Natchus MG, Patti J, Vernachio J, Plemper RK (2022) Orally efficacious lead of the AVG inhibitor series targeting a dynamic interface in the respiratory syncytial virus polymerase. Sci Adv 8(25):eabo2236. https://doi.org/10.1126/sciadv.abo2236
13. Kleiner VA, Fischmann TO, Howe JA, Beshore DC, Eddins MJ, Hou Y, Mayhood T, Klein D, Nahas DD, Lucas BJ, Xi H, Murray E, Ma DY, Getty K, Fearns R (2023) Conserved allosteric inhibitory site on the respiratory syncytial virus and human metapneumovirus RNA-dependent RNA polymerases. Commun Biol 6(1):649. https://doi.org/10.1038/s42003-023-04990-0
14. Cao D, Gao Y, Chen Z, Gooneratne I, Roesler C, Mera C, D'Cunha P, Antonova A, Katta D, Romanelli S, Wang Q, Rice S, Lemons W, Ramanathan A, Liang B (2024) Structures of the promoter-bound respiratory syncytial virus polymerase. Nature 625(7995): 611–617. https://doi.org/10.1038/s41586-023-06867-y
15. Cao D, Gooneratne I, Mera C, Vy J, Royal M, Huang B, Park Y, Manjunath A, Liang B (2022) Analysis of template variations on RNA synthesis by respiratory syncytial virus polymerase. Viruses 15(1). https://doi.org/10.3390/v15010047
16. Cao D, Gao Y, Roesler C, Rice S, D'Cunha P, Zhuang L, Slack J, Antonova A, Romanelli S, Liang B (2020) In vitro primer-based RNA elongation and promoter fine mapping of the respiratory syncytial virus. J Virol 95:1. https://doi.org/10.1128/JVI.01897-20
17. Braun MR, Deflube LR, Noton SL, Mawhorter ME, Tremaglio CZ, Fearns R (2017) RNA elongation by respiratory syncytial virus polymerase is calibrated by conserved region V. PLoS Pathog 13(12):e1006803. https://doi.org/10.1371/journal.ppat.1006803
18. Noton SL, Aljabr W, Hiscox JA, Matthews DA, Fearns R (2014) Factors affecting de novo RNA synthesis and back-priming by the respiratory syncytial virus polymerase. Virology 462–463:318–327. https://doi.org/10.1016/j.virol.2014.05.032
19. Cressey TN, Noton SL, Nagendra K, Braun MR, Fearns R (2018) Mechanism for de novo initiation at two sites in the respiratory syncytial virus promoter. Nucleic Acids Res 46(13): 6785–6796. https://doi.org/10.1093/nar/gky480
20. Tremaglio CZ, Noton SL, Deflube LR, Fearns R (2013) Respiratory syncytial virus polymerase can initiate transcription from position 3 of the leader promoter. J Virol 87(6):3196–3207. https://doi.org/10.1128/JVI.02862-12

21. Noton SL, Nagendra K, Dunn EF, Mawhorter ME, Yu Q, Fearns R (2015) Respiratory syncytial virus inhibitor AZ-27 differentially inhibits different polymerase activities at the promoter. J Virol 89(15):7786–7798. https://doi.org/10.1128/JVI.00530-15
22. Yoon JJ, Toots M, Lee S, Lee ME, Ludeke B, Luczo JM, Ganti K, Cox RM, Sticher ZM, Edpuganti V, Mitchell DG, Lockwood MA, Kolykhalov AA, Greninger AL, Moore ML, Painter GR, Lowen AC, Tompkins SM, Fearns R, Natchus MG, Plemper RK (2018) Orally efficacious broad-spectrum ribonucleoside analog inhibitor of influenza and respiratory syncytial viruses. Antimicrob Agents Chemother 62(8). https://doi.org/10.1128/AAC.00766-18
23. Cox RM, Toots M, Yoon JJ, Sourimant J, Ludeke B, Fearns R, Bourque E, Patti J, Lee E, Vernachio J, Plemper RK (2018) Development of an allosteric inhibitor class blocking RNA elongation by the respiratory syncytial virus polymerase complex. J Biol Chem 293(43):16761–16777. https://doi.org/10.1074/jbc.RA118.004862
24. Yu X, Abeywickrema P, Bonneux B, Behera I, Anson B, Jacoby E, Fung A, Adhikary S, Bhaumik A, Carbajo RJ, De Bruyn S, Miller R, Patrick A, Pham Q, Piassek M, Verheyen N, Shareef A, Sutto-Ortiz P, Ysebaert N, Van Vlijmen H, Jonckers THM, Herschke F, McLellan JS, Decroly E, Fearns R, Grosse S, Roymans D, Sharma S, Rigaux P, Jin Z (2023) Structural and mechanistic insights into the inhibition of respiratory syncytial virus polymerase by a non-nucleoside inhibitor. Commun Biol 6(1):1074. https://doi.org/10.1038/s42003-023-05451-4
25. Sourimant J, Lieber CM, Aggarwal M, Cox RM, Wolf JD, Yoon JJ, Toots M, Ye C, Sticher Z, Kolykhalov AA, Martinez-Sobrido L, Bluemling GR, Natchus MG, Painter GR, Plemper RK (2022) 4'-Fluorouridine is an oral antiviral that blocks respiratory syncytial virus and SARS-CoV-2 replication. Science 375(6577):161–167. https://doi.org/10.1126/science.abj5508
26. Beutler E, Guinto E (1976) A simple, rapid, efficient method for the preparation of gamma 32P-labeled guanosine triphosphate (GTP) and adenosine triphosphate (ATP). J Lab Clin Med 88(3):520–524
27. Ogino M, Gupta N, Green TJ, Ogino T (2019) A dual-functional priming-capping loop of rhabdoviral RNA polymerases directs terminal de novo initiation and capping intermediate formation. Nucleic Acids Res 47(1):299–309. https://doi.org/10.1093/nar/gky1058
28. Cressey TN, Shareef AM, Kleiner VA, Noton SL, Byrne PO, McLellan JS, Muhlberger E, Fearns R (2022) Distinctive features of the respiratory syncytial virus priming loop compared to other non-segmented negative strand RNA viruses. PLoS Pathog 18(6):e1010451. https://doi.org/10.1371/journal.ppat.1010451
29. Noton SL, Tremaglio CZ, Fearns R (2019) Killing two birds with one stone: how the respiratory syncytial virus polymerase initiates transcription and replication. PLoS Pathog 15(2): e1007548. https://doi.org/10.1371/journal.ppat.1007548
30. Shareef AM, Ludeke B, Jordan P, Deval J, Fearns R (2021) Comparison of RNA synthesis initiation properties of non-segmented negative strand RNA virus polymerases. PLoS Pathog 17(12):e1010151. https://doi.org/10.1371/journal.ppat.1010151

Chapter 13

Biophysical Characterization of the M_{2-1} Transcription Antiterminator with RNA

Ramon Peralta-Martinez, Ivana Molina, Sebastian Esperante, and Gonzalo de Prat-Gay

Abstract

The respiratory syncytial virus (RSV) encodes a singular transcription antiterminator or processivity factor M_{2-1}. This protein ensures the adequate expression of genes toward the 5′ end of the genome that results from transcription polarization, in which genes located near the 3′ genomic end are expressed at much higher levels than those at the 5′ end, resulting in a gradient of transcripts. Although its mechanism of action is not fully understood, it is based on its RNA-binding capacity. This chapter describes biophysical spectroscopic methods for quantitatively analyzing the RNA binding activity of M_{2-1} in vitro from the pure recombinant protein. By employing short fluorescently labeled RNA molecules, the chapter describes methodologies for characterizing physical interactions, determining binding stoichiometries and binding affinities, using a combination of electrophoretic mobility shift assays (EMSA) and equilibrium fluorescence spectroscopic titrations. In addition, the determination of dissociation constants, encompassing both independent single binding sites within the M_{2-1} core domain and cooperative RNA binding to the M_{2-1} tetramer, is described. Given the intricate nature and functional significance of RNA binding, the RSV M2–1 viral tetramer also constitutes a valuable model system for studying RNA-protein interactions among oligomeric proteins.

Key words RSV, Antiterminator, M2–1, RNA binding, Fluorescence

1 Introduction

Pneumoviruses code for the M_{2-1} transcription antiterminator factor, with the most clinically relevant members being the human respiratory syncytial virus (hRSV) and the human metapneumovirus (hMPV) [1, 2]. M_{2-1} forms a tight alpha-helical-rich homotetramer. Each promoter contains three domains separated by flexible linkers: an N-terminal zinc-binding motif (Cys3His1), a tetramerization domain, and a C-terminal core domain with RNA binding properties [3–6] (Fig. 1). Structural studies revealed that the zinc finger and core domains contribute to RNA binding independently of each other [7]. The VP30 protein of similar

Bo Liang and Rachel Fearns (eds.), *Human Respiratory Syncytial Virus: Methods and Protocols*, Methods in Molecular Biology, vol. 2948, https://doi.org/10.1007/978-1-0716-4666-3_13,

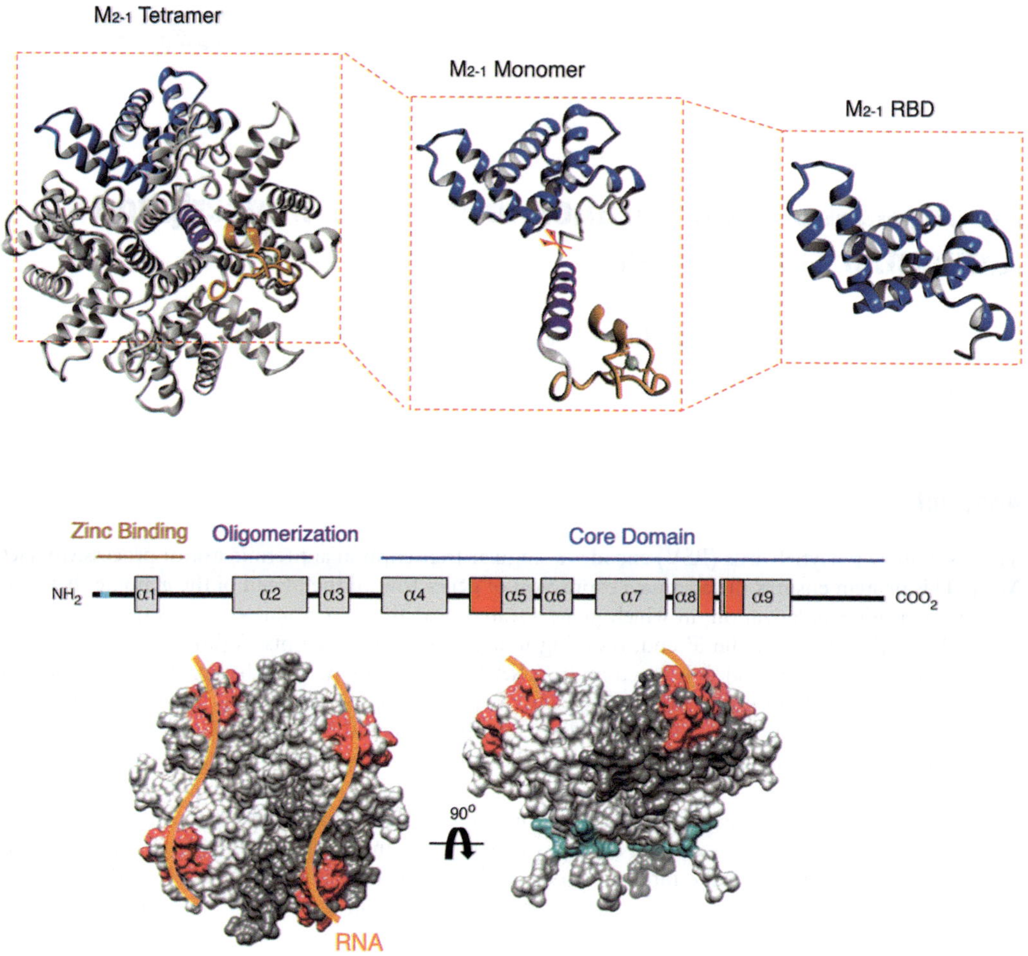

Fig. 1 M_{2-1} protein forms a stable tetramer in solution and binds RNA through a core domain: M_{2-1} is a basic protein of 194 residues that forms a stable tetramer ($K_D = 10^{-38}$ M^3) [5]. Each protomer consists of three regions: a Zinc Binding region (C7-H25—yellow), an Oligomerization Domain (P32-S49—purple), and an RNA Binding ("Core") Domain (RBD) (Y72-T184—light blue). RNA binding residues are highlighted in red (bottom). To obtain $M_{2\text{-}1\text{-RBD}}$, the DNA sequence is cleaved through molecular engineering (red scissors) and expressed. It can also be obtained by limited proteolysis of the full-length tetramer with chymotrypsin [20]

characteristics, i.e., RNA binding, role in transcription modulation, oligomerization, structural resemblance, and Cys3His1 zinc motif, is found in the closely related Ebola virus [8].

RSV M_{2-1} acts as a viral transcription elongation and antiterminator factor, which prevents both intragenic and intergenic premature transcription termination by the viral RNA-dependent RNA polymerase (RdRP) [9–11]. As a result, M_{2-1} is essential for the polymerase to efficiently synthesize long mRNAs and to transcribe sequentially downstream genes within the negative-sense viral RNA genome [1, 10]. This function is likely associated with its interaction with nascent viral mRNA [12]. After the synthesis of viral mRNAs, M_{2-1} transiently concentrates untranslated viral mRNAs into mRNP granules, forming sub-compartments that exhibit liquid droplet behavior [13, 14]. During the viral life cycle, M_{2-1} interacts specifically with positive-sense RSV mRNAs, but not negative-sense viral genomic RNA. M_{2-1} makes contacts along the entire length of each viral mRNA, and it has also been shown to bind cellular mRNAs [15]. In vitro studies indicated that the RNA:M_{2-1} interaction displays a nanomolar affinity for both short A-rich oligonucleotides and RNA molecules corresponding to positive-sense viral gene ends [6, 12]. Additional studies revealed that M_{2-1} protomers bind cooperatively to RNAs, allowing M_{2-1} tetramers to establish high-affinity interactions with low sequence specificity [16]. Cyclic M_{2-1} phosphorylation/dephosphorylation appears to regulate its binding affinity with viral protein partners and mRNAs [6, 17]. Dephosphorylated M_{2-1} displays a higher affinity for RNAs, in particular for poly(A) sequences [7].

The underlying biochemical basis of M_{2-1} 's activity is not yet fully understood; nevertheless, a critical step for investigating M_{2-1} function is RNA binding, for which a thorough understanding starts from structural information but requires a quantitative approach covering thermodynamics and kinetics. Although specific binding activities were thought to take place at gene end regulatory sites [6], preferences, even for RSV-unrelated sequences, are more likely to occur than specific binding sites [6, 16]. In the present chapter, we describe methods for investigating the RNA binding activity of M_{2-1}, by determining a minimal model based on stoichiometry and affinity [16].

Establishing a quantitative spectroscopic method in solution requires the validation that a saturable signal change corresponds to an actual binding event. A simple method to achieve this is an electrophoretic mobility shift assay (EMSA) [18, 19]. In this article, we present accessible, robust, and reproducible methods using simple techniques to investigate the most relevant parameters in a protein:RNA binding mechanism, e.g., stoichiometry and affinity, using recombinant protein and a model synthetic RNA oligonucleotide.

2 Materials

2.1 Proteins

1. Purified full-length M_{2-1} protein: The recombinant M_{2-1} protein is expressed in *E. coli* and purified as previously described [5] (*see* **Note 1**).
2. Purified monomeric M_{2-1}-RBD: The RNA binding "core domain" RBD consists of a 13.6-kDa monomeric fragment that contains the major RNA binding site (Fig. 1) [12]. This globular species is produced by limited proteolysis of the pure M_{2-1} tetramer by chymotrypsin and further purified by size exclusion chromatography as described in [20].

2.2 Unlabeled and 6-Carboxyfluorescein (6-FAM)-Labeled RNA Oligonucleotides

Several companies offer custom RNA oligonucleotide synthesis services and labeling, including Integrated DNA Technologies IDT, Genescript, ThermoFisher Scientifics, among others. Unlabeled desalted oligonucleotides meeting QC requirements usually yield products of enough quality for biophysical characterization (*see* **Note 2**). RNAs are ordered with 6-carboxyfluorescein (6-FAM) at their 5′ end and subjected to additional desalting and RP-HPLC purification (*see* **Note 3**). M_{2-1} is known to bind RNAs of various sequences, but the specific factors governing its binding preferences are not yet understood [6, 16]. We use a model sequence consisting of repetitions of the pentamer AGUUG referred to as RNA_{10} (AGUUGAGUUG), RNA_{15} (AGUUGAGUUGAGUUG), and RNA_{20} (AGUUGAGUUGAGUUGAGUUG), described to bind M_{2-1} with nanomolar affinity [16]. The affinity of a 20mer corresponding to the gene end (GE) of the SH gene of RSV is indistinguishable from the repetitive 20mer ([16], Molina, PhD thesis).

1. RNA_{10} (AGUUGAGUUG).
2. RNA_{15} (AGUUGAGUUGAGUUG).
3. RNA_{20} (AGUUGAGUUGAGUUGAGUUG).

2.3 Reagents

1. Tris–HCl (spectroscopic quality).
2. Sodium phosphate (spectroscopic quality).
3. HEPES (spectroscopic quality).
4. Sodium chloride (spectroscopic quality).
5. DTT (spectroscopic quality).
6. Glycerol.
7. Ultrapure water for buffer preparation.

2.4 Materials and Solutions

1. 5× Tris-glycine buffer (pH 9.0): Dissolve 15.1 g Tris–HCl (FW: 121.14 g/mol) and 72.1 g glycine (FW 71.07 g/mol) in 700 mL of ultrapure water: After solid is dissolved, adjust pH to 9.0 with sodium hydroxide and bring the volume to 1 L with ultrapure water. Store at 4 °C.
2. Thirty percent acrylamide/bis solution (29.2:0.8). Dissolve in 40 mL of water 29.2 g of acrylamide and 0.8 g bisacrilamide and adjust the volume to 100 mL in a graduated cylinder. Filter through a 0.45 μm filter and store at 4 °C, in a bottle wrapped with aluminum foil.
3. 50% glycerol: Combine 50 mL of water and 50 mL of glycerol 100% in a glass flask. Shake the flask until the liquids are mixed. Sterilize by autoclaving at 15 lbs. pressure (121 °C) for 15 min.
4. Ammonium persulfate: 10% solution in water.
5. N,N,N,N′-Tetramethyl-ethylenediamine.

6. 0,2 M HEPES pH 7,4, weigh 4.76 g of HEPES (FW: 238.3 g/mol) and transfer to a 100 mL glass beaker containing 70 mL of water. Stir until completely dissolved and adjust pH to 7.4 with 10 N NaOH.
7. Native 4% polyacrylamide gel: Mix 1.33 mL of 30% acrylamide/bisacrylamide, 1 mL of 5× Tris-glycine buffer, 1 mL of 50% glycerol, and 6.6 mL of water in a 50 mL conical flask. Add 80 μL of 10% ammonium persulfate and 8 μL of TEMED and cast gel within a 7.25 cm × 10 cm × 1.5 mm gel cassette. Insert a 10-well gel comb immediately.
8. Binding reaction solution: 20 mM HEPES (pH 7.4), 80 mM NaCl, 10% glycerol.

2.5 Equipment

1. Electrophoresis: Standard acrylamide electrophoresis equipment and methods are used.
2. Spectrofluorimeter: Equipment of most known brands can fulfill the requirements for monitoring fluorescence intensity changes upon titration. It must operate under controlled temperature conditions either by an attached circulating bath or by a Peltier.
3. Storm Imager 840 (Molecular Dynamics Storm 840).
4. ImageJ 1.48 V software.
5. Pipettes: quality calibrated micropipettes are strongly recommended (*see* **Note 4**).

3 Methods

3.1 Defining Physical Interaction and Stoichiometry by EMSA

3.1.1 Preparation

1. Perform a pre-run of the native 4% polyacrylamide gel (2.4.7) for 30 min at 60 volts before loading the samples.
2. Prepare the binding reaction solution (2.4.8) and 6-FAM-labeled RNA at the desired final concentration. The mixture (100 μL) is divided into 10 tubes, and increasing concentrations of the M_{2-1} protein are added, leaving aside one tube with RNA alone, to a final volume of 20 μL per well (5% final glycerol).
3. Incubate the samples for 20 min at room temperature to allow complex formation.
4. Perform the electrophoretic run at 40 volts for 20 min to allow the samples to enter the gel, followed by 60 volts for 1 h. Perform all experiments in the 0.5× Tris-Glycine run buffer (2.4.1) at 20 °C.
5. Detect RNA fluorescence by densitometry in Storm Imager 840. Experiments should be performed at total fixed RNA concentrations ranging from 200 nM to 4 μM. In cases where

the RNA concentration is greater than 200 nM, use an RNA concentration of 6-FAM 200 nM, and add unlabeled RNA to reach the final RNA concentration (*see* **Note 5**).

3.1.2 Densitometric Analysis of the EMSA Experiments

Digital imaging of the EMSA gel is used to quantify the fraction of free 6-FAM RNA, using the ImageJ 1.48 V software. The strong quenching of the probe's fluorescence upon binding precludes an accurate estimation of the fraction of bound versus free species by densitometry, unlike the EMSAS, where radioactively labeled oligonucleotides are used. Therefore, we estimate the percentage of free probe species. The boundaries of the first free probe band (the first tube that does not contain unlabeled protein) are determined and used as a guide to analyze the rest of the bands. The program generates a value that corresponds to the signal strength within the selected area. The values obtained by increasing the concentrations of unlabeled protein are normalized to the intensity value of the first band (which corresponds to 100% of the free probe) and expressed as the percentage ' of the unbound probe, which can be plotted as a function of the molar ratio of M_{2-1}:RNA to determine the binding stoichiometry, or as a function of the concentration of M_{2-1} to estimate the dissociation constant (K_D).

3.1.3 Examples of Stoichiometry Determination of M_{2-1}-RNA Complexes by EMSA

Proceed carefully, step by step, the points detailed in 3.1.1 using the following parameters and concentrations:

1. Prepare and pre-run native 4% polyacrylamide gels (2.4.7).
2. In separate tubes, prepare RNA solutions in binding reaction solution (2.4.8) with a final volume of 100 μL each, containing: (a) 4 μM of 10-mer RNA, (b) 200 nM of 20-mer RNA, or (c) 4 μM of 20-mer RNA. As was mentioned above (3.1.1.5), add 200 nM of 6-FAM-labeled RNAs, and add unlabeled RNA to reach the final RNA concentrations.
3. Divide each mixture into 10 tubes by adding 10 μL of the corresponding RNA solution.
4. Add 10 μL of serial dilutions of M_{2-1} proteins prepared in binding reaction solution, ranging from 0 to 2 molar equivalents of M2–1 tetramer in binding reactions a and b or from 0 to 5 molar equivalents of M_{2-1} RBD in binding reaction c. Add 10 μL of binding buffer to one tube as a negative control (RNA only).
5. Incubate the samples for 20 min at room temperature to allow complex formation and run the electrophoresis as detailed in Subheading 3.1.1.
6. Detect RNA fluorescence by densitometry and perform the densitometric analysis of the EMSA experiments as described in Subheading 3.1.2. Results of these experiments are shown in Fig. 2 and further discussed in **Note 6**.

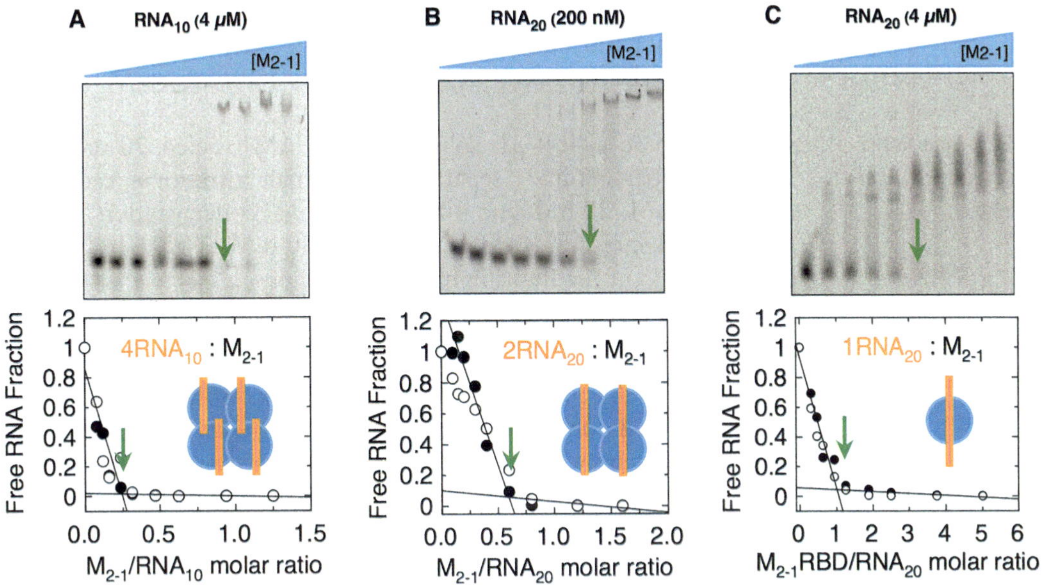

Fig. 2 M_{2-1} tetramer-RNA-binding stoichiometry depends on RNA length: Binding stoichiometry was determined by EMSA assays (top), where the fluorescent-RNA fixed concentration was five-fold higher than the K_D. The concentration of free RNA was estimated by image analysis and plotted (bottom) as a function of M_{2-1}/RNA ratio. The binding stoichiometry is obtained at the M_{2-1}/RNA ratio, where the curve has a pronounced change in the slope (green arrow). For M_{2-1}, the stoichiometry depends on RNA length, where 4 RNA_{10} molecules bind one tetramer and 2 RNA_{20} molecules bind one tetramer (**A** and **B** respectively). Finally, a 1:1 stoichiometry is obtained for the $M_{2-1-RBD}$ independently of RNA length **C**. (Modified from Ref. [16] with permission)

3.2 RNA Binding by Fluorescence Spectroscopy Titration

3.2.1 Titration Assay Setup

1. RNA binding by fluorescence spectroscopy titration assays requires a fixed concentration of fluorescent-labeled RNA to which increasing concentrations of protein yield a change in fluorescence intensity. The concentration of the fixed fluorescent-labeled RNA will depend on the equipment sensitivity and the affinity (K_D). Choose an RNA concentration near the expected K_D if known, if not, exploratory experiments should be done to define this concentration.
2. Fluorescence emission spectrum should be done for each pair of RNA:Protein concentrations. Use an excitation wavelength of 495 nm and record the emission from 505 to 600 nm. Change the emission slits as sensitivity is required; 5 nm of both excitation and emission slits is a good starting point (*see* **Note 7**).
3. As a starting point, add a two-fold excess of M_{2-1} to a fixed amount of 6-FAM-RNA and monitor the fluorescence intensity change. Add an additional excess of M_{2-1} until the fluorescence signal reaches a plateau, reflecting RNA:protein binding

saturation. Determining the free labeled RNA signal (starting point) and binding saturation (endpoint) is crucial for a successful titration.

3.2.2 Examples of Stoichiometric Titration Assays

1. Prepare a 400 μL solution of 6-FAM-labeled 20-mer RNA (prepare three separate sequential titrations containing 10 nM, 20 nM, and 40 nM RNA final concentration) in Tris–HCl (pH 7.5), 0.1 M NaCl, and 1.0 mM DTT. Equilibrate at 25 °C in the fluorimeter cuvette holder.
2. Record the fluorescence spectrum of free 6-FAM RNA as the starting point of the titration.
3. Gradually add increasing concentrations of M_{2-1} tetramer (*see* **Note 8**) and incubate at 25 °C in the fluorimeter cuvette holder between 5 and 10 min prior next spectrum measurement (*see* **Note 9**).
4. Repeat **step 3** until signal saturation at 525 nm.
5. Plot the values of fluorescence intensity at the wavelength where the maximum difference is observed versus total protein concentration. Results of these experiments are shown in Fig. 3 and explained in *see* **Note 10**.
6. Proceed with the analysis as described in the next Subheading 3.3.

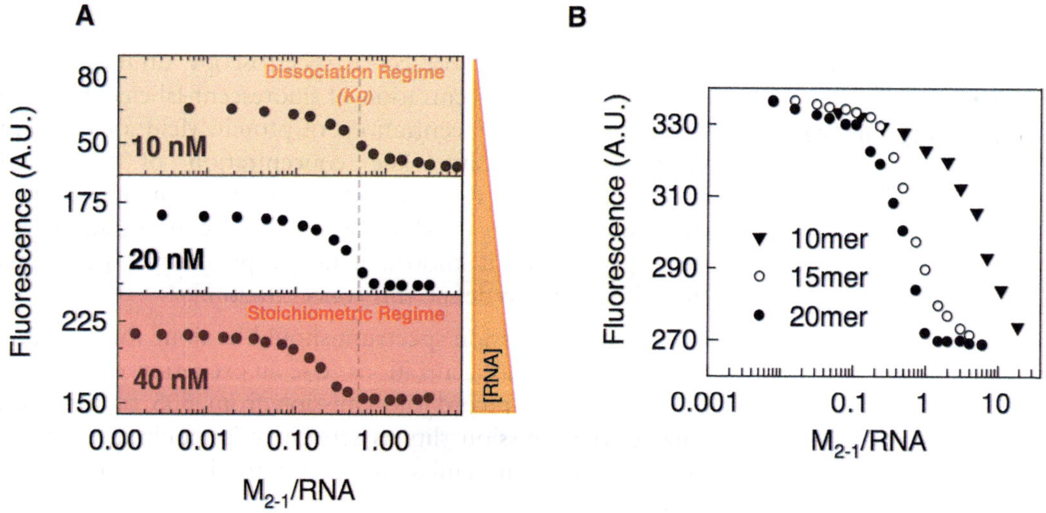

Fig. 3 Different fixed RNA concentrations allow exploration of different binding regimes: The fixed concentration of ligand in a binding titration experiment can be adjusted to obtain a K_D (Dissociation Regime: [RNA] ~ K_D) (**A**—top) or the binding stoichiometry (Stoichiometric Regime: [RNA] > > K_D) (**A**—bottom). As the binding affinity decreases, more titrant (M_{2-1} in this case) is needed for signal saturation (**B**)

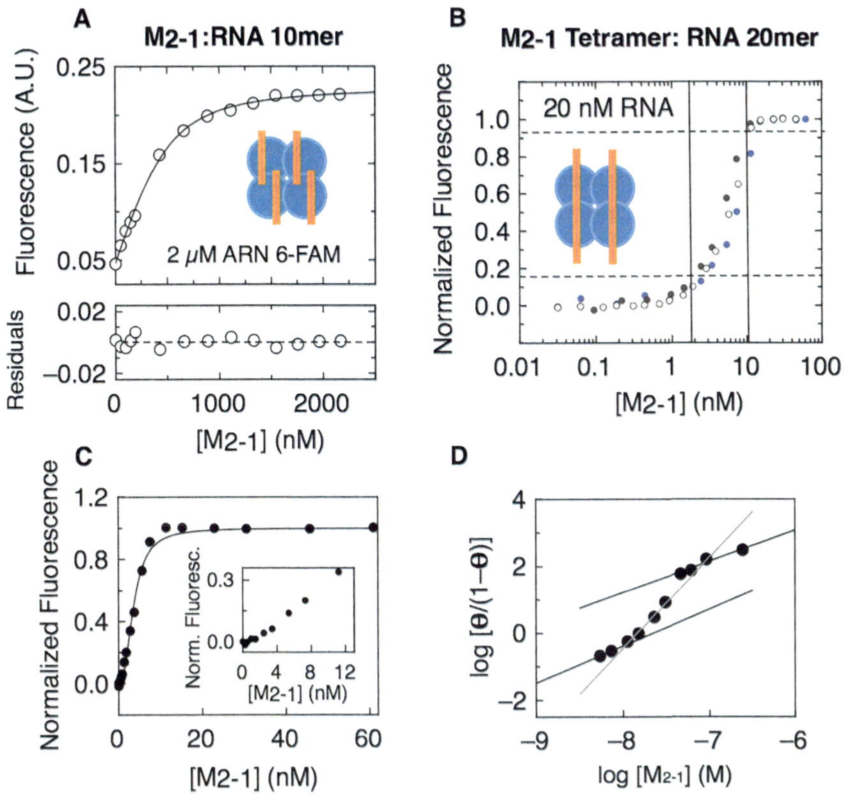

Fig. 4 M_{2-1} binding to RNA is cooperative: For a given RNA, the binding titration experiment is fitted with a 1:1 binding model (Eq. 1) and the quality of the fitting is verified with the residuals. **(A)** Binding saturation is necessary for best fitting and is better visualized when $[M_{2-1}]$ is expressed in logarithmic scale. **(B)** Cooperativity is suggested in **(C)** and verified in **(B)** where less than 2 log units are necessary for the transition of the 0.1–0.9 saturation fraction, which is indicative of positive cooperativity. Finally, apparent K_D and Hill coefficient (n_H) can be estimated with the transformation of the Hill equation **(D)**. (Modified from Ref. [16] with permission)

3.3 Determination of Dissociation Constants

3.3.1 Independent Single Binding Site

First, we analyze complexes with 1:1 stoichiometry per RNA binding site, which comprises the M_{2-1} tetramer binding to RNA_{10} (determined from Fig. 2A). Fig. 4A shows the binding titration of the M_{2-1}-RNA_{10} complex, where the data are fitted to an independent single binding site [16, 21]:

$$F = F_{\text{Free}} + \frac{\Delta F_{\text{int}}}{2} \cdot \left[(x + [\text{RNA}] + K_D) - \left((x + [\text{RNA}] + K_D) - 4[\text{RNA}]_x \right)^{0.5} \right] \tag{1}$$

The distribution of the residuals to the fitted data confirms the application of this simple binding model (Fig. 4A, bottom) (*see* **Note 11**). The K_D obtained is 0.29 ± 0.17 μM. Analysis of monomeric M_{2-1}RBD with RNA_{20} (1:1 binding event determined from Fig. 2C) yields a K_D of 0.24 ± 0.11 μM [16]. Moreover, the K_D values recovered from the combination of electrophoretic and spectroscopic techniques are similar [16]. Based on these results, it can be concluded that RNA_{10} saturates the binding to the primary site located in the core domain of individual M_{2-1} protomers assembled into the tetramer, without significant additional interactions contributed by longer RNAs. The RNA binds like a minimal and independent binding site, whether the RNA-binding site is within the context of the M_{2-1} tetramer or the M_{2-1}RBD core monomer.

3.3.2 Cooperative Binding of M_{2-1} Tetramer

Analysis of the M_{2-1}-RNA_{20} complex reveals a 2 RNA:1 M_{2-1} stoichiometry (Figs. 2B and 3A) and close inspection of the raw and normalized binding data reveals a sigmoidal shape of the binding curves at sub-stoichiometric concentrations with a pronounced increase in the fraction of ligand bound after the addition of M_{2-1}, suggestive of cooperativity (Fig. 4C).

One approach to assessing cooperativity is to quantify the log units of ligand required to produce a transition of bound fractions from 0.1 to 0.9 [22]. For this, the normalized fluorescence values are plotted as a function of the ligand concentration with a logarithmic x-axis. While non-cooperative interactions usually span 2 log units of ligand concentration, negative cooperativity is characterized by shallower transitions spanning more than 2 log units and positive cooperativity by sharper transitions spanning less than 2 log units. The M_{2-1} tetramer:RNA_{20} binding isotherm transition spans less than two log units of protein concentration, indicative of positive cooperativity (Fig. 4D). Based on this, we used the Hill formula to assess cooperativity [23].

$$\Theta = \frac{[M_{2-1}]^{n_H}}{[M_{2-1}]^{n_H} + [K_D]^{n_H}} \tag{2}$$

where θ is the degree of saturation, $[M_{2-1}]$ is the free ligand concentration, K_D is the apparent dissociation constant, and n_H is the Hill coefficient. From the linear transformation of the Hill equation, it is possible to estimate the n_H from the slope of the linear plot of log θ as a function of log $[M_{2-1}]$, and the apparent binding constant from the ordinate.

$$\log \frac{\Theta}{1-\Theta} = n_H.\log[M_{2-1}] - n_H.\log K_D \tag{3}$$

Reproducible fits are obtained at different RNA_{20} concentrations [16], with average values of $n_H = 2.3 \pm 0.5$, confirming positive cooperativity (Table 1). The average apparent binding constant recovered (K_{Dapp}) was 14 ± 6 nM (Table 1), revealing a 15-fold increase over the K_D values of the single-site interactions observed with the core domain monomer and the non-cooperative interactions of the M_{2-1} tetramer with the shorter RNA_{10} (Fig. 4A). Thus, the occurrence of two adjacent sites from different protomers within the M_{2-1} tetramer led to the observed

Table 1
Analysis of binding cooperativity (Hill) of the RNA_{20} (DEN 20 nt) repetitive sequence described in this chapter, compared to the sequence corresponding to a SH RSV-Gene End 20mer sequence (Reproduced from Ref. [16] with permission)

'	n_H	K_{Dapp} (nM)	Corr. coefficient
DEN 20 nt RNA association to M_{2-1} tetramer			
10 nM	2.0 ± 0.1	14.6 ± 2.7	0.968
10 nM II	1.4 ± 0.1	5.6 ± 1.4	0.974
20 nM	2.0 ± 0.2	12.0 ± 3.2	0.973
20 nM II	1.7 ± 0.4	13.4 ± 6.6	0.936
30 nM	1.9 ± 0.3	16.1 ± 5.1	0.963
40 nM	1.4 ± 0.2	22.8 ± 5.4	0.980
RSV 20 nt RNA association to M_{2-1} tetramer			
10 nM	1.7 ± 0.1	14.1 ± 1.7	0.994
10 nM II	1.9 ± 0.2	13.5 ± 3.2	0.979
10 nM III	2.0 ± 0.2	9.5 ± 3.0	0.963
20 nM	1.6 ± 0.3	18.0 ± 5.0	0.975
20 nM II	1.6 ± 0.3	18.0 ± 6.8	0.950
20 nM III	1.8 ± 0.2	15.3 ± 3.4	0.984
40 nM	2.0 ± 0.1	24.1 ± 4.4	0.983

cooperativity and determined its high affinity. As our experiments are performed with a monomeric M_{2-1}-RBD lacking the N-terminal region containing the zinc-binding motif, we cannot rule out that the observed increase in binding affinity is also due in part to additional interactions of the RNA with the N-terminal region, previously shown to contribute to RNA association (R3 and R4) [7, 16]. The minimum RNA length that yields cooperative high-affinity binding is 4 bases [6, 16].

4 Notes

1. Storage of protein in 20 mM sodium phosphate, pH 7.0, 0.3 M NaCl, 1.0 mM DTT. The 260/280 nm ratio between 0.5 and 0.6 is indicative of minimal nucleic acid presence. After concentration in the last step, the protein can be stored in the same buffer. It is a good practice to check oligos the first time

resuspended and from time to time by PAGE to ensure integrity.

2. The concentration of reagents should be carefully determined. Protein concentration should be determined from its extinction coefficient 13,200 M^{-1} cm^{-1} for the M_{2-1} monomer and 4470 M^{-1} cm^{-1} for M_{2-1}RBD [5, 20]. Protein quantification by Bradford assay consistently yields comparable results, but absorbance measurement at 280 nm is usually the preferred method. Always centrifuge for 10 min at 12,000 rpm at 4 °C in a microcentrifuge after thawing and prior to measuring the concentration of the protein. Resuspended RNA can be diluted and measured by absorbance at 260 nm using the extinction coefficient according to its nucleotide content [16].
3. Other strong and stable fluorophores can be used, provided they yield an appropriate signal change, which is largely empirical.
4. Hamilton pipettes can also be used (*see* **Note 9**).
5. Two hundred nanomolar is the lowest probe concentration (RNA 6-FAM) that produces a detectable signal in our Storm equipment. The fluorescent RNA concentration could be lower, depending on the specific fluorescent gel scanner employed. For instance, the ChemiDoc Touch imaging system from Bio-Rad can detect RNA 6 FAM ranging from 5 to 10 nM. The ratio between labeled and unlabeled RNA can be adjusted according to the equipment sensitivity requirements. Depending on the dye used, fluorescence anisotropy can be a parameter of choice. However, if there is a large fluorescence intensity change, this should be the method of choice since, besides the sensitivity, intensity corrections must be applied when measuring anisotropy changes.
6. The decrease in intensity of the free labeled RNA band indicates the formation of the complex as the protein concentration is increased. The band disappears at a 2 RNA20:1 M_{2-1} tetramer ratio, indicating the stoichiometric titration (Fig. 2A bottom, arrows). When a 10-mer oligonucleotide is used, the ratio is 4 RNA10:M_{2-1} tetramer. (Fig. 2B). The monomeric M_{2-1}-RBD shows a 1:1 stoichiometry with RNA20 (Fig. 2C). The requirement of higher fixed RNA concentrations to observe saturation anticipates a lower affinity, as is the case with M_{2-1}-RNA10 (Fig. 2B) or M_{2-1}RBD-RNA20 (Fig. 2C). High-affinity interactions saturate M2–1 binding using low concentrations of RNA (200 nM), as is the case with M_{2-1}-RNA20 (Fig. 2A).
7. An alternative method to a conventional spectrofluorometer is to use a fluorescence multi-well plate reader. The use of this type of instrument has several advantages, including lower

error due to higher volume pipetted and rapid measurement of multiple curves in parallel. A possible downside is the amount of sample to be used in each experiment.

8. Recommended volume can range from 400 to 1000 μL, where saving reagents without compromising sensitivity is always a goal. Volumes to be added should not be lower than 3–5 μL for best accuracy, using P10 or P20 pipettes, and neither the working solution of M_{2-1} nor the pipette should be changed along the titration. If Hamilton pipettes are used, the same volume ranges apply. The fluorescence intensity values should be corrected for the corresponding dilution at each titration point, and overall dilution should not exceed 10%, so that the concentration of RNA can be assumed to have remained constant throughout the titration.
9. Most protein-nucleic acid interactions are electrostatically driven and occur near diffusion limits. This means that reactions are completed in seconds, and a 5 min incubation period ensures a completed equilibrium reaction.
10. Experiments at 10, 20, and 40 nM RNA show different regimes, where stoichiometry is best assessed at the highest concentration, whereas the dissociation constant (K_D) is more accurately determined at lower concentrations (Fig. 3A). At high concentrations, every molecule added forms a complex, which precludes a correct determination of the ratio of free and complexed molecules for accurate K_D determination. The lower the concentration, every time saturation is also observed, the better the K_D is defined. Ideally, the concentration range of the added ligand should be 1/5 to five-fold the K_D, but often three-fold excess is sufficient. In any case, spreading data points yields more accurate data than a larger number of points within a reduced concentration range.
11. Titrations at sub-stoichiometric RNA concentrations revealed changes in binding transitions as a function of RNA length, with transitions occurring at lower M_{2-1} concentrations for longer RNAs, indicative of a gradient of relative affinities, where RNA10 shows the lowest affinity (Fig. 3B).

References

1. Collins PL, Hill MG, Cristina J, Grosfeld H (1996) Transcription elongation factor of respiratory syncytial virus, a nonsegmented negative-strand RNA virus. Proc Natl Acad Sci USA 93(1):81–85. https://doi.org/10.1073/pnas.93.1.81
2. Buchholz UJ, Biacchesi S, Pham QN, Tran KC, Yang L, Luongo CL et al (2005) Deletion of M2 gene open reading frames 1 and 2 of human metapneumovirus: effects on RNA synthesis, attenuation, and immunogenicity. J Virol 79(11):6588–6597. https://doi.org/10.1128/JVI.79.11.6588-6597.2005
3. Hardy RW, Wertz GW (2000) The Cys(3)-His(1) motif of the respiratory syncytial virus M2-1 protein is essential for protein function. J Virol 74(13):5880–5885. https://doi.org/10.1128/jvi.74.13.5880-5885.2000

4. Tran TL, Castagne N, Dubosclard V, Noinville S, Koch E, Moudjou M et al (2009) The respiratory syncytial virus M2-1 protein forms tetramers and interacts with RNA and P in a competitive manner. J Virol 83(13): 6363–6374. https://doi.org/10.1128/JVI.00335-09
5. Esperante SA, Chemes LB, Sanchez IE, de Prat-Gay G (2011) The respiratory syncytial virus transcription antiterminator M(2-1) is a highly stable, zinc binding tetramer with strong pH-dependent dissociation and a monomeric unfolding intermediate. Biochemistry 50(40):8529–8539. https://doi.org/10.1021/bi200661k
6. Tanner SJ, Ariza A, Richard CA, Kyle HF, Dods RL, Blondot ML et al (2014) Crystal structure of the essential transcription antiterminator M2-1 protein of human respiratory syncytial virus and implications of its phosphorylation. Proc Natl Acad Sci USA 111(4): 1580–1585. https://doi.org/10.1073/pnas.1317262111
7. Gao Y, Cao D, Pawnikar S, John KP, Ahn HM, Hill S et al (2020) Structure of the human respiratory syncytial virus M2-1 protein in complex with a short positive-sense gene-end RNA. Structure 28(9):979–90 e4. https://doi.org/10.1016/j.str.2020.07.001
8. John SP, Wang T, Steffen S, Longhi S, Schmaljohn CS, Jonsson CB (2007) Ebola virus VP30 is an RNA binding protein. J Virol 81(17): 8967–8976. https://doi.org/10.1128/JVI.02523-06
9. Collins PL, Hill MG, Camargo E, Grosfeld H, Chanock RM, Murphy BR (1995) Production of infectious human respiratory syncytial virus from cloned cDNA confirms an essential role for the transcription elongation factor from the 5′ proximal open reading frame of the M2 mRNA in gene expression and provides a capability for vaccine development. Proc Natl Acad Sci USA 92(25):11563–11567. https://doi.org/10.1073/pnas.92.25.11563
10. Fearns R, Collins PL (1999) Role of the M2-1 transcription antitermination protein of respiratory syncytial virus in sequential transcription. J Virol 73(7):5852–5864. https://doi.org/10.1128/JVI.73.7.5852-5864.1999
11. Noton SL, Fearns R (2015) Initiation and regulation of paramyxovirus transcription and replication. Virology 479–480C:545–554. S0042-6822(15)00017-3 [pii]. https://doi.org/10.1016/j.virol.2015.01.014
12. Blondot ML, Dubosclard V, Fix J, Lassoued S, Aumont-Nicaise M, Bontems F et al (2012) Structure and functional analysis of the RNA- and viral phosphoprotein-binding domain of respiratory syncytial virus M2-1 protein. PLoS Pathog 8(5):e1002734. https://doi.org/10.1371/journal.ppat.1002734
13. Rincheval V, Lelek M, Gault E, Bouillier C, Sitterlin D, Blouquit-Laye S et al (2017) Functional organization of cytoplasmic inclusion bodies in cells infected by respiratory syncytial virus. Nat Commun 8(1):563. https://doi.org/10.1038/s41467-017-00655-9
14. Visentin A, Demitroff N, Salgueiro M, Borkosky SS, Uversky VN, Camporeale G et al (2023) Assembly of the tripartite and RNA condensates of the respiratory syncytial virus factory proteins in vitro: role of the transcription antiterminator M(2-1). Viruses 15(6). https://doi.org/10.3390/v15061329
15. Braun MR, Noton SL, Blanchard EL, Shareef A, Santangelo PJ, Johnson WE et al (2021) Respiratory syncytial virus M2-1 protein associates non-specifically with viral messenger RNA and with specific cellular messenger RNA transcripts. PLoS Pathog 17(5):e1009589. https://doi.org/10.1371/journal.ppat.1009589
16. Molina IG, Esperante SA, Marino-Buslje C, Chemes LB, de Prat-Gay G (2018) Cooperative RNA recognition by a viral transcription antiterminator. J Mol Biol 430(6):777–792. https://doi.org/10.1016/j.jmb.2018.01.018
17. Durrer J, Polackova J (1978) Frequency analysis of the vibrations of the temporal bone due to head injury. Sb Ved Pr Lek Fak Karlovy Univerzity Hradci Kralove Suppl 21(2): 127–131
18. Hellman LM, Fried MG (2007) Electrophoretic mobility shift assay (EMSA) for detecting protein-nucleic acid interactions. Nat Protoc 2(8):1849–1861. https://doi.org/10.1038/nprot.2007.249
19. Stowell JAW, Tang TTL, Seidel M, Passmore LA (2021) Gel-based analysis of protein-nucleic acid interactions. Methods Mol Biol 2263:321–339. https://doi.org/10.1007/978-1-0716-1197-5_14
20. Molina IG, Josts I, Almeida Hernandez Y, Esperante S, Salgueiro M, Garcia Alai MM et al (2018) Structure and stability of the human respiratory syncytial virus M(2-1) RNA-binding core domain reveals a compact and cooperative folding unit. Acta Crystallogr F Struct Biol Commun 74(Pt 1):23–30. https://doi.org/10.1107/S2053230X17017381
21. Ferreiro DU, Lima LM, Nadra AD, Alonso LG, Goldbaum FA, de Prat-Gay G (2000) Distinctive cognate sequence discrimination, bound DNA conformation, and binding modes in the E2 C-terminal domains from

prototype human and bovine papillomaviruses. Biochemistry 39(47):14692–14701. bi001 694r [pii]

22. Cattoni DI, Chara O, Kaufman SB, Gonzalez Flecha FL (2015) Cooperativity in binding processes: new insights from phenomenological modeling. PLoS One 10(12):e0146043. https://doi.org/10.1371/journal.pone.0146043

23. Ames JB, Porumb T, Tanaka T, Ikura M, Stryer L (1995) Amino-terminal myristoylation induces cooperative calcium binding to recoverin. J Biol Chem 270(9):4526–4533. https://doi.org/10.1074/jbc.270.9.4526

Chapter 14

Biophysical Characterization of Protein-Protein Interactions in Respiratory Syncytial Virus Infections

Nicole D. Wagner, David A. Price, and Daisy W. Leung

Abstract

Human respiratory syncytial virus (RSV) encodes multifunctional viral proteins that form interactions with each other and with host proteins during different stages of the viral replication cycle to facilitate viral replication and pathogenesis. Biochemical and biophysical characterization of protein-protein interactions is important for validation and for providing insights into the regulatory mechanisms used by viral proteins during RSV infections. In this chapter, we describe a process that can be used to validate direct protein-protein interactions by in vitro pulldown assay and biophysical characterization using size-exclusion chromatography-multiangle light scattering (SEC-MALS) and hydrogen-deuterium exchange-mass spectrometry (HDX-MS).

Key words RSV, Protein-protein interactions, Pulldown assay, SEC-MALS, HDX-MS

1 Introduction

The human respiratory syncytial virus (RSV) genome encodes for 10 genes and 11 proteins that include nonstructural protein 1 (NS1), nonstructural protein 2 (NS2), nucleocapsid protein (N), phosphoprotein (P), matrix (M), small hydrophobic (SH), glycoprotein (G), fusion (F), replication/transcription cofactors M2–1 and M2–2, and the large polymerase (L). Given the limited size of the RSV genome, many of these RSV proteins are multifunctional, occupying different roles during various stages of the viral replication cycle.

RSV proteins interact with a multitude of host factors and cellular pathways to antagonize host immune responses and to facilitate viral replication and pathogenesis. Recent proteomics-based efforts identified relevant host factors interacting with RSV proteins during infection, which can provide potential therapeutic

Nicole D. Wagner and David A. Price contributed equally.

Bo Liang and Rachel Fearns (eds.), *Human Respiratory Syncytial Virus: Methods and Protocols*, Methods in Molecular Biology, vol. 2948, https://doi.org/10.1007/978-1-0716-4666-3_14,

targets [1–14]. With the generation of these datasets, it is critical that robust methods are used to validate protein complexes and to characterize these host-viral interactions to gain mechanistic insight into how RSV proteins function and are modulated by interactions with other viral or host proteins. Furthermore, the availability of an RSV reverse genetics system also offers opportunities to apply information from protein-protein interaction studies and generate specific mutations at critical protein interfaces that can modulate viral attenuation while maintaining immunogenicity. Multiple biochemical and biophysical techniques can be used to confirm protein-protein interactions, the conditions in which these interactions occur, and to define the interaction interface. These techniques range from the more qualitative assays that are performed initially, such as coimmunoprecipitation and in vitro pulldown assays, to more rigorous, quantitative analyses, including (but not limited to) chromatography coupled to light scattering, biolayer interferometry, isothermal titration calorimetry, mass spectrometry-based methods, and X-ray crystallography. Each of these techniques can be carried out independently, but in combination, these complementary methods will generate a more detailed description of a protein-protein interaction that can inform on the mechanism.

Here we describe a workflow for characterizing stable protein-protein interactions and defining the regions required for interaction once recombinantly expressed and purified proteins are available. In this chapter, we will focus on in vitro pulldown assays, size-exclusion chromatography coupled to multiangle light scattering (SEC-MALS), and hydrogen-deuterium exchange coupled to mass spectrometry (HDX-MS).

2 Materials

2.1 Chemicals and Solutions

All buffers are prepared using ultrapure water from the Milli-Q IQ 7000 water system unless otherwise noted. Solvents used for LC-MS are of at least LC-MS grade or HPLC grade. Prepare and store all reagents at 4 °C unless otherwise indicated.

2.1.1 In Vitro Pulldown Assay

1. Amylose resin slurry in 20% ethanol (*see* **Note 1**).
2. General buffer: 25 mM sodium phosphate (pH 7.0), 150 mM NaCl, 5 mM 2-mercaptoethanol (*see* **Note 2**).
3. Elution buffer: 25 mM sodium phosphate (pH 7.0), 150 mM NaCl, 5 mM 2-mercaptoethanol, 1% Maltose.
4. 100 μL of MBP-tagged protein (1 mg/mL) in general buffer.
5. 100 μL of binding protein partner (1 mg/mL).
6. 15% polyacrylamide gel for SDS-PAGE.

7. 2× SDS loading buffer: 4% SDS, 20% glycerol, 0.004% bromophenol blue, 0.125 M Tris–HCl, pH 6.8, 10% 2-mercaptoethanol.
8. 10× SDS running buffer: 0.25 M Tris base, 1.9 M Glycine, 0.035 M sodium dodecyl sulfate.
9. Staining solution: 45% ethanol, 10% acetic acid (v/v), 2% Coomassie brilliant blue dye (w/v).
10. Destaining solution: 45% ethanol, 10% acetic acid (v/v).

2.1.2 Size-Exclusion Chromatography Coupled to Multiangle Light Scattering (SEC-MALS)

1. General buffer (*see* Subheading 2.1.1).
2. 50 μL Bovine serum albumin (BSA) (2 mg/mL).
3. 50 μL of protein or protein complex (2 mg/mL).

2.1.3 Hydrogen/Deuterium Exchange-Mass Spectrometry (HDX-MS)

1. H_2O buffer (for mapping and undeuterated control): 25 mM sodium phosphate (pH 7.0), 150 mM NaCl, 2 mM Tris(2-carboxyethyl)phosphine hydrochloride (TCEP-HCl) in water (*see* **Note 3**). Aliquot and store at −20 °C. Thaw as needed (*see* **Note 4**).
2. D_2O buffer: 25 mM sodium phosphate (pD 7.0; pD = $pH_{measured}$ + 0.41), 150 mM NaCl, 2 mM Tris(2-carboxyethyl) phosphine hydrochloride (TCEP-HCl) in 99.9% deuterium oxide. Adjust pH using DCl and NaOD to maintain deuterium concentrations. Aliquot and store at −20 °C. Thaw as needed (*see* **Note 4**).
3. PBS buffer (pH 2.5): 100 mM phosphate buffer (pH 2.5), 2.7 mM KCl, 137 mM NaCl.
4. Quench buffer: 3 M urea in PBS buffer (pH 2.5), pH adjusted with HCl to pH 2.5. Aliquot and store at −20 °C. Thaw as needed (*see* **Notes 4–6**).
5. FP-XIII: For in-solution digestion, 10 mg/mL protease from *Aspergillus saitoi* type XIII in PBS buffer (pH 2.5). Store at 4 °C.
6. H_2O/FA or LC-A: 99.9% water, 0.1% formic acid (% v/v).
7. ACN/FA: 99.9% acetonitrile, 0.1% formic acid (% v/v).
8. LC-B: 79.95% acetonitrile, 19.95% water, 0.1% formic acid (% v/v/v).

2.2 Instrumentation and Equipment

2.2.1 In Vitro Pulldown Assay

1. 1.5-mL microcentrifuge tubes.
2. Microcentrifuge.
3. Electrophoresis cell and power supply.
4. Protein standard.

2.2.2 SEC-MALS

1. Cytiva Superdex™ 200 Increase 5/150 GL 3 mL column.
2. Analytical HPLC system.
3. Wyatt Technologies DAWN HELEOS-II light scattering detector with an Optilab T-rEX refractive index detector.

2.2.3 HDX-MS

1. Isocratic pump capable of 100 μL/min flow.
2. Binary gradient pump capable of 100 μL/min flow.
3. Mass spectrometer with electrospray ionization (ESI), capable of MS and tandem MS/MS with CID fragmentation.
4. In-lab coupled and packed pepsin column (*see* **Note** 7).
5. C8 trap (Agilent ZORBAX Eclipse XDB C8 column, 2.1 mm x 15 mm).
6. C18 analytical column (Waters XSelect CSH C18 XP, 130 Å, 2.5 μm, 2.1 × 50 mm).
7. Two valves (one capable of ≥2000 psi and the other ≥15,000 psi).

3 Methods

3.1 In Vitro Pulldown Assay

The in vitro pulldown assay is used to detect physical binding between two (or more) proteins. In a pulldown assay, one protein is immobilized to a stationary phase through an affinity tag and incubated with an untagged binding protein partner. After washing of the stationary phase to eliminate non-specific binders, interaction is qualitatively assessed via SDS-PAGE followed by Coomassie staining. Generally, interactions confirmed by in vitro pulldown have a binding affinity in the low μM or nM range. Proper controls must be done in parallel to ensure that non-specific binding to the stationary phase or precipitation does not provide false-positive results.

1. Equilibrate 100 μL of amylose resin in a 1.5 mL tube with buffer. Spin down at 2000 × *g* at room temperature for 10 s. Carefully remove supernatant and discard, leaving the resin at the bottom of the tube. Add 100 μL of general buffer to the resin. Repeat two more times (*see* **Note 8**).
2. Add 100 μL of MBP-tagged protein (1 mg/mL) to equilibrated resin and incubate at room temperature for 5 min.
3. Spin down at 2000 × *g* at room temperature for 10 s. Remove the supernatant and add 100 μL of the general buffer. Immediately remove 10 μL of resin slurry for an "initial beads" gel sample.
4. Spin down the remaining 90 μL resin slurry at 2000 × *g* at room temperature for 10 s. Carefully remove the supernatant and discard it.

5. Add 100 μL of binding protein partner (1 mg/mL) and immediately remove 10 μL for an "input" gel sample.
6. Incubate at room temperature for 5 min.
7. Spin down at 2000 × *g* at room temperature for 10 s. Carefully remove the supernatant and discard it.
8. Add 100 μL of the general buffer. Spin at 2000 × *g* at room temperature for 10 s. Remove supernatant and reserve 10 μL for a "wash 1" gel sample.
9. Repeat **step 8** two more times, labeling consecutive washes as "wash 2" and "wash 3".
10. Add 80 μL of general buffer to the resin and immediately remove 10 μL for a "final beads" gel sample.
11. Spin down at 2000 × *g* at room temperature for 10 s. Remove the supernatant and add 70 μL of elution buffer. Incubate for 5 min.
12. Spin down at 2000 × *g* at room temperature for 10 s. Remove supernatant and reserve 10 μL for an "elution" gel sample.
13. For each gel sample collected, add 10 μL of 2× SDS loading buffer. Boil samples briefly.
14. Load gel samples on a 15% polyacrylamide SDS-PAGE gel along with a protein standard.
15. Run the gel at 240 V for 35 min at room temperature.
16. Stain the gel in 10 mL of staining solution (*see* **Note 9**).
17. Discard the staining solution in the proper waste receptacle and place the gel in 10 mL of destaining solution.
18. Destain until protein bands are visible. Representative pulldown gels are shown in Fig. 1.

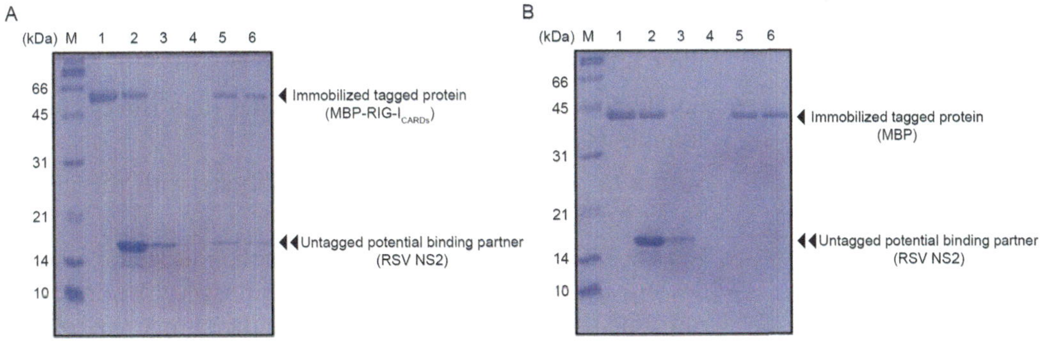

Fig. 1 SDS-PAGE analysis of in vitro pulldown assays. Lanes correspond to the following samples: M—marker, 1—initial beads, 2—input, 3—wash 1, 4—final wash, 5—final beads, 6—elution. (**A**) Analysis of a positive pulldown assay between MBP-RIG-I CARDs and RSV NS2. (**B**) Analysis of a negative pulldown assay between MBP and RSV NS2. (Adapted from Ref. [10])

3.2 SEC-MALS

SEC-MALS is the combination of size-exclusion chromatography (SEC) with multiangle light scattering (MALS) used for the biophysical characterization of macromolecules [15, 16]. This method provides users with the ability to determine physical properties such as molar mass, size, and conformation [17]. Since these properties are determined in solution, the user must ensure that their sample is stable in the desired buffer conditions and limit transient species, as they may interfere with the analysis. Samples are first separated based on hydrodynamic volume using an SEC column and are then illuminated by a laser beam in the MALS detector, where light scattering is used to determine protein properties [15]. Here, we describe the setup of a SEC-MALS experiment using a Wyatt Technologies MALS instrument and data analysis using the VISION CSH 2.0.2.9 software suite [17]. Additional information can be obtained from the manufacturer.

1. Place 1 L of ultrapure water in pump position A and 1 L of filtered general buffer in pump position B. Prime pump A by setting the pump position A method to 0.1 mL/min with an upper pressure limit of 40 bar. Open the pump purge valve to ensure there are no bubbles in the line of HPLC. Repeat for pump position B.
2. Switch flow back to pump position A. Close the pump purge valve and let pressure stabilize for 5 min.
3. Place the pump in "standby" mode.
4. Attach the SEC column (Cytiva Superdex 200 Increase 5/150 GL) stored in 20% ethanol by disconnecting the bypass fitting and connecting the column while the instrument is flowing at 0.1 mL/min. Try to minimize the introduction of air into the line/column when installing the column. Ensure that there is no leakage.
5. Allow pressure to stabilize for 10 min at 0.1 mL/min flowrate. Gradually increase the flowrate to 0.2 mL/min and allow pressure to stabilize for 10 min. Repeat at 0.3 mL/min and allow pressure to stabilize for 10 min.
6. Equilibrate column in ultrapure water for three column volumes (CV).
7. Switch to pump position B and equilibrate with the general buffer for 3 h (*see* **Note 10**).
8. Next, a 19-min VOYAGER method with an isocratic gradient flowing at 0.3 mL/min is used to encompass a 14-min sample run-time and 5-min COMET run after the sample run to clean the flow cell. Each series of experiments should begin with a validation standard or isotropic scatterer such as BSA.
9. Load 10 μL BSA (2 mg/mL) into the autosampler in position 1 (*see* **Note 11**).

10. Load or create a sequence in the VOYAGER software. Voyager sequences are designated for use with an autosampler and can automate multiple experiments with independent methods. Here, we describe a sequence for a single sample that contains one voyager and ASTRA method for simplicity.
11. Edit the sequence to include the 19-min VOYAGER method and an ASTRA method for BSA (*see* **Note 12**). Edit the sequence fields to the following settings: autosampler location to position 1, injection volume to 10 μL, concentration to 2.0 mg/mL, dn/dc to 0.185 mL/g, and UV extinction to 0.667 mL/(mg*cm).
12. Save and run the sequence.
13. After completion of the run, VOYAGER software will indicate all modules "ready", and the previous run can be analyzed in ASTRA.
14. Open ASTRA software and open the results from the BSA experiment for analysis. Begin analysis by navigating to the "peaks" tab and selecting the main BSA peak for normalization. Click "OK."
15. Navigate to the "baselines" tab and ensure proper baselines for light scattering detectors 2–18 (LS 2–18), UV, differential refractive index (dRI), and count rate (CR) of the Quasi-elastic light scattering (QELS) detector. Baselines are auto-generated based on the collected signal and can be corrected via snap-dragging with mouse or through entering a user defined function. Click "OK" when complete.
16. Navigate to the "configuration" tab, right-click and select "normalization". Normalization is performed to relate the various detector signals to the 90° detector signal and the instrument calibration constant.
17. Ensure that all detectors except the 90° detector (LS 11) and the QELS detector are different than 1.000. Click "normalize" followed by "OK."
18. Navigate to the "configuration" tab, right-click and select "alignment". The alignment function corrects for the delay volume between the UV, LS, and dRI detectors.
19. Select main peaks from UV, LS, and dRI signals by dragging the box to encompass these peaks. Click "align signals" followed by "OK."
20. Navigate to the "configuration" tab, right-click and select "band broadening". The band broadening function corrects for sample peak dilution between transfers between detectors, which can cause mismatched shapes between signals.

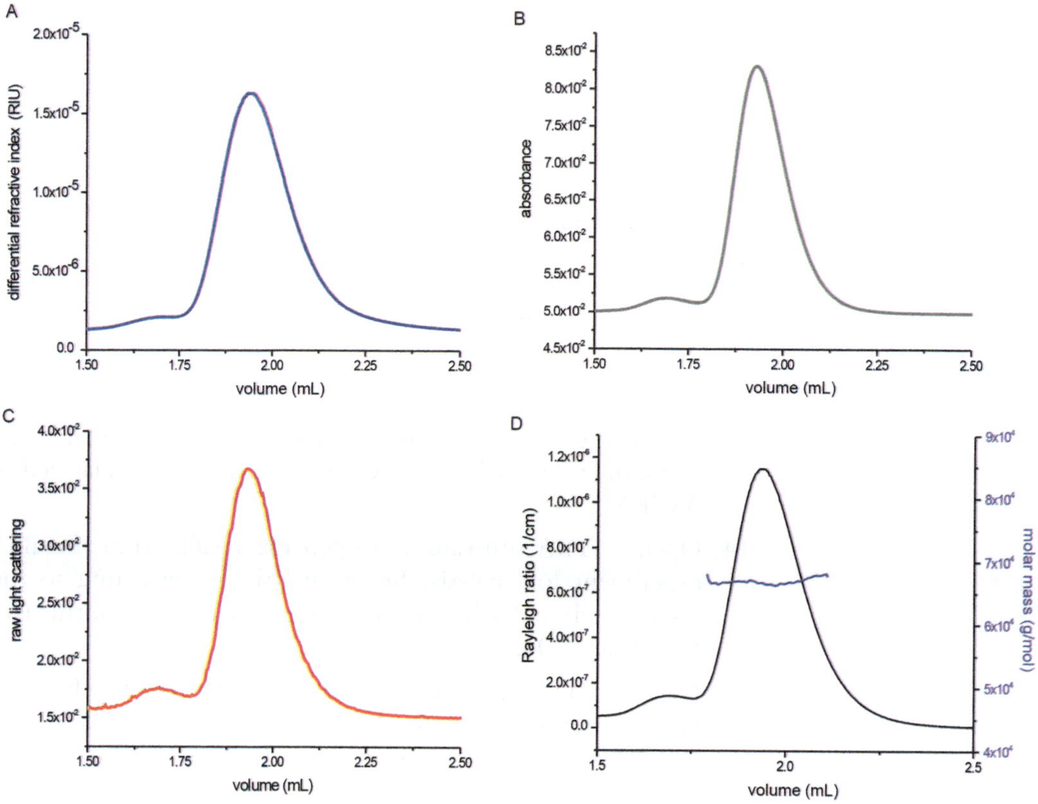

Fig. 2 Size-exclusion chromatography coupled to multiangle light scattering analysis of BSA. Representative raw (**A**) differential refractive index, (**B**) absorbance, and (**C**) light scattering data are shown, and (**D**) analysis demonstrates an experimentally determined molecular weight of 66.7 kDa with a polydispersity of 1.00. The theoretical molecular weight of BSA is 66.5 kDa

21. Select main peaks from UV, LS, and dRI signals by dragging the box to encompass these peaks. Click "perform fit" followed by "OK."

22. Navigate to the "results" tab and open the summary. The molecular weight of monomeric BSA is 66 kDa. Check that the recovered mass is near 200 μg (10 μL injection of 2 mg/mL sample). Ensure that the molar mass moments are near 6.6×10^4 and that the polydispersity is near 1.0 (Fig. 2).

23. Save the experiment as an ASTRA method and then save the experiment itself. This will create two separate files.

24. The instrument is now ready to run a new protein sample. Repeat **steps 10–24** with the new protein sample in position 1 using the ASTRA method created in **step 23**.

25. When the run is complete, navigate to the "baselines" tab and ensure proper baselines for light scattering detectors 2–18 (LS 2–18), UV, dRI, and CR. Baselines are auto-generated

based on the collected signal and can be corrected via snap-dragging with the mouse or through entering a user-defined function. Click "OK" when complete.

26. Navigate to the "results" tab and open the summary. It is not necessary to repeat normalization, band broadening, and alignment because these corrections are accounted for in the newly created ASTRA method that was used for the sample run.
27. The summary provides mass values and associated error for number average molar mass (M_n), molar mass at the apex of the concentration (RI or UV) peak (M_p), viscosity-average molar mass (M_v), weight-average molar mass (M_w), z-average molar mass (M_z), and uncertainty-weighted average molar mass (M_{avg}). The summary also includes values and errors for polydispersity (M_w/M_n), mean square radius, and hydrodynamic radius. In typical SEC-MALS experiments, users report M_w, polydispersity, and hydrodynamic radius to characterize the physical properties of the protein sample in solution.

3.3 HDX-MS

Hydrogen-deuterium exchange-mass spectrometry (HDX-MS) is a sensitive biophysical method that provides structural and conformational dynamics information on individual proteins and interactions in protein complexes. After the exchange of labile protein backbone amide hydrogens with deuterium, the protein is denatured, digested, and submitted to liquid chromatography (LC)-MS analysis to examine mass differences (Fig. 3); thus, the rate of HDX can provide information on the folded state of the protein, backbone hydrogen bonding, and solvent accessibility. Furthermore, a

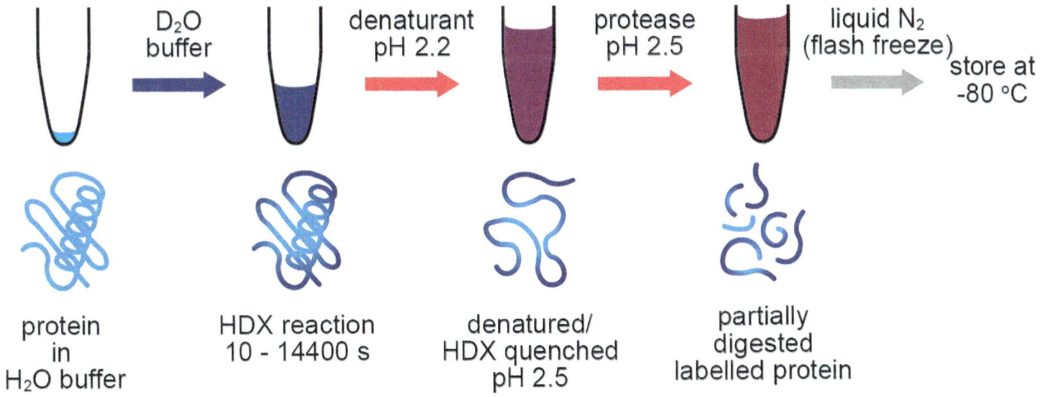

Fig. 3 Sample preparation steps for an HDX experiment. From left to right, the protein in H_2O (light blue) is diluted with D_2O (dark blue), starting the HDX incubation time. The HDX reaction is quenched with low pH and the protein denatured (purple). Optionally, the protein is partially digested in solution using a protease functional under acidic conditions (magenta). Timings for all these steps should be performed as precisely as possible. The prepared sample is flash frozen and stored at –80 °C for later LC-MS analysis

differential HDX-MS experiment, comparing the HDX rates of a protein with its protein complex, can be used to localize direct interactions between proteins within a protein complex, as well as any induced changes in the protein's conformational dynamics as a function of binding [18, 19]. While the steps and buffer solutions covered here are an example of the steps used for manual HDX-MS analysis of RSV NS2 binding to RIG-I CARDs (*see* **Note 13**), we advise also referring to the recommendations made by the broader HDX community [20].

3.3.1 Prepare Protein Samples for HDX-MS

A differential HDX-MS experiment for localizing binding interactions and the effects of binding on protein structure requires the comparison of HDX-MS results from at least two protein states, an "unbound" state and a "bound" state. Besides the addition of the binding partner(s), all other experimental variables must be kept as constant as possible. Buffer conditions that might affect HDX rates include salt identity, salt concentration, pH, and other solution additives [21]. Ideally, both the unbound and bound protein states should be dialyzed into an identical buffer solution to minimize deviations in buffer composition. This same buffer composition will be used for all HDX dilutions, H_2O buffer and D_2O buffer (*see* **Note 3**).

1. Inject both proteins into individual dialysis devices with an appropriate molecular weight cut-off (MWCO) (Thermo-Fisher Scientific, Slide-A-Lyzer™ MINI Dialysis Device, 3.5 K MWCO, 2 mL).
2. Dialyze in >1 L of the chosen buffer for ~1 h at 4 °C.
3. Dialyze in >1 L of fresh buffer overnight at 4 °C.
4. Recover the sample and determine protein concentration (*see* **Note 14**).

3.3.2 Peptide Mapping: Digestion Optimization and LC-MS/MS Analysis

Successful identification of a protein-protein interaction site using HDX-MS is contingent on the peptides identified and tracked throughout the rest of the experiment. Factors such as coverage (% of the protein sequence covered by peptides), the length of peptides, and overlapping redundancy define the possible resolution of an HDX-MS experiment. The conditions producing an effective set of peptides differ for each protein system; therefore, peptide mapping experiments varying quench conditions, protease choices/order, the LC gradient, and MS/MS conditions should be tested to identify appropriate conditions. Additional suggestions to improve pre-HDX mapping optimization are described elsewhere [22].

1. Prepare the following non-deuterated samples for mapping in at least duplicate (**steps 2–4**).

2. To 8 μL of 12.5 μM unbound protein in H_2O buffer, add 72 μL of H_2O buffer (to mimic adding D_2O buffer in later acquisitions), and then 120 μL of a chosen quench solution (2: 3 dilution) (*see* **Note 15**).
3. Add 10 μL of FP-XIII solution and incubate at 25 °C for 3 min (*see* **Note 16**).
4. Flash freeze the prepared samples in liquid nitrogen (*see* **Note 17**).
5. Two valves should be plumbed as shown in Fig. 4. All tubing and columns should be submerged in an ice-water bath (except for some protease columns, which lose activity at low temperature) to minimize back-exchange during LC-MS analysis. When prepared for LC-MS/MS analysis, ensure the valve flow-path is prepared for sample load (Fig. 4a).
6. Wash the loop: 3× with H_2O/FA, 3× with ACN/FA, 3× H_2O/FA.

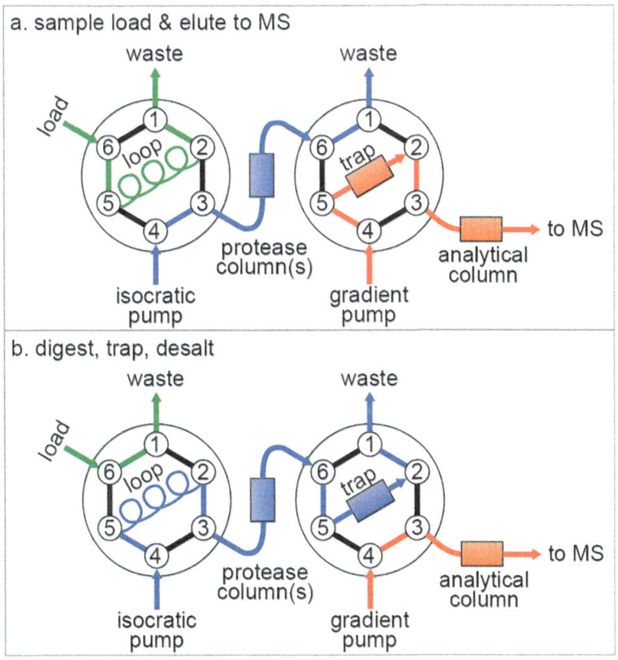

Fig. 4 Valve setup for LC-MS/MS (mapping) and LC-MS (HDX) analysis. (**A**) Flow path for sample loading (green) and elution of trapped/desalted peptides into the MS (orange). (**B**) Flow path for sample injection, on-column digestion, peptide trapping, and desalting (blue). Gradient pump flow path continues pushing a mixture of H_2O/FA and ACN/FA through the analytical column and into the MS (orange). Optionally, another valve between the analytical column and MS can be used to divert eluents from the analytical column during washing steps to limit introducing the eluents into the ESI source and front-end of the MS

7. Name the MS acquisition, including the protein state and "0 s" in the filename.
8. Thaw the sample in an incubator set to 37 °C (~30 s) and load the sample into the sample loop.
9. Turn both valves to push the sample through the protease column and to load the peptide products onto the trap (Fig. 4b). Desalt for long enough to push all of the sample volume out of the loop, plus at least three trap column volumes (for this trap, 3 min total, *see* **Note 18**).
10. 30 s before **step 11** (i.e., after loading and desalting for 150 s), start MS/MS acquisition.
11. Turn both valves to elute the trapped peptides onto the analytical column for separation and ESI into the mass spectrometer for MS/MS analysis (Fig. 4a).
12. Using proteomics software such as Byos® (Protein Metrics Inc.), identify peptide digestion fragments for all proteins of interest using appropriate parameters. To start, we suggest the default unless otherwise noted:
    (a) Sequences: enter .fasta file for the protein construct(s).
    (b) Instrument parameters: appropriate for the mass spectrometer; for an LTQ-FTMS, we used a precursor mass tolerance of 5 ppm and fragment mass tolerance of 10 ppm.
    (c) Digestion specificity: fully non-specific.
    (d) Modifications: for this system, there were no post-translational modifications. Adjust accordingly for your protein system.
    (e) Glycans: none. Special considerations are needed for HDX-MS analysis of glycosylated proteins [23–25].
13. Repeat mapping steps with varied quench, protease, LC gradient, and MS/MS conditions (Table 1) (*see* **Note 16**).
14. Create a master peptide list using the "best" mapping conditions for the protein correlating peptide sequence, charge state, m/z, and LC retention time.
15. Once the master peptide list is assembled, any changes to quench conditions, digestion, and LC gradient necessitate repeating the mapping experiment.

3.3.3 HDX-MS Sample Preparation and LC-MS Analysis

1. For each protein state, prepare non-deuterated controls (i.e., "0 s" samples) in at least triplicate (**steps 2–4**, *see* **Note 19**).
2. To 8 μL of 12.5 μM unbound protein in H_2O buffer, add 72 μL of H_2O buffer (or other 1:9 dilution, to mimic adding

Table 1
Example quench and digestion workflow and mapping optimization results (*see* Note 16)

Quench and digestion conditions					Peptide mapping results				
Condition	Denaturant	Protease column	In-sol'n protease	Post-Quench Incubation	Peptide Count	Coverage (%)	Redundancy	Weighted Avg Pep Length	Avg Abun / peptide
1	3 M urea	Pepsin	None	None	81	100	9.8	16.5	1.84E+06
2	3 M urea	2xpepsin	None	None	34	100	3.4	11.6	3.31E+06
3	4 M GndHCl	Pepsin	None	None	21	72.8	3.5	15.3	1.38E+06
4	3 M urea	Pepsin	10 μL FP-XIII	1 min	66	100	5.2	10.7	2.86E+06
5	3 M urea	Pepsin	10 μL FP-XIII	3 min	66	99.2	4.9	9.9	2.66E+06
6	3 M urea	Pepsin	10 μL FP-XIII	5 min	73	100	5.1	9.4	2.44E+06
7	3 M urea	Pepsin	20 μL FP-XIII	1 min	48	98.4	3.6	10.2	3.55E+06
8	3 M urea	Pepsin	20 μL FP-XIII	3 min	47	92.8	3.3	8.6	2.74E+06
9	3 M urea	Pepsin	20 μL FP-XIII	5 min	46	87.2	3.4	8.3	3.13E+06
10	4 M GndHCl	Pepsin	10 μL FP-XIII	3 min	9	64	1.6	14.6	1.53E+06

D_2O buffer in later acquisitions), and then 120 μL of Quench buffer (2:3 dilution) (*see* **Note 15**).

3. Add 10 μL of FP-XIII solution and incubate at 25 °C for 3 min (*see* **Note 16**).
4. Flash freeze prepared samples in liquid nitrogen (*see* **Note 17**).
5. For each protein state, prepare deuterated samples (**steps 6–10**) (*see* **Note 20**).
6. To 8 μL of 12.5 μM protein in H_2O buffer, add 72 μL of D_2O buffer, and start the HDX timer.
7. After the HDX time expires, add 120 μL of quench solution (2: 3 dilution) (*see* **Note 21**).
8. Add 10 μL of FP-XIII solution and incubate at 25 °C for 3 min (*see* **Note 16**).
9. Flash freeze prepared samples in liquid nitrogen (*see* **Note 17**).
10. Repeat **steps 6–9** in at least triplicate and with different deuteration incubation times. Typical timepoints include 10 s, 30 s, 300 s (5 min), 3600 s (1 h), and 14,400 s (4 h).
11. When prepared for LC-MS analysis (*see* **Note 22**), ensure the valve flow path is prepared for sample loading (Fig. 4a).
12. Wash the loop with: 3× with H_2O/FA, 3× with ACN/FA, 3× H_2O/FA.
13. Name the MS acquisition, including the protein state and HDX time.
14. Thaw the sample in an incubator set to 37 °C (30 s, *see* **Note 22**) and load the sample into the sample loop.
15. Turn both valves to push the sample through the protease column and to load the peptide products onto the trap (Fig. 4b). Desalt for at least three trap column volumes (3 min for this trap column, *see* **Note 18**).
16. Start MS acquisition 30 s before **step 17** (i.e., after loading and desalting for 150 s) (*see* **Note 18**).
17. Turn both valves to elute the trapped peptides onto the analytical column for separation and direct infusion into the mass spectrometer for MS analysis (Fig. 4a).
18. Repeat **steps 12–18** for all prepared and frozen samples.

3.3.4 HDX-MS Data Processing, Statistical Validation, and Data Representation

There are many HDX-MS data processing software available to curate and compile LC-MS results and to calculate the deuterium uptake for each peptide as a function of protein state and HDX time. We used and will describe processing with HDExaminer (Sierra Analytics). For more detailed instructions, refer to the HDExaminer User Manual found in the software under "Help" > "Documentation." We also use and briefly describe statistical analyses to confirm true positive differences in HDX results [26].

1. After opening HDExaminer, create a new project file and select the .fasta file for the protein. You should then see the protein sequence in the "Protein" tab.
2. Right-click the newly created "Protein State 1" and "Edit Protein State." Rename the protein state appropriately (often "Unbound"). If the fasta sequence does not start at amino acid 1 of the protein, right-click and adjust the numbering using "residue offset."
3. Add a protein state for each set of HDX-MS samples (very often two states, "Unbound" and "Bound").
4. Switch to the "Peptides" tab. Right-click "Peptide Sources" and "Add Peptide Source". Select your master peptide list from the mapping experiments to import the peptide list. Check that columns are automatically assigned appropriately.
5. Click "Add Marked Peptides to Pool."
6. Under "Tools" > "Options," check that the calculation parameters are appropriate for the experiment (*see* **Note 23**). In particular, we specify that the non-deuterated and deuterated RT windows must be within 1 min and 0.1 min of the expected RT (these parameters may vary depending upon the reproducibility of your peptide LC retention times), and that both ends of the RT window must be within the specific range (e.g. strict mode). These restrictions reduce the risk of using an LC-MS result for an incorrect peptide.
7. Switch to the "Analysis" tab. Drag and drop the data files for the non-deuterated controls into the window and ensure in the pop-up window that the protein state is correct and that "undeuterated" is selected. Once you finish with this window and select "OK", the software will begin searching the data for LC-MS results consistent with your master peptide list. Wait for the software to finish this processing step (*see* **Note 24**).
8. In the "Peptides" tab, go through and adjust LC windows for each peptide, charge state, and protein state to:
   (a) Use the same LC window for all protein states (*see* **Note 25**).
   (b) Minimize contaminant m/z peaks that could affect the deuteration calculation.
   (c) Remove peptides with poor signal or extensive contaminant interference.

9. Once satisfied with the LC windows for the non-deuterated controls, switch to the "Analysis" tab and drag the rest of the data into the window. Ensure that the appropriate values for each sample are chosen including protein state, HDX time ("D time"), and % of deuterium in the HDX solution ("D_2O %", 90% for the steps here). The software will begin processing the data as soon as you select "OK". We advise waiting for all processing to finish before proceeding.
10. While the "strict" limits to the RT search windows should limit false identifications, in the "Peptides" tab, go through and check/adjust LC windows for each peptide, charge state, and protein state to:
    (a) Use the same LC window for all protein states (*see* **Note 25**).
    (b) Minimize contaminant m/z peaks that could affect the deuteration calculation.
    (c) Remove peptides with poor signal or extensive contaminant interference.
11. When satisfied with data processing, right-click "Pool 1" (unless it has been renamed) and select "Export Peptide Pool Results." The results will be summarized in this master file.
12. Use the exported peptide pool results to produce kinetic plots: plot the average and standard deviations of the deuterium uptake for every peptide at every time point for both states as a function of $\log_{10}$ of the deuterium incubation time (Fig. 5a).
13. Calculate the difference in deuterium uptake (delta D) for every peptide at every time point between two protein states. Calculate the *p*-value for each difference.
14. Calculate the global significance limit for the HDX-MS dataset using $\alpha < 0.01$ [26].
15. For each difference in deuterium uptake between the two states, those data points with *p*-values <0.01 and absolute delta *D* values greater than the calculated global significance limit are considered a true positive difference in deuterium uptake [26].
16. Compile the significant HDX results, showing per timepoint significant differences in deuterium uptake between two protein states. Many options for data representation are possible. We prefer coloring peptides in a "coverage map" with respect to significant differences in deuterium uptake (Fig. 5b).
17. If the structure of the protein is available, recolor significant differences in HDX onto the structure of the protein (Fig. 5c).

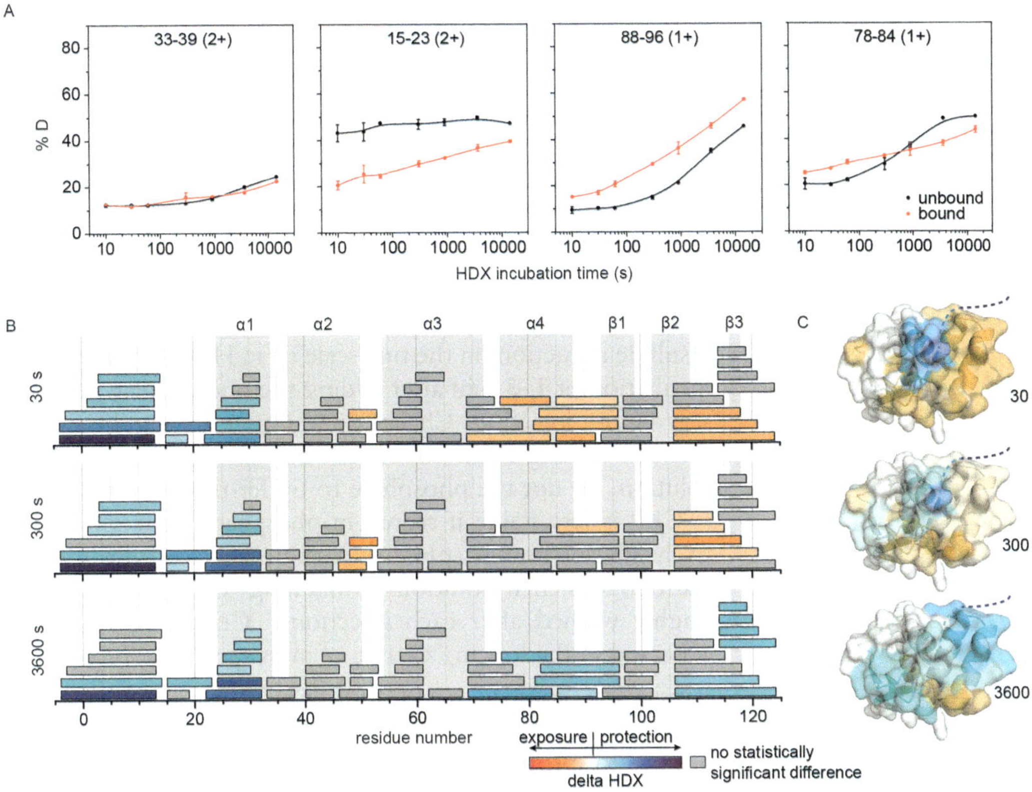

Fig. 5 Compilation of HDX-MS results and mapping results onto an available structure. (**A**) Representative deuterium uptake curves for select peptides showing possible HDX differences between NS2 (black) and the NS2-RIG-I CARDs complex (red), including no difference in HDX, binding-induced decreased HDX, binding-induced increased HDX, and a mixed character curve with both early increased HDX and late decreased HDX. (**B**) Peptide-level difference maps showing absolute differences in HDX for all identified peptides after 30, 300, and 3600 s (top to bottom, respectively). Peptides are represented as bars covering their peptide sequence. HDX changes are indicated by a color gradient (warm colors, increased HDX; cool colors, decreased HDX) and (**C**) mapped onto an available NS2 structure. (Adapted from Ref. [10])

4 Notes

1. Resin can vary by experiment depending on the affinity tag of proteins to be used in the assay.
2. Buffers can vary by experiment. It is recommended to buffer match the storage buffer of the proteins to be used in the assay.
3. Use any buffer conditions that stabilize both proteins. Prepare an equivalent buffer in deuterium oxide. It is especially critical that the starting buffer and the deuterated buffer conditions are identical in salt concentrations and pH.
4. It is advisable to prepare excess H_2O buffer, D_2O buffer, and quench solution and to aliquot and freeze the excess solutions

for future additions of protein states, replicates, or extra HDX time points to ensure no deviation in the solution preparations.

5. Check the pH for a test sample to ensure that the H_2O buffer and quench mixture is pH 2.5. This is especially critical if the starting buffer conditions differ from pH 7.0. At pH 2.5, backbone amide proton exchange and back-exchange are minimized, while other labile (e.g. side-chain) deuterium exchange back to hydrogen.
6. While this protein did not have disulfide bonds, a high concentration of TCEP (i.e. ~500 mM) is often necessary for efficient disulfide reduction on the timescale of an HDX experiment; an incubation of 1–3 min after adding the quench may be necessary before injection. Check the pH after adding TCEP to a quench. If using TCEP to reduce disulfide bonds in the quench solution, do not use phosphate to buffer the quench solution (TCEP is less stable if stored in phosphate buffers); alternatively, glycine buffers (pH 2.4) have been successfully used for buffering quench solutions containing TCEP. If not thoroughly washed after each injection, TCEP clusters may be observed in the low m/z range of mass spectra.
7. Pepsin was coupled to POROS AL-20 beads as previously described and packed into a 2 × 20 mm column [27]. Many options for proteolytic columns active in acidic conditions are available. Alternatively, digestion can be done in solution using a protease active in acidic conditions and limited handling time (e.g., fungal protease *Aspergillus satoi* XIII, nepenthesin, and others). A combination of in-solution and on-column digestion could yield better peptide coverage for some proteins.
8. When adding solution to resin, place pipette tip against the wall of the 1.5 mL tube near the top of the resin and dispense solution down the side of the tube to gently resuspend the resin.
9. Gel can be destained faster in a microwave on high power for 1 min and placed on a nutator for 5–10 min.
10. Column equilibration can be done overnight, but ensure there is enough buffer available so that the reservoir does not run dry. At least 3 h of equilibration time at the experimental flow rate is recommended to reduce column shedding during the run that occurs with changes in flow rate and pressure. This can increase baseline noise in the light scattering detector during the experiment.
11. Here, the experimental setup includes an HPLC equipped with an autosampler, necessitating the use of a VOYAGER sequence. Alternatively, the user can load the sample into a loop for manual injection.

12. The ASTRA method can be from a previous set of experiments or the default method from instrument installation. While the workflow remains the same across Wyatt MALS instruments, newer software packages incorporate HPLC and MALS instrument control solely into ASTRA.
13. Automated systems are available for HDX-MS. All solutions can be used for an automated HDX platform as described here; load all solutions into their appropriate positions within the autosampler.
14. Some protein loss after dialysis is expected.
15. The usage of these exact volumes for an HDX experiment is not necessary. More important is to use enough protein for detection, as large a D_2O dilution as is feasible for injection (ideally >90% (% v/v), but in this case a 1:9 dilution), and a large dilution with the quench solution to yield a final pH of 2.5.
16. The most effective quench and digestion conditions, and the workflow to identify these conditions, are highly protein dependent. For example, NS2 does not have any disulfide bonds, so the addition of a reducing agent to the quench was unnecessary. We also found it necessary to use in-solution partial digestion to increase the peptide coverage of NS2. A good mapping experiment is a balance of observing many peptides covering most of the protein sequence, with emphasis on shorter peptides (> 5 amino acids and <20 amino acids), while limiting sample handling time to minimize back-exchange. Variables monitored to evaluate these digestion conditions include the number of peptides identified using LC-MS/MS (peptide count), the percentage of the protein sequence covered by identified peptides (coverage), the average number of peptides reporting on each amino acid (redundancy), the weighted average peptide length scaled relative to their observed abundance (weighted average peptide length), and the observed average abundance of each peptide. The quench and digestion conditions chosen for HDX of this system are indicated as condition 5 in Table 1.
17. Flash frozen non-deuterated and deuterated samples can be stored for later analysis for up to 1 month without the risk of degradation and/or deuterium back-exchange.
18. Keep desalting time consistent between acquisitions to minimize drifting LC retention times. Using the flow rate of the loading pump and the volume of each component, ensure that the "desalting time" is long enough to empty the sample loop and travel the entire volume through to the trap to ensure that the entire volume of the sample passes through the protease

column(s) and is loaded onto the trap. Failure to do so will result in sample losses.

19. The non-deuterated control, or time 0 s, is used to establish LC window limits to follow each peptide after deuteration. If data is acquired on a different day, we suggest acquiring another non-deuterated control to determine if the LC window for each peptide has shifted.

20. Timings for the preparation of deuterated samples are critical and should be as precise as possible to minimize deviations in deuterium uptake.

21. Any sample handling time after quench, especially at higher temperatures, will allow for increased back-exchange of the deuterated backbone amide protons, thus decreasing HDX information and dynamic range. All timings should be precise.

22. Here we are performing peptide-level HDX-MS. While mapping experiments acquire LC-MS/MS analyses to identify peptides and correlate their identity to retention times and m/z values, acquisitions for the HDX-MS analysis should be done with LC-MS. Collision-induced dissociation (CID) can induce deuterium scrambling and should not be used. Alternatively, other fragmentation methods such as electron-transfer dissociation (ETD) or electron-capture dissociation (ECD) might be used (with care) to localize deuterium uptake to the residue level. More recently, an HDX-MS experiment using data-independent acquisition methodologies has been proposed, using CID and predictable deuterium scrambling for fast auto-curation of peptides and MS-based determination of deuteration extents [28].

23. These parameters are simplest to change now; if these parameters are changed after calculations have been started or completed, the software will recalculate all the results.

24. HDExaminer uses the master peptide list to search for peptides in the non-deuterated control (or 0 s sample) LC-MS results. The LC window selected for each peptide will then be used to extract the m/z of the deuterated peptides and to calculate their deuterium uptake. If you change the LC window for the non-deuterated control sample, the software will redo the analysis for the related deuterated samples. Optimization of the LC windows before importing deuterated sample LC-MS data decreases processing time.

25. The isotopic distribution of a peptide will vary over different LC windows. Deuterium very slightly affects LC retention times. Keeping the LC window consistent for each peptide across all protein states decreases these deviations.

Acknowledgments

We thank G.K. Amarasinghe, M.L. Gross, and members of the Leung, Amarasinghe, and Gross labs for support. This work was supported in part by NIH grants R01AI40758, P01AI120943, and R01AI159678 to D.W.L. The content is solely the responsibility of the authors and does not necessarily represent the official views of the National Institutes of Health.

References

1. Bancroft-Hunt JB, Weston JF, Munday JS (2010) Necrotising fasciitis in a bull due to infection with Arcanobacterium haemolyticum. N Z Vet J 58(5):260–264. https://doi.org/10.1080/00480169.2010.69300
2. Brasier AR, Spratt H, Wu Z, Boldogh I, Zhang Y, Garofalo RP et al (2004) Nuclear heat shock response and novel nuclear domain 10 reorganization in respiratory syncytial virus-infected a549 cells identified by high-resolution two-dimensional gel electrophoresis. J Virol 78(21):11461–11476. https://doi.org/10.1128/JVI.78.21.11461-11476.2004
3. Dave KA, Norris EL, Bukreyev AA, Headlam MJ, Buchholz UJ, Singh T et al (2014) A comprehensive proteomic view of responses of A549 type II alveolar epithelial cells to human respiratory syncytial virus infection. Mol Cell Proteomics 13(12):3250–3269. https://doi.org/10.1074/mcp.M114.041129
4. Hastie ML, Headlam MJ, Patel NB, Bukreyev AA, Buchholz UJ, Dave KA et al (2012) The human respiratory syncytial virus nonstructural protein 1 regulates type I and type II interferon pathways. Mol Cell Proteomics 11(5):108–127. https://doi.org/10.1074/mcp.M111.015909
5. Jamaluddin M, Wiktorowicz JE, Soman KV, Boldogh I, Forbus JD, Spratt H et al (2010) Role of peroxiredoxin 1 and peroxiredoxin 4 in protection of respiratory syncytial virus-induced cysteinyl oxidation of nuclear cytoskeletal proteins. J Virol 84(18):9533–9545. https://doi.org/10.1128/JVI.01005-10
6. Kipper S, Hamad S, Caly L, Avrahami D, Bacharach E, Jans DA et al (2015) New host factors important for respiratory syncytial virus (RSV) replication revealed by a novel microfluidics screen for interactors of matrix (M) protein. Mol Cell Proteomics 14(3):532–543. https://doi.org/10.1074/mcp.M114.044107
7. Munday DC, Emmott E, Surtees R, Lardeau CH, Wu W, Duprex WP et al (2010) Quantitative proteomic analysis of A549 cells infected with human respiratory syncytial virus. Mol Cell Proteomics 9(11):2438–2459. https://doi.org/10.1074/mcp.M110.001859
8. Munday DC, Hiscox JA, Barr JN (2010) Quantitative proteomic analysis of A549 cells infected with human respiratory syncytial virus subgroup B using SILAC coupled to LC-MS/MS. Proteomics 10(23):4320–4334. https://doi.org/10.1002/pmic.201000228
9. Pei J, Beri NR, Zou AJ, Hubel P, Dorando HK, Bergant V et al (2021) Nuclear-localized human respiratory syncytial virus NS1 protein modulates host gene transcription. Cell Rep 37(2):109803. https://doi.org/10.1016/j.celrep.2021.109803
10. Pei J, Wagner ND, Zou AJ, Chatterjee S, Borek D, Cole AR et al (2021) Structural basis for IFN antagonism by human respiratory syncytial virus nonstructural protein 2. Proc Natl Acad Sci USA 118(10). https://doi.org/10.1073/pnas.2020587118
11. Ternette N, Wright C, Kramer HB, Altun M, Kessler BM (2011) Label-free quantitative proteomics reveals regulation of interferon-induced protein with tetratricopeptide repeats 3 (IFIT3) and 5′-3′-exoribonuclease 2 (XRN2) during respiratory syncytial virus infection. Virol J 8(1):442. https://doi.org/10.1186/1743-422X-8-442
12. van Diepen A, Brand HK, de Waal L, Bijl M, Jong VL, Kuiken T et al (2015) Host proteome correlates of vaccine-mediated enhanced disease in a mouse model of respiratory syncytial virus infection. J Virol 89(9):5022–5031. https://doi.org/10.1128/JVI.03630-14
13. van Diepen A, Brand HK, Sama I, Lambooy LH, van den Heuvel LP, van der Well L et al (2010) Quantitative proteome profiling of respiratory virus-infected lung epithelial cells. J Proteome 73(9):1680–1693. https://doi.org/10.1016/j.jprot.2010.04.008

14. Van Royen T, Sedeyn K, Moschonas GD, Toussaint W, Vuylsteke M, Van Haver D et al (2022) An unexpected encounter: respiratory syncytial virus nonstructural protein 1 interacts with mediator subunit MED25. J Virol 96(19): e0129722. https://doi.org/10.1128/jvi.01297-22
15. Minton AP (2016) Recent applications of light scattering measurement in the biological and biopharmaceutical sciences. Anal Biochem 501:4–22. https://doi.org/10.1016/j.ab.2016.02.007
16. Stetefeld J, McKenna SA, Patel TR (2016) Dynamic light scattering: a practical guide and applications in biomedical sciences. Biophys Rev 8(4):409–427. https://doi.org/10.1007/s12551-016-0218-6
17. Wyatt PJ (1993) Light scattering and the absolute characterization of macromolecules. Anal Chim Acta 272(1):1–40. https://doi.org/10.1016/0003-2670(93)80373-S
18. Chalmers MJ, Busby SA, Pascal BD, West GM, Griffin PR (2011) Differential hydrogen/deuterium exchange mass spectrometry analysis of protein-ligand interactions. Expert Rev Proteomics 8(1):43–59. https://doi.org/10.1586/epr.10.109
19. Englander JJ, Del Mar C, Li W, Englander SW, Kim JS, Stranz DD et al (2003) Protein structure change studied by hydrogen-deuterium exchange, functional labeling, and mass spectrometry. Proc Natl Acad Sci USA 100(12): 7057–7062. https://doi.org/10.1073/pnas.1232301100
20. Masson GR, Burke JE, Ahn NG, Anand GS, Borchers C, Brier S et al (2019) Recommendations for performing, interpreting and reporting hydrogen deuterium exchange mass spectrometry (HDX-MS) experiments. Nat Methods 16(7):595–602. https://doi.org/10.1038/s41592-019-0459-y
21. Walters BT, Ricciuti A, Mayne L, Englander SW (2012) Minimizing back exchange in the hydrogen exchange-mass spectrometry experiment. J Am Soc Mass Spectrom 23(12): 2132–2139. https://doi.org/10.1007/s13361-012-0476-x
22. Hamuro Y, Coales SJ (2018) Optimization of feasibility stage for hydrogen/deuterium exchange mass spectrometry. J Am Soc Mass Spectrom 29(3):623–629. https://doi.org/10.1007/s13361-017-1860-3
23. Guttman M, Scian M, Lee KK (2011) Tracking hydrogen/deuterium exchange at glycan sites in glycoproteins by mass spectrometry. Anal Chem 83(19):7492–7499. https://doi.org/10.1021/ac201729v
24. Lambert T, Gramlich M, Stutzke L, Smith L, Deng D, Kaiser PD et al (2023) Development of a PNGase Rc column for online deglycosylation of complex glycoproteins during HDX-MS. J Am Soc Mass Spectrom 34(11): 2556–2566. https://doi.org/10.1021/jasms.3c00268
25. Wagner ND, Huang Y, Liu T, Gross ML (2021) Post-HDX deglycosylation of Fc gamma receptor IIIa glycoprotein enables HDX characterization of its binding interface with IgG. J Am Soc Mass Spectrom 32(7): 1638–1643. https://doi.org/10.1021/jasms.1c00003
26. Hageman TS, Weis DD (2019) Reliable identification of significant differences in differential hydrogen exchange-mass spectrometry measurements using a hybrid significance testing approach. Anal Chem 91(13):8008–8016. https://doi.org/10.1021/acs.analchem.9b01325
27. Wang L, Pan H, Smith DL (2002) Hydrogen exchange-mass spectrometry: optimization of digestion conditions. Mol Cell Proteomics 1(2):132–138. https://doi.org/10.1074/mcp.m100009-mcp200
28. Filandr F, Sarpe V, Raval S, Crowder DA, Khan MF, Douglas P et al (2024) Automating data analysis for hydrogen/deuterium exchange mass spectrometry using data-independent acquisition methodology. Nat Commun 15(1):2200. https://doi.org/10.1038/s41467-024-46610-3

Chapter 15

Bisulfite Sequencing to Map 5-Methylcytosine Sites in Human Respiratory Syncytial Virus RNA

Li-Sheng Zhang, Yuexiu Zhang, Amit Sharma, Chuan He, and Jianrong Li

Abstract

Viral epigenetic modification is an emerging field. As an obligate intracellular parasite, viral genetic material is exposed to host RNA-modifying enzymes. Currently, more than 180 types of RNA modifications have been identified. Like all non-segmented negative-sense RNA viruses, human respiratory syncytial virus (RSV) infection produces three types of RNA: genome and antigenome that are encapsidated by viral nucleocapsid and 10 mRNAs that are capped, methylated, and polyadenylated. Recently, the RSV genome, antigenome, and mRNAs have been shown to be modified by *N*6-methyladenosine (m^6A), which plays critical roles in replication, gene expression, and innate immune responses. However, whether RSV RNAs contain other types of modification remains unclear. Here, we describe an optimized bisulfite sequencing method and rigorous data analysis procedure to quantitatively map 5-methylcytosine (m^5C), one of the most prevalent RNA modifications, in RNA isolated from A549 cells infected by RSV and vesicular stomatitis virus (VSV) at single-base resolution. This method identified 20 m^5C sites in RSV genome but no m^5C sites were found in RSV antigenome, RSV mRNAs, or VSV RNAs. This protocol can be easily adapted for m^5C mapping of the RNAs of other viruses.

Key words Respiratory syncytial virus (RSV), RNA modification, 5-methylcytosine, Bisulfite sequencing

1 Introduction

Chemical modification of RNA bases within RNA is a widespread posttranslational modification that epigenetically regulates numerous biological processes [1–3]. To date, more than 180 types of RNA modifications have been identified. The most abundant RNA modifications include N^6-methyladenosine (m^6A), 5-methylcytosine (m^5C), pseudouridine (ψ), N^1-methyladenosine (m^1A), N^7-methylguanosine (m^7G), and $2'$-*O*-methylation (N_m) [4]. Among these modifications, m^5C, methylation on the fifth carbon of cytosine, is one of the most prevalent RNA modifications. This modification was initially identified in DNA [5, 6] and later found in RNA in 1958 [7]. The detection of m^5C in polyadenylated host RNA was

Bo Liang and Rachel Fearns (eds.), *Human Respiratory Syncytial Virus: Methods and Protocols*, Methods in Molecular Biology, vol. 2948, https://doi.org/10.1007/978-1-0716-4666-3_15,

demonstrated in 1970 for the first time [8, 9]. RNA m^5C modification is catalyzed by the host m^5C methyltransferase (MTase) family (also called m^5C writers), which includes seven members of the NOL1/NOP2/SUN domain (NSUN) family (NSUN1–7) and DNA methyltransferase-like 2 (DNMT2) [10, 11]. Evidence suggests that m^5C modification of DNA can be reversibly removed by m^5C demethylases (also called m^5C erasers) [12, 13] although conversion of m^5C back to C in RNA molecules has not been confirmed. Potential m^5C erasers include members of the ten-eleven translocation (TET) family of demethylases, TET1, TET2, and TET3, which catalyze m^5C demethylation and generate 5-hydroxymethylcytosine (hm^5C) [12–15]. The biological function of m^5C methylation can be mediated by m^5C-binding proteins (m^5C readers), such as ALYREF and YBX1, which bind preferentially to m^5C sites in RNA [14, 16].

Viruses use DNA or RNA as their genetic materials. During the replication cycle, viral genetic materials are exposed to the host m^5C MTases thereby possibly acquiring m^5C modifications. The m^5C methylation was found in a viral RNA, the Sindbis virus 26S RNA in 1975 [17]. The following year, the mRNAs of adenovirus, a DNA virus, were shown to be m^5C modified [18]. These studies mainly used DEAE cellulose chromatography, thin-layer chromatography, or liquid chromatography. However, these methods are not able to map the exact location of the m^5C sites in RNA. Subsequent functional studies of viral m^5C methylation were hampered for decades due to the lack of a reliable method to precisely map the m^5C sites in viral RNA at a single-base level and the lack of knowledge of the enzymes involved. In recent years, the development of next-generation sequencing methods such as RNA bisulfite sequencing, m^5C RNA immunoprecipitation sequencing (m^5C-RIP-seq), 5-azacytidine–mediated RNA immunoprecipitation sequencing (Aza-IP-seq), and methylation individual-nucleotide-resolution crosslinking and immunoprecipitation sequencing (miCLIP-seq) have enabled transcriptome-wide mapping of m^5C methylation in host and viral RNAs [19–25]. Recent studies suggest that m^5C sites in viral RNA modulate viral replication and gene expression via numerous mechanisms including regulation of viral mRNA translation, RNA alternative splicing, RNA stability, RNA export, viral reverse transcription, and particle production [26–30].

An immunoprecipitation-based method (photo-crosslinking-assisted m^5C sequencing, PA-m^5C-seq) was first used to detect m^5C sites in human immunodeficiency virus-1 (HIV-1) RNA [26]. This method identified 19 distinct m^5C methylation peaks spread throughout the HIV-1 genome RNA [26]. Using the same method, the same group found that murine leukemia virus (MLV) RNAs bear an exceptionally high level of m^5C modifications, which were especially concentrated in the gag/pol region [27]. This method also identified several m^5C sites in the hepatitis B virus (HBV) pregenomic RNA (pgRNA) [28]. The limitations of IP-based methodology include its high dependence on specific antibodies, the risk of nonspecific binding to RNA, the inability to accurately identify the location at single-nucleotide resolution, and inefficient detection of m^5C sites located in the regions with RNA secondary structures.

Nanopore direct RNA sequencing (DRS) has been used for m^5C detection in coronavirus RNA. DRS analysis of human coronavirus 229E (HCoV-229E) RNA suggested numerous m^5C sites distributed across viral RNAs [31]. However, these sites need furthermore validation due to the lack of proper controls in the study. Using a similar Nanopore DRS, Kim et al. reported 41 modifications sites in SARS-CoV-2 RNA although the types of these modifications could not be determined [32]. In the DRS analysis, Kim et al. also found numerous signal level changes corresponding to m^5C modification, even with SARS-CoV-2 RNA synthesized in vitro which should completely lack m^5C sites [32]. They argued that the m^5C sites detected without proper controls are likely to be false positives.

Bisulfite sequencing (BS) is the gold standard method for detecting m^5C modification sites, allowing quantitative mapping of m^5C sites in RNA at a single-base resolution [11, 19]. Bisulfite treatment chemically deaminates unmethylated cytosine in single-stranded RNA into uracil. However, methylated cytosine at the carbon 5 position is protected from this reaction. Therefore, the m^5C sites on RNA can be identified by constructing a library followed by next-generation sequencing. Using this technique, m^5C sites have been proposed in EBER1, the Epstein-Barr virus (EBV) noncoding RNA [33], HBV mRNA [29], enterovirus 71 (EV71) RNA [34], hepatitis C virus (HCV) RNA [35], and SARS-CoV-2 RNA [36]. However, a major caveat of this method is the potential for incomplete deamination of unmodified cytidine after bisulfite treatment, which often occurs in double-stranded RNA regions, RNA stems, RNA secondary structures, and RNA duplexes (e.g., between sense and antisense RNA), resulting in false-positive m^5C identifications [11, 37, 38]. In recent years, many controversial issues regarding m^5C sites in viral RNA have been raised, particularly for SARS-CoV-2 RNA. Soon after SARS-CoV-2 appeared, Nanopore DRS could hardly detect confident m^5C sites in SARS-CoV-2 RNA with confidence [32]. However, Lin et al. detected 555 high-confidence m^5C sites in SARS-CoV-2 RNA isolated from Vero E6 cells [39]. Similarly, Wang et al. identified a large number of m^5C-methylated sites in SARS-CoV-2 RNA purified from Caco2 cells [36]. To address these inconsistent m^5C sequencing data, Alexandra Lusser's group employed a rigorous bisulfite-seq protocol and stringent data analysis to determine the m^5C methylome of SARS-CoV-2 and reexamine the reported m^5C sites in HIV and MLV [37]. To prevent false positives, they optimized the deamination and desulfonation methods, improved RNA double strand melting, and included additional stringency filters for analysis. Surprisingly, there is no evidence for the existence of m^5C sites in SARS-CoV-2 RNA isolated from Vero E6-TMPRSS2-ACE2 or Caco2 cells [37]. No confident m^5C sites were found in HIV-1 or in MLV RNA [37]. Thus, it is critical to optimize the conditions for bisulfite sequencing and carefully analyze the m^5C sequencing data.

Human respiratory syncytial virus (RSV) is a member of the *Pneumoviridae* family of non-segmented negative strand (NNS) RNA viruses. Like all NNS RNA viruses, the RSV genome is rapidly encapsidated by its nucleocapsid (N) protein as it is being synthesized, preventing secondary structures from forming. Its nucleocapsid complex serves as the active template for replication and transcription [40, 41]. During RSV replication, the RNA-dependent RNA polymerase (RdRp) synthesizes full-length antigenomic RNA, which is also encapsidated by the N protein and functions as a template for synthesis of progeny genomic RNA. During RSV transcription, the RdRp transcribes the viral genome into ten 5′ capped and methylated and 3′ polyadenylated mRNAs in a sequential and gradient manner, such that 3′ proximal genes are transcribed more abundantly than the distal genes.

Recent studies have shown that genome, antigenome, and mRNAs of many NNS RNA viruses (such as RSV, vesicular stomatitis virus, human metapneumovirus, measles virus, and Sendai virus) are epigenetically modified by m^6A [3, 42–44]. Since viral genome and antigenome are completely encapsidated by N protein, it is unclear how host m^6A methyltransferases would gain access to viral genomic and antigenomic RNA. However, a recent study found that host m^6A methyltransferases co-localize with the viral N protein in inclusion bodies where new ribonucleoprotein (RNP) is assembled and active viral replication occurs [3]. Therefore, it is likely that m^6A methylation in genomic and antigenomic RNA occurs prior to or concomitant with encapsidation. It also suggests the possibility that other host RNA-modifying enzymes could similarly modify the RSV RNA. Functionally, these m^6A sites in the viral RNA promote RSV replication and gene expression, regulate innate and adaptive immune responses and pathogenesis [3, 42–44]. Whether genome, antigenome, and mRNAs of other NNS RNA viruses contain other types of internal RNA modification is currently poorly understood.

We recently conducted bisulfite sequencing to detect m^5C in RSV and vesicular stomatitis virus (VSV), representative members of the *Pneumoviridae* and *Rhabdoviridae*, respectively [45]. We identified a total of 20 m^5C sites in RSV genome with high confidence [45]. However, no m^5C sites were found in the RSV antigenome (the positive sense replication intermediate) or mRNAs

[45]. In addition, no m^{5}C sites were found in the genome, antigenome, or mRNAs of VSV [45]. Several studies have proposed that NSUN2 may add m^{5}C methylation in HIV-1 [27], HBV [28, 29], EV71 [34], and SARS-CoV-2 [36] viral RNA. We also conducted bisulfite sequencing of RNA isolated from RSV-infected A549 cells and NSUN2 knockout A549 cells. Our result showed that none of these 20 m^{5}C sites in RSV genome were notably affected by NSUN2 depletion [45], suggesting that NSUN2 is not the enzyme responsible for catalyzing m^{5}C methylation in RSV genome and that m^{5}C methylation in RSV genome must be added by a different m^{5}C methyltransferase.

This protocol in this review aims to provide a clear procedure to conduct bisulfite sequencing to map m^{5}C sites in host and viral RNAs at a single-base resolution. This includes a detailed description of each step in m^{5}C mapping: preparation of viral RNA, bisulfite treatment, library construction, next-generation sequencing, and m^{5}C data analysis. In addition, we provide a detailed procedure to optimize bisulfite sequencing and use rigorous data analysis to avoid false-positive m^{5}C calls. This protocol can be easily adapted for m^{5}C mapping of the RNAs of other viruses.

2 Materials

2.1 Materials Required for Preparation of Total RNA for m^{5}C Sequencing

1. A549 cells (ATCC CCL-185).
2. NSUN2$^{+/+}$ A549 cells (transduced with control sgRNA).
3. NSUN2$^{-/-}$ A549 cells (NSUN2 knockout).
4. Recombinant human respiratory syncytial virus (RSV) A2 strain expressing GFP (rgRSV).
5. Recombinant vesicular stomatitis virus (VSV) Indiana strain expressing GFP (rVSV-GFP).
6. TRIzol reagent (Invitrogen).
7. Chloroform (Invitrogen).
8. Isopropanol.
9. 75% ethanol.
10. RiboMinus™ Eukaryote System v2 (Invitrogen).
11. Dynabeads™ mRNA DIRECT™ Purification Kit (Invitrogen).

2.2 Materials Required for m^{5}C Bisulfite Sequencing

1. rRNA-depleted total RNA from rgRSV- or rVSV-GFP-infected A549 cells.
2. rRNA-depleted polyadenylated mRNA from rgRSV- or rVSV-GFP-infected A549 cells.
3. $Na_2CO_3/NaHCO_3$ buffer (pH 9.2).

4. Oligo Clean & Concentrator (Zymo Research).
5. RNA Clean & Concentrator (Zymo Research).
6. DNA Clean & Concentrator (Zymo Research).
7. EZ RNA Methylation Kit (Zymo Research).
8. T4 Polynucleotide Kinase (EK0032, Thermo Fisher Scientific).
9. 10× T4 Polynucleotide Kinase Reaction Buffer (NEB).
10. 3′-Adaptor (5′App-NNNNN ATCACGAGATCGGAAGAG CACACGTCT-3SpC3).
11. 5′-Adaptor (5′-GUUCAGAGUUCUACAGUCCGACGAUC NNNNN-3′).
12. NEBNext® Small RNA Library Prep Set for Illumina® includes 3′ Ligation Reaction Buffer (2×), 3′ Ligation Enzyme Mix, and SR RT Primer.
13. 10 mM dNTP Solution Mix (NEB).
14. RNaseOUT Recombinant Ribonuclease Inhibitor (Thermo Scientific).
15. SuperScript IV Reverse Transcriptase (Thermo Scientific), 100 mM DTT and 5× SSIV Buffer included.
16. RNase H (NEB, M0297L).
17. NEBNext® Multiplex Oligos for Illumina (Index Primers Set 1).
18. NEBNext® Multiplex Oligos for Illumina (Index Primers Set 2).

2.3 Software Required for Analysis of Bisulfite Sequencing Data

1. Cutadapt tool.
2. BBMap tool (https://sourceforge.net/projects/bbmap/).
3. Bismark-0.22.3 tool, https://github.com/FelixKrueger/Bismark/releases using Tophat2 (version 2.1.1) and bowtie2 (version 2.3.5.1).
4. Samtools (http://samtools.sourceforge.net/).
5. Fine-tuned bam-readcount (https://github.com/genome/bam-readcount).
6. IGV software (https://software.broadinstitute.org/software/igv/).

3 Methods

3.1 Total RNA Isolation

Our goals are to map m^5C sites in both viral and host RNAs and to determine whether these m^5C sites are regulated by NSUN2. Briefly, wild-type A549 cells, NSUN2$^{+/+}$ cells, and NSUN2$^{-/-}$ A549 cells were infected with rgRSV or rVSV-GFP at an MOI of

1.0. When maximum cytopathic effects (CPE) were observed (48 h for rgRSV and 18 h for rVSV-GFP), total RNA was extracted using TRIzol reagent (Invitrogen). Subsequently, ribosomal RNA (rRNA) was removed from the total RNA by using RiboMinus™ Eukaryote System v2 (Invitrogen). To isolate polyadenylated (polyA$^+$) RNA for m^{5}C sequencing, polyA$^+$ RNA was enriched by using Dynabeads™ mRNA DIRECT™ Purification Kit (Invitrogen).

Steps:

1. Seed the NSUN2$^{+/+}$ A549 cells or NSUN2$^{-/-}$ A549 cells in a 10-cm plate the day before infection.
2. Infect cells with rgRSV or rVSV-GFP at an MOI of 0.1.
3. Incubate the cells at 37 °C for 1 h, remove the virus inoculum, and replace it with fresh DMEM with 2% FBS.
4. Incubate infected cells at 37 °C for 48 h for rgRSV or 18 h for rVSV-GFP.
5. Remove media.
6. To lyse the cells, add 1 mL TRIzol™ Reagent to the plate, incubate at room temperature for 5 min, and collect the lysate in a 2-mL tube.
7. Add 0.2 mL of chloroform to the lysate, vortex the samples briefly, and incubate for 2–3 min.
8. Centrifuge at 10,000 × *g* for 15 min at 4 °C. The mixture separates into a lower phenol-chloroform, an interphase, and a colorless upper aqueous phase.
9. Transfer the aqueous phase containing the RNA to a new tube.
10. Add 0.5 mL isopropanol to the aqueous phase and incubate at room temperature for 10 min.
11. Centrifuge at 10,000 × *g* for 10 min at 4 °C. Total RNA precipitate forms a white gel-like pellet at the bottom of the tube.
12. Discard the supernatant with a micropipettor.
13. Resuspend the RNA pellet in 1 mL of 75% ethanol. Centrifuge at 10,000 × *g* for 5 min at 4 °C.
14. Discard the supernatant with a micropipettor.
15. Air dry the RNA pellet for 5–10 min.
16. Resuspend the pellet in 50 μL of RNase-free water.
17. Measure RNA concentration and purify using a Nanodrop device.

3.2 Bisulfite Sequencing for m⁵C Site Detection

In bisulfite-based m^5C sequencing, the RNA will be first treated with bisulfite, which converts cytosine residues to uracil but leaves 5-methylcytosine residues unaffected. Bisulfite treatment introduces specific changes in the RNA sequence that depend on the methylation status of individual cytosine residues. Therefore, RNA has been treated with bisulfite retains only 5-methylcytosine, allowing the mapping of m^5C sites in a single-nucleotide resolution.

Steps:

1. Aliquot 200 ng of rRNA-depleted total RNA or mRNA, and fragment into 50–70 nt length using $Na_2CO_3/NaHCO_3$ buffer (pH 9.2).
2. Heat the RNA samples at 95 °C for 3 min.
3. Purify the RNA with Oligo Clean & Concentrator (Zymo Research). Elute with 10 μL RNase-free water.
4. Freeze 1.5 μL in -80 °C freezer as "input" RNA.
5. Use the remaining RNA for bisulfite treatment using EZ RNA Methylation Kit (Zymo Research). Elute with 10 μL RNase-free water.
6. Repair 3′- and 5′-end for "input" RNA and bisulfite-treated RNA by T4 Polynucleotide Kinase (EK0032, Thermo Fisher Scientific). Dilute 'input' RNA to 10 μL. Prepare each reaction according to Table 1.
7. Purify RNA by Oligo Clean & Concentrator (Zymo Research). Elute with 7 μL RNase-free water.
8. Perform 3′-adaptor ligation: incubate 6 μL repaired RNA fragments with 1.0 μL of 3′-Adaptor (5′App-NNNNN ATCAC GAGATCGGAAGAGCACACGTCT -3SpC3) at 70 °C for 2 min, and immediately placed on ice.

Table 1
Repair 3′- and 5′-end of RNA by T4 polynucleotide kinase

Reagent	Volume (μL per reaction)	Note
RNA	10	
10× T4 polynucleotide kinase reaction buffer	3	
T4 PNK	3	
RNase-free water	14	
Total volume/reaction	30	Incubated at 37 °C for 1 h
T4 PNK	1.5	
10 mM ATP	1.5	Incubation 37 °C for 30 min

9. To the RNA-adapter mixture, add 10 μL of 3′ Ligation Reaction Buffer (2×) (NEB) and 3 μL of 3′ Ligation Enzyme Mix (NEB).
10. Incubation of the reaction at 25 °C for 2 h, followed by incubation at 16 °C for 10 h.
11. Add 1.0 μL of SR RT Primer to the mixture for Illumina (NEB), followed by 5 min at 75 °C, 15 min at 37 °C, 15 min at 25 °C, and then on ice.
12. Add 1.0 μL of the denatured 5′-Adaptor (5′-GUUCAGAGUUCUACAGUCCGACGAUCNNNNN-3′), 1.0 μL of 5′ Ligation Reaction Buffer (10×) (NEB), and 2.5 μL of 5′ Ligation Enzyme Mix (NEB) to the mixture. Incubate the reaction at 25 °C for 4 h.
13. Purify RNA by RNA Clean & Concentrator (Zymo Research). Elute with 11 μL RNase-free water.
14. Set up reverse transcription (RT) step: mix the purified RNA with 4 μL 5× SSIV Buffer (Thermo Scientific), 2 μL of 10 mM dNTP Solution Mix (NEB), 1 μL 100 mM DTT, 1 μL of RNaseOUT Recombinant Ribonuclease Inhibitor (Thermo Scientific), 1 μL of SuperScript IV Reverse Transcriptase (Thermo Scientific), and 11 μL of eluted RNA a total of 20 μL reaction. Incubate the reaction at 50 °C for 30 min. 1.0 μL of RNase H (NEB, M0297L) was then added, and the reaction was incubated at 37 °C for 20 min.
15. Purify the cDNAs as ssDNA by DNA Clean & Concentrator (Zymo Research). Use 7 volumes of DNA Binding Buffer for each sample.
16. Amplify the libraries using the SR Primer for Illumina (NEB) and indexed primers (from NEBNext Multiplex Oligos for Illumina).
17. Subject all libraries to sequencing on Illumina NovaSeq 6000 with single-end 100 bp read length.

3.3 Identification of Bisulfite-Induced Misincorporation

Steps:

1. Trim the sequencing data with cutadapt tool to remove adapters and low-quality reads.
2. Remove PCR duplicates with BBMap tool (https://sourceforge.net/projects/bbmap/).
3. Trim the 5-mer random barcodes at reads' ends.
4. Remove low-quality or short reads (less than 30 nt) using cutadapt tool.
5. Align the remaining reads to the converted human genome (hg38, C- > T or G- > A converted, by Bismark-0.22.3 tool, https://github.com/FelixKrueger/Bismark/releases) using

Tophat2 (version 2.1.1) and bowtie2 (version 2.3.5.1), allowing at most three mismatches per read.

6. Split the generated .bam files into positive and negative strands and sort using Samtools (http://samtools.sourceforge.net/).
7. Identify sequence variants by measuring the base composition at each position using fine-tuned bam-readcount (https://github.com/genome/bam-readcount).
8. Parse the generated bam-readcount results and analyze by in-house scripts.
9. Convert all cytidine sites in RNA into uridine after bisulfite treatment, and the sites showing T→C mutations are the m^5C candidate sites.
10. Calculate the T to C mutation ratio at each m^5C candidate site by data output from bam-readcount pipeline.
11. Confirm the T to C mutation ratio by direct visualization through IGV software (https://software.broadinstitute.org/software/igv/).
12. Identify an m^5C candidate site that meet the criterion in its misincorporation profile: (1) the T→C mutation count of this candidate site should be above 5; (2) the mutation ratio (C→A or G or T) of this candidate site needs to be zero in "input" library; (3) the T→C mutation ratio of this candidate site in "bisulfite treated" library is at least 20% and the candidate site does not appear within a cluster of multiple successive T→C mutations; (4) all T→C misincorporation signatures must appear at the internal positions within reads, instead of occurring at reads ends. All confident sites with T→C mutation above 20% will be used for downstream study, like investigating NSUN2-hypomethylated sites.

4 Notes

4.1 Bisulfite Sequencing Results and Analysis

Both RSV and VSV are NNS RNA viruses. In virus-infected cells, RSV produces three viral RNA species: genome (negative-sense), antigenome (replicative intermediate), and 10 mRNAs. Similarly, VSV produces genome, antigenome, and 6 mRNAs. The bisulfite sequencing identified 20 m^5C sites with T to C mutation ratio at 45 ~ 99%, spread in NS2, intergenic sequence (*IG*) between NS2 and P, P, M, *IG* between P and M, M, G, and M2 genes (Fig. 1 and Table 2). However, no m^5C site was identified in the RSV antigenome or 10 mRNAs. For VSV, no m^5C site was observed in genome, antigenome, or 6 mRNAs. To validate the lack of m^5C sites in RSV and VSV mRNAs, poly(+)A RNA was isolated and subjected to bisulfite sequencing; no m^5C site was identified in RSV or VSV mRNAs.

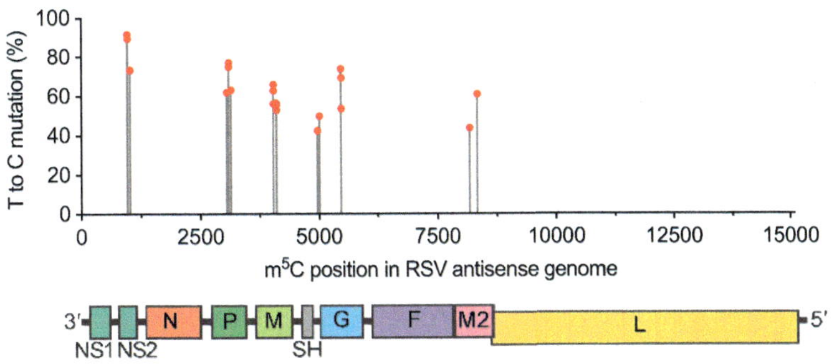

Fig. 1 Locations of 20 m^5C sites within RSV genome RNA. The x-axis shows the m^5C site position in the RSV genome; the y-axis shows the m^5C site mutation level

Table 2
Number and location of m^5C sites in RSV genome

Gene/position	Number of m^5C sites	Position in RSV genome
NS2	2	958, 962
NS2-N intergenic	1	1019
P	2	3050, 3085
P-M intergenic	2	3038, 3133
M	6	4032, 4034, 4036, 4098, 4100, 4101
G	5	4965, 5000, 5459, 5463, 5466
M2	2	8183, 8339
Total	20	

Next, the T to C mutation ratio in the RSV genome isolated from $NSUN2^{+/+}$ and $NSUN2^{-/-}$ A549 cells was compared. The T to C mutation ratio is highly consistent within two replicates. There was no significant reduction in T to C mutation ratio in the RSV genome isolated from $NSUN2^{-/-}$ A549 cells compared to those isolated from $NSUN2^{+/+}$ A549 cells (Table 3), demonstrating that NSUN2 regulates none of these m^5C sites in the RSV genome.

4.2 Types of RNA Used for Bisulfite Sequencing

Both virion RNA (only containing genome RNA) and total RNA (containing genome, antigenome, and mRNAs) isolated from virus-infected cells could be used for bisulfite sequencing. Polyadenylated mRNAs can be enriched using Dynabeads™ mRNA DIRECT™ Purification Kit for bisulfite sequencing.

Table 3
m^{5}C sites in RSV genome are not regulated by NSUN2

m^{5}C site position	NSUN2$^{+/+}$ A549 cells T to C mutation % (rep1)	NSUN2$^{+/+}$ A549 cells T to C mutation % (rep2)	NSUN2$^{-/-}$ A549 cells T to C mutation % (rep1)	NSUN2$^{-/-}$ A549 cells T to C mutation % (rep2)	Regulated by NSUN2
m^{5}C 958	86.7	97.14	53.85	93.33	No
m^{5}C 962	83.8	95.74	60	94.12	No
m^{5}C 1019	60	86.84	83.33	77.78	No
m^{5}C 3050	68.2	56	66.67	75	No
m^{5}C 3085	71.9	82.61	69.23	74.07	No
m^{5}C 3088	71.9	78	81.48	72.41	No
m^{5}C 3133	63.6	62.96	58.33	44.44	No
m^{5}C 4032	64	48.39	70	57.89	No
m^{5}C 4034	70.4	61.76	72.73	63.64	No
m^{5}C 4036	67.7	58.33	76.92	68.18	No
m^{5}C 4098	46.2	67.21	59.09	60	No
m^{5}C 4100	43.6	67.86	57.14	57.14	No
m^{5}C 4101	44.4	61.82	57.14	55	No
m^{5}C 4965	40.5	44.68	52.63	57.14	No
m^{5}C 5000	48.1	51.61	66.67	92.86	No
m^{5}C 5459	68	80	63.64	78.57	No
m^{5}C 5463	64	75	61.54	64.29	No
m^{5}C 5466	46.2	61.54	50	72.73	No
m^{5}C 8183	49	38.64	33.33	47.06	No
m^{5}C 8339	58.3	63.89	68.18	55.17	No

4.3 Time for Harvesting RNA in Virus-Infected Cells

Total RNA should be harvested when maximum cytopathic effects (CPEs) are observed in virus-infected cells. Abundant viral RNA will be present in the RNA preparation.

4.4 Method for RNA Purification

Because the size of the RSV genome and antigenome is considerably large (15.2 kb), the TRIzol extraction method is recommended for RNA extraction to recover most of the full-length genome and antigenome. Host rRNA should be depleted from total RNA prior to bisulfite sequencing.

4.5 Cell Contamination

Mycoplasma contamination can affect cellular gene expression and may further disturb m^{5}C profiles detected from bisulfite sequencing. Check cells for mycoplasma contamination before any downstream experiments. Use a fresh frozen cell stock or treat with an appropriate antibiotic to eliminate mycoplasma before conducting the virus infection.

4.6 RNase Contamination

All procedures should avoid contaminating RNase. RNase inhibitors should be added during RNA extraction and RT steps, and RNA degradation should be avoided.

4.7 Optimization of Bisulfite Sequencing

One major issue in conventional RNA m^{5}C bisulfite sequencing is the high false-positive rates caused by incomplete C-to-U conversion. Researchers often used relatively low temperatures (54 °C or 75 °C) and less incubation time to avoid severe RNA degradation. These conditions are insufficient to denature local secondary structures of highly structured RNAs, leading to high background and false positives (overestimating m^{5}C level). Our protocol used an optimized condition to denature RNA samples at 95 °C for 3 min, accelerating the reaction and achieving complete bisulfite conversion.

Acknowledgments

The method developed in this chapter was supported in part by grants from the NIH (1R01AI185617-01A1, R01AI090060 and P01AI175399). The content is solely the responsibility of the authors and does not necessarily represent the official views of the National Institutes of Health. Y.Z. was supported by the C. Glenn Barber Fund Trust. We thank Dr. Qing Dai for the critical reading of the chapter.

References

1. Zhao BS, Roundtree IA, He C (2017) Post-transcriptional gene regulation by mRNA modifications. Nat Rev Mol Cell Biol 18(1): 31–42. https://doi.org/10.1038/nrm.2016.132
2. Fu Y, Dominissini D, Rechavi G, He C (2014) Gene expression regulation mediated through reversible m(6)A RNA methylation. Nat Rev Genet 15(5):293–306. https://doi.org/10.1038/nrg3724
3. Lu MJ, Zhang ZJ, Xue MG, Zhao BS, Harder O, Li AZ et al (2020) N^6-methyladenosine modification enables viral RNA to escape recognition by RNA sensor RIG-I. Nat Microbiol 5:584. https://doi.org/10.1038/s41564-019-0653-9
4. Roundtree IA, Evans ME, Pan T, He C (2017) Dynamic RNA modifications in gene expression regulation. Cell 169(7):1187–1200. https://doi.org/10.1016/j.cell.2017.05.045
5. Hotchkiss RD (1948) The quantitative separation of purines, pyrimidines, and nucleosides by paper chromatography. J Biol Chem 175(1):315–332
6. Wyatt GR (1950) Occurrence of 5-methylcytosine in nucleic acids. Nature 166(4214):237–238. https://doi.org/10.1038/166237b0
7. Amos H, Korn M (1958) 5-methyl cytosine in the RNA of Escherichia coli. Biochim Biophys Acta 29(2):444–445. https://doi.org/10.1016/0006-3002(58)90214-2

8. Desrosiers R, Friderici K, Rottman F (1974) Identification of methylated nucleosides in messenger RNA from Novikoff hepatoma cells. Proc Natl Acad Sci USA 71(10): 3971–3975. https://doi.org/10.1073/pnas.71.10.3971
9. Dubin DT, Taylor RH (1975) The methylation state of poly A-containing messenger RNA from cultured hamster cells. Nucleic Acids Res 2(10):1653–1668. https://doi.org/10.1093/nar/2.10.1653
10. Bohnsack KE, Hobartner C, Bohnsack MT (2019) Eukaryotic 5-methylcytosine (m(5)C) RNA Methyltransferases: mechanisms, cellular functions, and links to disease. Genes (Basel) 10(2). https://doi.org/10.3390/genes10020102
11. Trixl L, Lusser A (2019) The dynamic RNA modification 5-methylcytosine and its emerging role as an epitranscriptomic mark. Wiley Interdiscip Rev RNA 10(1):e1510. https://doi.org/10.1002/wrna.1510
12. Fu L, Guerrero CR, Zhong N, Amato NJ, Liu Y, Liu S et al (2014) Tet-mediated formation of 5-hydroxymethylcytosine in RNA. J Am Chem Soc 136(33):11582–11585. https://doi.org/10.1021/ja505305z
13. Shen Q, Zhang Q, Shi Y, Shi Q, Jiang Y, Gu Y et al (2018) Tet2 promotes pathogen infection-induced myelopoiesis through mRNA oxidation. Nature 554(7690): 123–127. https://doi.org/10.1038/nature25434
14. Yang X, Yang Y, Sun BF, Chen YS, Xu JW, Lai WY et al (2017) 5-methylcytosine promotes mRNA export—NSUN2 as the methyltransferase and ALYREF as an m(5)C reader. Cell Res 27(5):606–625. https://doi.org/10.1038/cr.2017.55
15. Zhang X, Zhang Y, Wang C, Wang X (2023) TET (ten-eleven translocation) family proteins: structure, biological functions and applications. Signal Transduct Target Ther 8(1):297. https://doi.org/10.1038/s41392-023-01537-x
16. Chen X, Li A, Sun BF, Yang Y, Han YN, Yuan X et al (2019) 5-methylcytosine promotes pathogenesis of bladder cancer through stabilizing mRNAs. Nat Cell Biol 21(8):978–990. https://doi.org/10.1038/s41556-019-0361-y
17. Dubin DT, Stollar V (1975) Methylation of Sindbis virus "26S" messenger RNA. Biochem Biophys Res Commun 66(4):1373–1379. https://doi.org/10.1016/0006-291x(75)90511-2
18. Sommer S, Salditt-Georgieff M, Bachenheimer S, Darnell JE, Furuichi Y, Morgan M et al (1976) The methylation of adenovirus-specific nuclear and cytoplasmic RNA. Nucleic Acids Res 3(3):749–765. https://doi.org/10.1093/nar/3.3.749
19. Guo G, Pan K, Fang S, Ye L, Tong X, Wang Z et al (2021) Advances in mRNA 5-methylcytosine modifications: detection, effectors, biological functions, and clinical relevance. Mol Ther Nucleic Acids 26:575–593. https://doi.org/10.1016/j.omtn.2021.08.020
20. Khoddami V, Yerra A, Mosbruger TL, Fleming AM, Burrows CJ, Cairns BR (2019) Transcriptome-wide profiling of multiple RNA modifications simultaneously at single-base resolution. Proc Natl Acad Sci USA 116(14):6784–6789. https://doi.org/10.1073/pnas.1817334116
21. Amort T, Lusser A (2017) Detection of 5-methylcytosine in specific poly(a) RNAs by bisulfite sequencing. Methods Mol Biol 1562: 107–121. https://doi.org/10.1007/978-1-4939-6807-7_8
22. Liu J, Jia G (2014) Methylation modifications in eukaryotic messenger RNA. J Genet Genomics 41(1):21–33. https://doi.org/10.1016/j.jgg.2013.10.002
23. Hussain S, Sajini AA, Blanco S, Dietmann S, Lombard P, Sugimoto Y et al (2013) NSun2-mediated cytosine-5 methylation of vault noncoding RNA determines its processing into regulatory small RNAs. Cell Rep 4(2): 255–261. https://doi.org/10.1016/j.celrep.2013.06.029
24. Khoddami V, Cairns BR (2013) Identification of direct targets and modified bases of RNA cytosine methyltransferases. Nat Biotechnol 31(5):458–464. https://doi.org/10.1038/nbt.2566
25. Squires JE, Patel HR, Nousch M, Sibbritt T, Humphreys DT, Parker BJ et al (2012) Widespread occurrence of 5-methylcytosine in human coding and non-coding RNA. Nucleic Acids Res 40(11):5023–5033. https://doi.org/10.1093/nar/gks144
26. Courtney DG, Tsai K, Bogerd HP, Kennedy EM, Law BA, Emery A et al (2019) Epitranscriptomic addition of m(5)C to HIV-1 transcripts regulates viral gene expression. Cell Host Microbe 26(2):217–27 e6. https://doi.org/10.1016/j.chom.2019.07.005
27. Courtney DG, Chalem A, Bogerd HP, Law BA, Kennedy EM, Holley CL et al (2019) Extensive epitranscriptomic methylation of A

and C residues on murine leukemia virus transcripts enhances viral gene expression. mBio 10(3). https://doi.org/10.1128/mBio.01209-19

28. Su PA, Chang CH, Yen SB, Wu HY, Tung WJ, Hu YP et al (2024) Epitranscriptomic cytidine methylation of the hepatitis B viral RNA is essential for viral reverse transcription and particle production. Proc Natl Acad Sci USA 121(24):e2400378121. https://doi.org/10.1073/pnas.2400378121
29. Ding S, Liu H, Liu L, Ma L, Chen Z, Zhu M et al (2024) Epigenetic addition of m(5)C to HBV transcripts promotes viral replication and evasion of innate antiviral responses. Cell Death Dis 15(1):39. https://doi.org/10.1038/s41419-023-06412-9
30. Eckwahl M, Xu R, Michalkiewicz J, Zhang W, Patel P, Cai Z et al (2020) 5-Methylcytosine RNA modifications promote retrovirus replication in an ALYREF reader protein-dependent manner. J Virol 94(13). https://doi.org/10.1128/JVI.00544-20
31. Viehweger A, Krautwurst S, Lamkiewicz K, Madhugiri R, Ziebuhr J, Holzer M et al (2019) Direct RNA nanopore sequencing of full-length coronavirus genomes provides novel insights into structural variants and enables modification analysis. Genome Res 29(9):1545–1554. https://doi.org/10.1101/gr.247064.118
32. Kim D, Lee JY, Yang JS, Kim JW, Kim VN, Chang H (2020) The architecture of SARS-CoV-2 transcriptome. Cell 181(4):914–21 e10. https://doi.org/10.1016/j.cell.2020.04.011
33. Henry BA, Kanarek JP, Kotter A, Helm M, Lee N (2020) 5-methylcytosine modification of an Epstein-Barr virus noncoding RNA decreases its stability. RNA 26(8):1038–1048. https://doi.org/10.1261/rna.075275.120
34. Liu LS, Chen Z, Zhang K, Hao HJ, Ma L, Liu HZ et al (2024) NSUN2 mediates distinct pathways to regulate enterovirus 71 replication. Virol Sin 39(4):574–586. https://doi.org/10.1016/j.virs.2024.05.002
35. Li ZL, Xie Y, Xie YK, Chen HL, Zhou X, Liu M et al (2024) HCV 5-methylcytosine enhances viral RNA replication through interaction with m5C reader YBX1. ACS Chem Biol 19(7):1648–1660. https://doi.org/10.1021/acschembio.4c00322
36. Wang H, Feng J, Fu Z, Xu T, Liu J, Yang S et al (2024) Epitranscriptomic m(5)C methylation of SARS-CoV-2 RNA regulates viral replication and the virulence of progeny viruses in the new infection. Sci Adv 10(32):eadn9519. https://doi.org/10.1126/sciadv.adn9519
37. Huang A, Riepler L, Rieder D, Kimpel J, Lusser A (2023) No evidence for epitranscriptomic m(5)C modification of SARS-CoV-2. HIV and MLV viral RNA RNA 29(6):756–763. https://doi.org/10.1261/rna.079549.122
38. Dai Q, Ye C, Irkliyenko I, Wang Y, Sun HL, Gao Y et al (2024) Ultrafast bisulfite sequencing detection of 5-methylcytosine in DNA and RNA. Nat Biotechnol 42(10):1559–1570. https://doi.org/10.1038/s41587-023-02034-w
39. Lin X, Ke XL, Jian XQ, Xia L, Yang Y, Zhang TY et al (2022) Azacytidine targeting SARS-CoV-2 viral RNA as a potential treatment for COVID-19. Sci Bull 67(10):1022–1025. https://doi.org/10.1016/j.scib.2022.02.002
40. Cowton VM, McGivern DR, Fearns R (2006) Unravelling the complexities of respiratory syncytial virus RNA synthesis. J Gen Virol 87:1805–1821. https://doi.org/10.1099/vir.0.81786-0
41. Whelan SPJ, Barr JN, Wertz GW (2004) Transcription and replication of nonsegmented negative-strand RNA viruses. Curr Top Microbiol 283:61–119
42. Xue M, Zhao BS, Zhang Z, Lu M, Harder O, Chen P et al (2019) Viral N(6)-methyladenosine upregulates replication and pathogenesis of human respiratory syncytial virus. Nat Commun 10(1):4595. https://doi.org/10.1038/s41467-019-12504-y
43. Lu M, Xue M, Wang HT, Kairis EL, Ahmad S, Wei J et al (2021) Nonsegmented negative-sense RNA viruses utilize N(6)-methyladenosine (m(6)A) as a common strategy to evade host innate immunity. J Virol 95(9). https://doi.org/10.1128/JVI.01939-20
44. Xue M, Zhang Y, Wang H, Kairis EL, Lu M, Ahmad S et al (2021) Viral RNA N6-methyladenosine modification modulates both innate and adaptive immune responses of human respiratory syncytial virus. PLoS Pathog 17(12):e1010142. https://doi.org/10.1371/journal.ppat.1010142
45. Zhang Y, Zhang LS, Dai Q, Chen P, Lu M, Kairis EL et al (2022) 5-methylcytosine (m(5)C) RNA modification controls the innate immune response to virus infection by regulating type I interferons. Proc Natl Acad Sci USA 119(42):e2123338119. https://doi.org/10.1073/pnas.2123338119

Chapter 16

Well-Differentiated Primary Human Airway Epithelial Cultures for Respiratory Syncytial Virus (RSV) Infection Studies

Phylip Chen, Mahesh K. C., and Mark E. Peeples

Abstract

Human viruses have traditionally been studied in immortalized, usually tumor-derived, cell lines because of their availability, low cost, ease of expansion, and ease of care. Although most viruses infect, replicate, and spread in these cells, all aspects of in vivo virus-host interaction are not accurately reflected in immortalized cells. The isolation, storage, and differentiation of human airway epithelial basal cells have enabled ex vivo studies of RSV infection in a near natural setting. Here we provide our rationale for using well-differentiated primary human bronchial and nasal epithelial cultures, and a current protocol for generating these fully differentiated airway cultures from donor tissue for ex vivo studies of RSV and other respiratory virus infections.

Key words Well-differentiated primary human bronchial epithelial (HBE) or nasal epithelial (HNE) cultures, Basal cells, Columnar cells, Ciliated cells, Goblet cells, Respiratory syncytial virus (RSV)

1 Introduction

In vivo, respiratory syncytial virus (RSV) infects the epithelial cells lining the airways: the nasal, tracheal, bronchial, and bronchiolar (small) airway epithelium. The ability to produce ex vivo well-differentiated primary human bronchial epithelial (HBE) cultures, or similar nasal (HNE) cultures, from fresh human tissue has enabled the identification of multiple critical differences between RSV infection of cell lines and infection of HBE cultures. Nearly all immortalized cell lines can be infected with RSV, but only the ciliated cells in HBE cultures are infected by RSV, not the neighboring goblet cells [1]. The RSV attachment glycoprotein (G) in virions is not needed for infection of immortal cell lines, although it increases infectivity two-fold [2]. In contrast, the G protein in virions is essential for infection of HBE cultures [2]. The entry receptor for RSV in immortalized cell lines are the heparan sulfate

Bo Liang and Rachel Fearns (eds.), *Human Respiratory Syncytial Virus: Methods and Protocols*, Methods in Molecular Biology, vol. 2948, https://doi.org/10.1007/978-1-0716-4666-3_16,

proteoglycans [3, 4] while the RSV receptor in HBE cultures is a unique protein, CX3CR1 [5] or possibly CCR3 [6], which have both been found to bind the G protein. And finally, the glycan modifications on the G protein are much more extensive in virions produced from HBE cultures, and these glycans interfere with infection of immortalized cells during the initial round of infection but not in subsequent rounds of infection [7]. The loss of this interference after a single passage in HEp-2 cells indicates that the inhibition was due to specific modification of the G protein that inhibited its attachment to the cell line, most likely the additional glycans, rather than the G protein itself. For these reasons, and perhaps others that have yet to be discovered, using HBE or HNE cultures is essential for confirming all aspects of the RSV lifecycle, including the epithelial cell response to infection.

Methods for isolating and differentiating respiratory epithelial basal cells in vitro were pioneered and codified by Scott Randell at the University of North Carolina [8, 9], with the support of the Cystic Fibrosis Foundation. Airway basal cells are isolated from nasal brushings and lung airway tissue from organ donors, most often smokers making the lungs ineligible for transplantation, or less frequently from surgical specimens. With currently available protocols, culture medium, and technology, primary airway cultures can be established at the air-liquid interface, resting on a filter with perforations that allow the basal surface of the cells to be exposed to the cell culture medium in the chamber below while the apical surface is exposed to the air, as they are in vivo. Although producing these cultures is time consuming and expensive, studying RSV infection of these ex vivo cultures has been valuable and will likely continue to be.

Well-differentiated primary human nasal epithelial (HNE) and human bronchial epithelial (HBE) cultures can be generated from isolated basal cells, by the protocol described below. This maturation process takes 3 to 4 weeks. The resulting cultures appear in cross section to be structurally similar to the bronchial epithelium (Fig. 1). Small basal (progenitor) cells line the top of the supporting filter. The basal cells differentiate into tall columnar cells that have a foot on the filter and extend to the apical surface of the culture. Many of these cells are ciliated. Other, goblet cells, produce mucus and secrete it apically, providing a protective coating on the surface of the culture. Minor cell types that are also represented in these cultures include club cells, tuft cells, and ionocytes.

This protocol aims to provide a clear and relatively easy entry point for those interested in producing and using primary airway cultures. We have tried to enhance and simplify earlier methods to expedite success in processing the airway tissue, isolating the basal cells, and storing, recovering, and differentiating them into primary well-differentiated HBE cultures ready for virus inoculation and study. We begin with a short description of the flow of the process:

the source material, the culture media, expansion and storage of the basal cells, and their differentiation at the air-liquid interface (ALI), followed by a detailed description of each step and the composition of solutions and the conditions needed to support the production of well-differentiated primary HBE and HNE cultures.

2 Materials

2.1 Source Material

Normal human donor lungs that are not acceptable for transplantation may be procured from a regional tissue bank. In our region, Lifeline of Ohio works with the Ohio State University Wexner Medical Center Comprehensive Transplant Center Biorepository to collect tissues that are not used for transplants and provides them to academic laboratories for research purposes. To obtain these tissues, investigators or a core facility must have the approval of their Institutional Review Board and a Materials Transfer Agreement with the Biorepository.

To develop HBE cultures, we obtain the dissected trachea and bronchi to their second branch point from the Biorepository. As a practical matter, the gross dissections are usually performed during the night, when the surgery suites are available, and the tissue is placed in ice-cold physiologic solution, such as Ringers, for delivery the following morning. If longer storage is required, the Perfadex solution can be used in place of Ringers to stabilize the tissue.

Nasal epithelial cells can be collected from nasal brushings of organ donors, volunteers, or nasal tissues removed during sinus surgeries. Nasal epithelial cells can be extracted from any of these sources and basal cells isolated by methods similar to bronchial basal cells and differentiated similarly to HBE cultures.

2.2 Medium and Isolation Solutions

As an overview, two types of media described in the seminal Fulcher and Randell paper are used, BEGM (Bronchial Epithelial Growth Medium) and ALI (Air-Liquid Interface medium) [8], and we have adopted them as an economical method of culturing the cells. Both have an extensive list of components, but are cost effective, relative to commercial products, and the list of components for both media can be found in that report. After the collection of the progenitor cells from the digested airway tissues, BEGM is used to expand the airway progenitor cells from the initial harvesting on collagen-coated tissue culture plates, and thereafter, frozen for future use. To generate fully differentiated pseudostratified cells from the frozen vials, as described here, we use the ALI medium, supplemented with ROCK inhibitor (Y-27632) [10] for seeding the cultures on the porous membrane transwell supports to enhance viability and to foster cell proliferation and prevent differentiation until they reach full confluency. Thereafter, to initiate differentiation and tight junction formation, the commercial medium from Stem Cell

Technologies, PneumaCult-ALI, assembled from their differentiation kit, is used.

As an alternative medium during the proliferative or expansion phases of culture growth to BEGM or even ALI supplemented with ROCK inhibitor, we have found that PneumaCult Ex Plus from Stem Cell Technologies [11] can be effective for expanding the primary cultures.

2.2.1 Reagents

Catalog numbers of the reagents to make the BEGM and ALI media and other solutions can be found in the Randell paper; however, the step-by-step methods to make the additives, other solutions, and media along with their expiration dates after making the solutions are listed below for convenience.

2.2.2 Reagent Setup

Additives for BEGM and ALI

1. *Bovine Serum Albumin (250X):* Make a 125 mg/mL solution by adding PBS directly to the BSA container. Allow powder to fully dissolve by gently rocking for >3 h or overnight at 4 °C. After fully dissolved, bring the concentration to 125 mg/mL. Aliquot 4 mL volumes and store at −20 °C for up to 2.5 years.
2. *Bovine Pituitary Extract:* The bovine pituitary extract is a solution that can be purchased, and the concentration can be found on the certificate of analysis. Typically, 1 L of medium is made. Add BPE to yield the final concentration of 10 μg/mL. This is a critical reagent. Follow the manufacturer's expiration date.
3. *Insulin (1000X):* Make a 5 μg/mL solution by adding 500 uL of glacial acetic acid to 9.5 mL of dH_2O then add 50 mg of insulin and mix to dissolve. Aliquot 1 mL volumes into vials and store at −20 °C for up to 12 months.
4. *Transferrin (1000X):* Make a 10 mg/mL solution by reconstituting transferrin in sterile PBS. Aliquot 1 mL into vials and store at −20 °C for up to 3 months.
5. *Hydrocortisone (1000X):* Reconstitute powder to 0.72 mg/mL in sterile water. Aliquot 1 mL volumes into vials and store at − 20 °C for up to 12 months. (Make 12–15 mL for 12–15 vials).
6. *Triiodothyronine (1000X):* Make a 0.0067 mg/mL solution by dissolving powder in 0.001 M NaOH. Aliquot 1 mL volumes into vials and store at −20 °C for up to 3 months.
7. *Epinephrine (1000X):* Make a 0.6 mg/ml solution by dissolving powder in 0.01 N HCl. Aliquot 1 mL vials and store at − 20 °C for up to 24 months.
8. *Epidermal Growth Factor (1000X for BEGM; 50,000X for ALI):* Dissolve 100 μg of powder into 4 mL of PBS. Aliquot 1 mL volumes into vials for BEGM and 20 μL volumes into vials for ALI and store at −20 °C for up to 6 months.

9. *Retinoic Acid (1000X):* Prepare 1×10^{-3} M concentrated stock by dissolving 12.0 mg in 40 mL of 100% ethanol. (RA is light sensitive; store at −20 °C wrapped in aluminum foil.) To prepare 1000X stock, sample the 1×10^{-3} M concentrated stock by first diluting in absolute ethanol. Assess the absorbance of a 1:100 dilution of the sample at 350 nm using a 1 cm light path quartz cuvette on a spectrophotometer against an ethanol blank. Since the molar extinction coefficient of RA in ethanol is 45,000 at 350 nm, the absorbance should equal 0.45. (Discard if values are <0.18.) If the absorbance value equals 0.45, add 3 mL of 1×10^{-3}/mL stock to 53 mL of PBS and add 4.0 mL of BSA 150 mg/mL stock (for absorbance values between 0.18 and 0.45, calculate as 1.35/absorbance and adjust the PBS volume appropriately). Aliquot 1 mL volumes of light-sensitive RA into vials and store at −20 °C for up to 12 months.
10. *Phosphoethanolamine (1000X):* Dissolve powder in PBS to make a 70 mg/ml solution. Aliquot 1 mL volumes into vials and store at −20 °C for up to 12 months.
11. *Ethanolamine (1000X):* Dissolve powder in PBS to make a 30 μg/mL solution. Aliquot 1 mL volumes into vials and store at −20 °C for up to 12 months.
12. *Stock 11 (1000X):* Dissolve zinc sulfate powder in dH_2O to make a 0.863 mg/mL solution. Aliquot 1 mL into vials and store at room temperature. No expiration date.
13. *Penicillin-Streptomycin (1000X):* Dissolve penicillin-G sodium (100,000 U/mL) and streptomycin sulfate (100 mg/mL) in sterile dH_2O. Aliquot 1 mL into vials and store at −20 °C for up to 24 months. Make 25 ml for 25 vials.
14. *Gentamicin (1000X):* Only used for BEGM. May be stored at room temperature; expires as indicated.
15. *Amphotericin B (1000X):* Only used for BEGM. Store at 4 °C; expires as indicated.
16. *Stock 4 (1000X):* Dissolve 0.42 g ferrous sulfate, 122.0 magnesium chloride, 16.17 g calcium chloride-dihydrate, and combine with 5.0 mL of 37% hydrochloric acid in water to 800 mL of dH_2O in a volumetric flask. Bring volume to 1 L. Aliquot 1 mL volumes into vials and store at room temperature. No expiration date.
17. *Trace Elements (1000X):* First, prepare separate 100 ml stock solutions of selenium (520 mg), manganese (20 mg), silicone (14.2 g), molybdenum (124 mg), vanadium (59 mg), nickel (26 mg), and tin (11 mg) [8]. Next, fill a volumetric 1-L flask with dH_2O. Remove 8 mL of dH_2O from the flask and then replace the volume with 1.0 mL of each stock solution and

1.0 mL of concentrated hydrochloric acid. Aliquot into vials and store at room temperature. No expiration date.

18. *25X Antibiotic Stock* (Ceftazidime, Tobramycin, Vancomycin, and Amphotericin B): Reconstitute 1 gram of Ceftazidime with 5 mL dH_2O. Reconstitute 1-gram equivalent Vancomycin with 20 mL of dH_2O. Make 300 mL by combining 229 mL of Minimal Essential Medium Eagle with 3.75 mL of reconstituted Ceftazidime, 15 mL of reconstituted Vancomycin, 15 mL of Tobramycin (40 mg/mL), 37.5 mL of Amphotericin B (250ug/mL). Aliquot 40 mL into 50 mL conical tubes and store at −20 °C. Stable for >6 months. (Final concentrations: Ceftazidime, 100 μg/mL; Vancomycin, 100 μg/mL; Tobramycin, 80 μg/mL; and Amphotericin B, 1.25 μg/mL.)

19. *ROCK Inhibitor Y-27632 (10 mM; 1000X):* Add 296 μL of dH_2O to 1 mg of Y-27632. Aliquot 50 μL per vial and store at −20 °C for up to 6 months. Used only for proliferative medium [9].

2.2.3 Preparation of Solutions and Medium

1. *Wash Medium:* Combine 500 mL of MEM with 5 mL of Glutamax, 0.40 mL Gentamicin, 0.5 mL Pen/Strep (1000X), and 40 ml of 25X antibiotics (see reagent setup). Store at 4 °C for up to 1 month.

2. *Soak Solution:* Combine 125 mL of Wash medium with 63 mg DTT and 1.3 mg DNase. Store at 4 °C for up to 1 month.

3. *2X Freezing Medium:* To make 250 mL, combine 5 ml of 1.5 M HEPES, 25 mL of FBS, 195 mL of F-12, and place the beaker on ice. When chilled, add 25 mL of DMSO gradually. Filter, sterilize, and store at 4C for up to 12 months.

4. *10X Protease XIV and DNase Stock Solution:* Combine 1 g Protease XIV powder and 10 mg of DNase powder with 90 mL 1X PBS. Stir with a magnetic stir bar without heat (>1 h). Bring volume up to 100 mL and stir for an additional 5 min. Filter with 0.2 μm filter in a biosafety cabinet and aliquot into 15 mL conical tubes.

5. *Declumping Solution:* Make 120 mL solution by combining 30 mg of collagenase and 1.2 mg DNase with 120 mL of PBS. Add 0.09 g of calcium chloride and 0.12 g of magnesium chloride and stir. Aliquot 12 mL volumes into 15 mL conical tubes. Store at −20 °C for up to 1 year. *For fresh preparations* prior to use, thaw 12 mL aliquot and combine with 6 mg of DTT and 0.120 mL of 0.2 M EDTA and sterilize using 0.2 um syringe filter. (Final concentrations: Collagenase, 0.25 mg/mL; DNase, 10 μg/mL; Calcium chloride, 0.75 mg/mL; Magnesium chloride, 1 mg/mL; Dithiothreitol, 0.5 mg/mL; EDTA, 2 mM.)

6. *0.1% Trypsin/EDTA:* Dilute concentrated 0.5% trypsin/EDTA. Store at 4 °C for up to 1 month.
7. *Soybean Trypsin Inhibitor (1 mg/mL):* Dissolve soybean trypsin inhibitor in Ham's F12. Sterilize with a 0.2um filtration unit and store at 4 °C for up to 1 year.
8. *PureCol Solution:* Dilute PureCol® Type I Collagen Solution 1: 60 in water to coat 100 mm tissue culture plates with 3 mL diluted solution per plate. Incubate plates overnight at 37 °C with solution.
9. *Collagen Type IV Solution (10X):* Dissolve by combining 10 mg in 20 mL of dH_2O with 40 μL of concentrated acetic acid and incubate at 37 °C for 30 min. Filter sterilize (0.2 μm) and store 500 μL aliquots at −20 °C for up to 6 months.
10. *BEGM:* Combine all of the previously prepared components mentioned (see reagent setup) in a 1 L LHC medium bottle also listed here: 4 mL of endotoxin-free BSA; bovine pituitary extract (volume based on the certificate of analysis to yield the final concentration of 10 μg/mL); 1 mL of all of the following: insulin, transferrin (be aware that this solution expires after 3 months); hydrocortisone; triiodothyronine (be aware that this solution expires after 3 months); epinephrine; EGF; retinoic acid; phosphoethanolamine; ethanolamine; pen/strep; amphotericin B; Stock 11; Stock 4; trace elements; Gentamicin. Use a 1-L bottle-top filter and add 25X antibiotics and Fluconazole. Store at 4 °C for up to 3 months covered in aluminum foil.
11. *ALI:* Combine all of the previously prepared components (see reagent setup) mentioned with 500 mL LHC medium and 500 mL DMEM medium also listed here: 4 mL of endotoxin-free BSA; bovine pituitary extract (volume based on the certificate of analysis to yield the final concentration of 10 μg/mL); 20 μL of EGF; 1 mL of all of the following: insulin, transferrin (this solution expires after 3 months); hydrocortisone; triiodothyronine (this solution also expires after 3 months); epinephrine; retinoic acid; phosphoethanolamine; ethanolamine; pen/strep; Stock 11; Stock 4; trace elements. Filter using a 1-L bottle-top filter and store at 4 °C for up to 3 months covered in aluminum foil.
12. *ROCK Inhibitor/ALI Seeding Medium:* Combine 50 mL of ALI medium with 50 μL of 10 mM (1000X) Y-27632. Store at 4 °C for up to 3 months.
13. *PneumaCult-Ex Plus:* Follow manufacturer's instructions. Thaw PneumaCult-Ex Plus 50X Supplement at room temperature (15–25 °C). Mix gently by inverting the vial; do not vortex. Aliquot 1 mL aliquots and store at −20 °C. To make 50 mL, add 1 ml of 50X supplement to 49 mL of basal

medium, also add 250 μL of Hydrocortisone (96 μg/mL). Store at 4 °C up to 4 weeks after assembling.

14. *PneumaCult-ALI:* Follow manufacturer's instructions. Thaw PneumaCult™-ALI 10X Supplement overnight at 2–8 °C. Mix gently by inverting the vial; do not vortex. Thaw PneumaCult™-ALI Maintenance Supplement at room temperature (15–25 °C). To make 50 mL, add 5 mL of 10X supplement to 45 mL of basal medium, also add 250 μL of Hydrocortisone (96 μg/mL) and 100 μL of 0.2% heparin solution. Store at 4 °C up to 2 weeks after assembling.

3 Methods

This section provides a step-by-step guide from the initial phases of basal progenitor cell extraction, culturing, and expansion of the airway cells for long-term storage. The collection of basal cells from the airway tissue from one human lung may provide 100 or more vials for future experimentation. One vial (2 X 10^6 basal cells in 1 ml freezing medium) typically seeds 48 transwells, with "passage 1" cultures that will fully differentiate into ciliated pseudostratified airway cultures, phenotypically similar to sections of the bronchial airway tissue. Notes on preparing cultures for viral infection are also provided.

3.1 Preparing the Airways for Epithelial Cell Extraction

Perform all the processing of the lung tissue, isolated cells, and culture plates in a biosafety cabinet wearing personal protective garb, including gloves and a lab coat with tight sleeve cuffs for the prevention of potential exposure to bloodborne pathogens. Instruments used: sterile rat-toothed forceps, heavy scissors, and curve-tipped scissors for the dissection.

1. Remove the extraneous parenchymal and connective tissue from the exterior of the airway (*see* **Note 1**).
2. Splay open the airways and section them into manageable pieces, which will enable easy handling and later scraping of the interior airway surface.
3. Place airway sections in a 50 mL conical tube on ice and cover them with sterile chilled physiologic solution.
4. Aspirate the solution. Fill the conical tube with the *"Wash medium"* (see reagent setup). Agitate the tissue sections with this solution in the closed tube(s). Aspirate, repeat 3X. (*see* **Note 2**).
5. Combine the tissue segments into 50 mL conical tubes with 30 mL of wash medium with 4 mL of *enzymatic Protease/ DNase solution* (*see* reagent setup and **Note 3**).

6. Place tubes on a rotating platform in a cold room at 4 °C, rocking at about 50–60 cycles/min and incubate for 48 h.

3.2 Coating Tissue Culture Plates for Cell Expansion

1. Under sterile conditions, prepare a sufficient volume of diluted *PureCol* in water (3 mg/mL collagen type I, bovine; 1:60) for 3 mL per 100 mm tissue culture plate (*see* **Note 4**).
2. Add 3 mL to each plate. Incubate overnight at room temperature or minimally for 2 h at room temperature.
3. Aspirate coating solution, and rinse with PBS and avoid scratching surfaces, and allow to dry in the biosafety cabinet (*see* **Note 5**).

3.3 Harvesting Airway Basal Cells

1. Following 48 h of tissue digestion, pour the contents of the conical tubes (lung tissues with solutions) into a 150 mm tissue culture plate and neutralize the enzymes by adding FBS to a final concentration of 10% (v/v).
2. Scrape the inner surfaces of the immersed airway tissue using a #10 surgical blade and forceps to loosen cells from the airway tissue.
3. Pool all solutions in 50 mL conical tubes (*see* **Note 6**).
4. Centrifuge at 500 g for 5 min at 4 °C.
5. Resuspend this P_0 cell pellet in F-12 medium (*see* **Note** 7).

3.4 Plating Airway Cells (P_0) for Selection and Expansion of Basal Cells

1. Count the cells in a cell counting machine or a hemocytometer.
2. Depending on the size of the trachea and bronchi, starting material, and the efficiency of extraction, the number of basal cells can be as high as 100 million or more.
3. Seed three million cells per collagen-coated 100 mm tissue culture plate with a proliferation medium supplemented with antibiotics.
4. The following day, replace the medium, and thereafter, change the medium every 2–3 days. The basal cells will attach and proliferate. Differentiated cells will not.

3.5 Harvesting Expanded Airway Basal Cells for Long-Term Storage

1. After basal cell cultures reach 90–95% confluency or 1-week post-seeding, whichever comes first, rinse each plate with 2 mL of room-temperature PBS and aspirate 2X (*see* **Note 8**).
2. Add 2 mL of fresh *0.1% Trypsin EDTA*, and place in a 37 °C incubator (*see* **Note 9**).
3. Briskly tap the plate to dislodge the cells from the tissue culture plate.
4. After most cells have detached from the surface, pipette up and down with a 5 mL pipette, without forming bubbles, to disperse any obvious cell clumps.

5. Neutralize the trypsin by adding 2 mL of chilled *soybean trypsin inhibitor* to each plate and pipette up and down 2–3 times (*see* **Note 10**).
6. Pipette the cells into 50 mL conical tubes and place on ice.
7. Centrifuge at 500 × *g* for 5 min at 4 °C (*see* **Note 11**).
8. Aspirate the supernatant from the pellet, resuspend each pellet in 5 mL of *Ham's F12 medium* (no additives), and pool the cells in one or more 50 mL conical tubes. Place the tubes of cells on ice (*see* **Note 12**).
9. Count the number of viable cells on a hemocytometer using a 0.4% Trypan Blue solution (*see* **Note 13**).
10. Adjust the live cell count to ~4×10^6 cells/ml with F12 medium.
11. Add an equal volume of 2X concentrated freezing medium.
12. Distribute 1 mL to each sterile freezer vial: ~2×10^6 cells per vial.
13. Place vials in a refrigerated "Mr. Frosty" (ThermoScientific) or a Styrofoam container with holes to keep the tubes upright and insulated. Place the box of tubes at −80 °C to freeze slowly. After 4–24 h, transfer the frozen tubes of cells to boxes in a liquid nitrogen freezer for long-term storage (*see* **Note 14**).

3.6 Generating Fully Differentiated Pseudostratified Airway Cultures

3.6.1 Coat the Transwells with Collagen Prior to Seeding with Basal Cells

1. Thaw a 10X collagen Type IV vial (500 μL) and add 4.5 mL sterile (autoclaved) distilled H_2O (see reagent setup 2.2.3).
2. To coat, add 100 μL per 6.5 mm insert or 200 μL to 12 mm transwell insert (*see* **Note 15**).
3. Allow the solution on the plate to dry with the lid off in the laminar flow hood overnight.
4. UV treat the surfaces of the plates for 30 min in the hood with the air flowing.

3.6.2 Seeding Cultures and their Proliferation

1. Prepare cells from actively dividing primary cells in culture or from frozen vials. If cells are from liquid nitrogen, quickly thaw the vials at 37 °C, removing the vial from the water bath just before the cell suspension is completely thawed.
2. Spray the exterior of the vial with 70% ethanol, wipe dry, transfer the contents to a 15 mL conical tube, and gradually add F12 medium to the cell suspension. Pellet cells at 500 g for 5 min.
3. Resuspend cells gently in proliferation medium (*ALI medium* supplemented with *ROCK Inhibitor or PneumaCult Ex Plus*; see reagent setup).
4. Count and assess viability.

5. Seed 40–50,000 cells per 6.5 mm transwell in 100 μL of the proliferation medium (density of ~1.5×10^5 cells/cm^2). Place 500 μL of this medium in the lower chamber, also (*see* **Note 15**).
6. On the following day, remove the non-adherent cells by suction and replace with fresh medium. Change the medium 3X/week (every 2–3 days).

3.6.3 Inducing Differentiation of Airway Cultures and Inoculating with RSV

1. Assess cultures for confluency and tight junction formation with an ohmmeter (tight junctions form 3–7 days after seeding) (*see* **Note 16**).
2. Once tight junctions have formed and cultures are confluent, transition to Air-Liquid Interface (ALI) differentiation medium, by removing the medium from the apical and basolateral sides of the cultures and adding differentiation medium (*PneumaCult ALI*; see reagent setup) below the transwells only, exposing the apical cell surface to the air.
3. Continue to change medium in the bottom chamber 3X/wk. Approximately 3 weeks after the initiation of ALI differentiation, cells will be fully differentiated. Beating cilia may be visible with an inverted microscope.
4. In cross-section, the structure of the differentiated airway cultures (Fig. 1B) is very similar to that of the bronchus (Fig. 1A). Basal (progenitor) cells cover the Transwell perforated filter that supports these cultures. Long, narrow columnar cells extend from the filter and basal cells to the apical surface. Most of these cells are either ciliated or goblet cells, with several minor cell types (club, brush, and ionocytes) interspersed.
5. The ciliated cell apical surface is the target for RSV attachment and infection initiation (Zhang, Peeples, et al., 2002, PMC13 7037). Following replication, RSV assembles under the apical

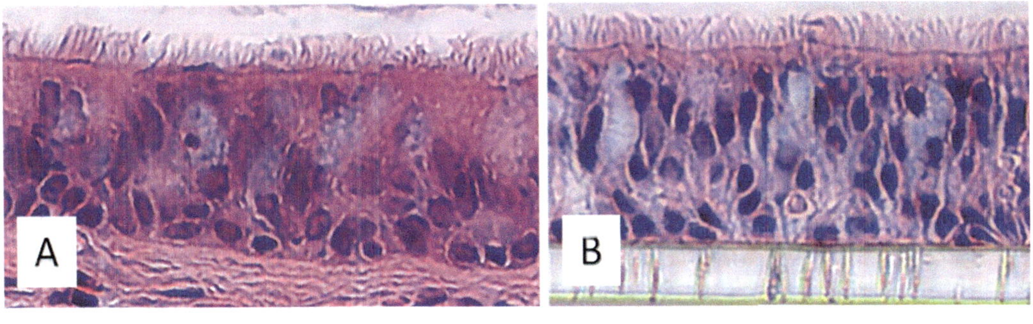

Fig. 1 Comparison of the structure of human bronchial epithelial tissue and HBE cultures. (**A**) Cross-section of the wall of a humanbronchus. (**B**) Cross-section of basal cells isolated from human bronchial tissue that were fully differentiated on a Transwell at the air-liquid interface over the course of 4 weeks

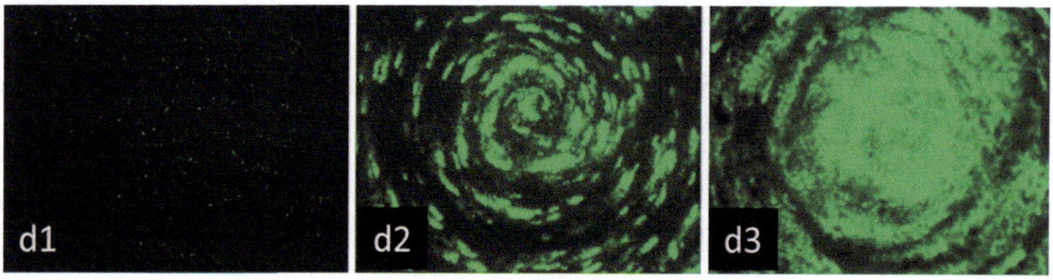

Fig. 2 Infection and spread of rgRSV in HBE cultures over 3 days. The culture was washed 3 times with DMEM and inoculated with 4 X 10^3 pfu recombinant green fluorescent protein expressing RSV (rgRSV) for 2 h and incubated at 37 °C. Cultures were photographed with an EVOS FL Auto 2 microscope using a 10X lens at daily intervals

membrane and buds through that membrane, releasing virions into the apical space which would be the lumen of the airways. The cilia, which beat rhythmically, spread RSV, often in a spiral pattern in these wells (Fig. 2). The RSV used here has been modified to express GFP, in addition to its own proteins in the cells it infects. In vivo, the beating cilia drive the mucus, with its trapped pathogens and debris, up and out of the airways.

3.6.4 Maintaining Differentiated Airway Cultures

1. Cultures may be maintained in the differentiated state for up to a month (*see* **Note 17**). During that time, it is best to remove mucus from the apical surface of the cultures when accumulation is apparent by adding 150 μL of pre-warmed DMEM to each 6.5 mm transwell insert culture and placing it in a 37 °C CO_2 incubator for 5 min.
2. Pipette up and down and completely and carefully remove all wash medium with a P200 pipetman without disturbing the cultures. Three such washes usually remove most of the mucus.
3. Maintaining cultures more than a month, results in deformation (Fig. 3) but the cultures retain their tight junctions.

4 Primary Human Nasal Epithelial (HNE) Cultures

Respiratory viruses, such as RSV and SARS-CoV-2, display different tropisms for HNE versus HBE cultures and these differences can be important for understanding their ecology. Cultures can be derived from nasal epithelial cells collected from volunteer or organ donors by nasal brushings or surgically resected nasal tissues.

Cells from brushings are collected, rinsed, and pelleted using ice-cold physiological solution such as DMEM. The declumping solution (see reagent setup) can be used when copious amounts of mucus are also collected with the cells. After this treatment, stop

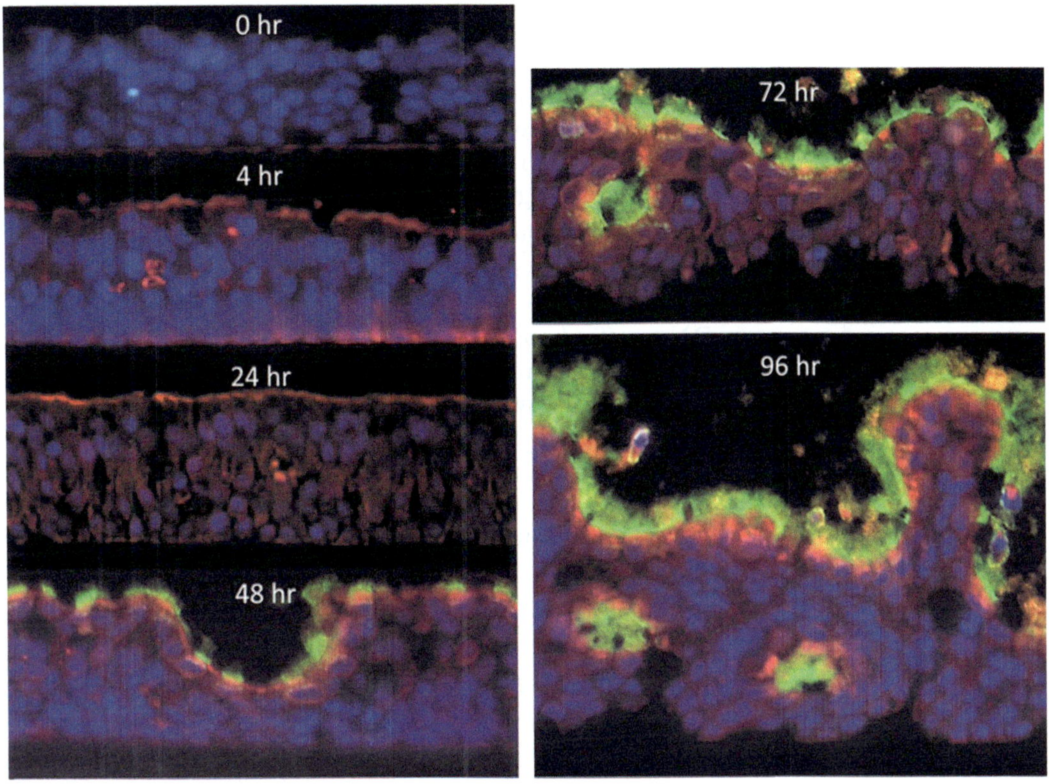

Fig. 3 Time course of RSV infection and spread in HBE cultures. Well-differentiated primary HBE cultures were inoculated with 2 X 10^4 pfu of RSV-A2 in 100 µl of DMEM. At the indicated times, one culture was fixed with 4% paraformaldehyde in PBS, embedded in paraffin and sliced in 5 µm sections, followed by 30 min of antigen retrieval. Sections were blocked then stained with L9, a mouse monoclonal antibody (mAb) against the RSV G protein followed by Dylight 488 Goat anti-mouse IgG (green), and Motavizumab, a humanized mAb against the RSV F protein and Cy3 (red) anti-human IgG. Nuclei were also stained with DAPI (blue) and photographed at 40X with an Axioscope for 15 msec (red and green) and 40 msec (blue)

the enzymatic reaction by adding FBS sufficient to reach 10%. Cells can also be filtered through a 70um filter, pelleted, and rinsed with the physiologic solution to remove mucus, DTT, and enzymes from the declumping solution.

Nasal tissue from surgical resections can be digested with the Protease XIV and DNase (see reagent setup) on a rocking device overnight at 4 °C. After this treatment, stop the enzymatic reaction with 10% FBS. Cells can be filtered through a 70 µm filter, pelleted, and rinsed with the physiologic solution used.

After cells have been collected and pelleted, they are best pro pagated by seeding onto collagen-coated plates with irradiated mouse 3T3 cells and fed with PneumaCult Ex Plus supplemented with 10 uM ROCK Inhibitor (Y-27632) along with anti-fungal and antibiotics. Cells are typically expanded at most, two weeks, and then harvested as described for the HBE cultures.

5 Notes

1. If a whole lung is obtained, dissect out the trachea and bronchus and process the bronchi in sections, while keeping the tissue moist with a chilled physiologic solution such as DMEM or Lactated Ringer's.
2. For a chronically infected lung with much mucus and debris (often the case with lungs from a person with cystic fibrosis, for instance), tissues can be soaked in wash medium with DTT and DNase *(Soak solution; see reagent setup)* twice prior to the regular washes mentioned above. This is done by adding the soak solution, swirling the contents, and allowing tissues to incubate for 5 min on ice before aspirating with a Pasteur pipette and repeating. Wash as above, three times with wash medium to ensure residual *DTT* and *DNase* is removed from the soak solution. Then proceed as above by placing tissues in enzyme solution for digestion for 48 h.
3. The dilution of the *Protease/DNase* solution in the wash medium and tissue should not be higher than 1:10.
4. The number of 100 mm plates coated can be adjusted by the amount of tissue collected. When full lung airway tissue has been collected, prepare 40–50 plates.
5. After coating, plates can be stored at 4°C for up to a month. Plates can then be UV-treated in the biosafety cabinet for 30 min to ensure sterilization before adding cells.
6. For more complete collection of cells and to maximize the number of cells, wash medium can be used to rinse the tissue sections in a 150-mm tissue culture plate.
7. If the pellet of cells has substantial amounts of collagenous and fibrous sediment, *a declumping solution* (see reagent setup) can be used. Aspirate supernatant and incubate with freshly prepared declumping solution for 5–15 min. Neutralize DTT with 10% FBS. Pellet cells at 500 g for 5 min. Wash once with F-12 medium, count the number of cells using a hemocytometer and plate for expansion.
8. To expand these P_0 cells to P_1, prepare proliferation medium with antibiotics. *BEGM* (Bronchial Epithelial Cell Growth Medium) (see reagent setup) can be made in-house and assembled using a bottle-top filter, or a commercial medium, *PneumaCult Ex Plus* from Stem Cell Technologies can be used. It is more expensive but enhances proliferation and expansion. It is assumed that the primary human tissues contain microbiological contaminants, for the viability of the cultures and future sterile experimental culture conditions, the addition of the *25X antibiotics* (see reagent setup) along with other antimicrobials

(such as Fluconazole) is necessary when growing and expanding airway cultures. (Passaging beyond P_3 not recommended. Additional passages will result in a decrease in the ability of the cells to divide and the loss of specialized populations. For RSV studies, loss of ciliated cell can dramatically reduce the susceptibility of the HBE cultures to RSV. For this reason, we do not passage the basal cells, unless there is something unique about the donor that we would like to study.)

9. Ensure that the trypsin is freshly made. If the trypsinization process is prolonged for more than 25–30 min, collect any cells that have detached along with the trypsin solution and neutralize trypsin with soybean trypsin inhibitor. Then replace with freshly made trypsin.
10. Rinse plates with additional rinses of soybean trypsin inhibitor to ensure more complete cell harvests. Check under the microscope after rinsing.
11. At this point, cells can be frozen or reseeded on the additional previously *PureCol type I* collagen-coated plates for further expansion to passage (P2) for a later harvest within 7 days, and fed with the addition of 10 ml of proliferation medium and changing medium every other day. Beyond passage 3, cell differentiation quickly diminishes.
12. Ensure all the cells from the tubes are collected by rinsing with small volumes and pooling into one conical tube. Triturate the cells to obtain a homogeneous cell suspension. Depending on the size of pellet, the cells are typically resuspended in 15–25 mL of F12 prior to counting.
13. Usually, the cell concentration is high and a 1:5 dilution using PBS (10 μL + 40 μl) is needed before counting. Triturate this mixture and then dilute 10 μL of this cell suspension with 10 μL of 0.4% Trypan blue. Count the number of unstained cells to assess the number of viable cells.
14. With this number of cells per vial, 48 transwells (6.5 mm) may be seeded with 40,000–45,000 cells/100 μL into each transwell for good culture production.
15. Transwells come in several sizes, depending on the needs of the study. We generally place 12 6.5 mm diameter wells in the two central rows of 24-well tissue culture dishes.
16. Upon reaching confluency, the culture should have a cobblestone appearance without spaces between them. Use an ohmmeter to determine complete confluency and tight junction formation of the cultures [12]. The transepithelial electrical resistance values should be at least in the range of 700 Ω*cm^2 or higher

17. Ideally, the fully differentiated cultures can be used for experimentation within 4–6 weeks after seeding. After 6 weeks, the cultures may still be viable with active ciliary beatings, however, they may show signs of stress and aging.
18. As the cultures mature and differentiate, mucus produced by goblet cells accumulates and may inhibit infection. Warm medium and the normal beating of the cilia will help loosen mucus from the culture surface. However, medium should not remain on cells for more than 4 h.

Acknowledgments

The authors thank Scott Randell, University of North Carolina, for training two of our lab members in the art of HBE culture preparation. Production of the HBE cultures was supported by the Cystic Fibrosis Foundation (CFF) through the Columbus Cures Cystic Fibrosis (C3) Research Development Program Grant (MCCOY17R2), which includes the Epithelial Production, Analysis, and Innovation Core (EPAIC) at Nationwide Children's Hospital (NCH) and The Ohio State University (OSU), by the NCH Division of Pediatric Pulmonary Medicine (MCCOY19Ro), and the OSU Center for Clinical and Translational Science (UL1TR002733). Source tissues for HBE cultures were provided by the Comprehensive Transplant Center Human Tissue Biorepository of The OSU Wexner Medical Center. Cells for HNE cultures were collected by nasal brushing of volunteers and organ donors. Other grants that support the EPAIC include: CFF006781R223, U19AI42733, R01AI185617, R01AI093848, R01AI090060, R01AI095684, R21HL165406, R01AI171027, R21OD032454, R56AI16658748, P01AI112524, P01AI175399, and P01HL051818. The content is solely the responsibility of the authors and does not necessarily represent the official views of the National Institutes of Health.

References

1. Zhang L et al (2002) Respiratory syncytial virus infection of human airway epithelial cells is polarized, specific to ciliated cells, and without obvious cytopathology. J Virol 76(11): 5654–5666
2. Techaarpornkul S, Barretto N, Peeples ME (2001) Functional analysis of recombinant respiratory syncytial virus deletion mutants lacking the small hydrophobic and/or attachment glycoprotein gene. J Virol 75(15): 6825–6834
3. Hallak LK et al (2000) Glycosaminoglycan sulfation requirements for respiratory syncytial virus infection. J Virol 74(22):10508–10513
4. Hallak LK et al (2000) Iduronic acid-containing glycosaminoglycans on target cells are required for efficient respiratory syncytial virus infection. Virology 271(2):264–275
5. Johnson SM et al (2015) Respiratory syncytial virus uses CX3CR1 as a receptor on primary human airway epithelial cultures. PLoS Pathog 11(12):e1005318

6. Wellemans V et al (2021) Role of CCR3 in respiratory syncytial virus infection of airway epithelial cells. iScience 24(12):103433
7. King T et al (2021) The larger attachment glycoprotein of respiratory syncytial virus produced in primary human bronchial epithelial cultures reduces infectivity for cell lines. PLoS Pathog 17(4):e1009469
8. Fulcher ML, Randell SH (2013) Human nasal and tracheo-bronchial respiratory epithelial cell culture. Methods Mol Biol 945:109–121
9. Fulcher ML et al (2005) Well-differentiated human airway epithelial cell cultures. Methods Mol Med 107:183–206
10. Horani A et al (2013) Rho-associated protein kinase inhibition enhances airway epithelial basal-cell proliferation and lentivirus transduction. Am J Respir Cell Mol Biol 49(3): 341–347
11. Rayner RE et al (2019) Optimization of normal human bronchial epithelial (NHBE) cell 3D cultures for in vitro lung model studies. Sci Rep 9(1):500
12. Srinivasan B et al (2015) TEER measurement techniques for in vitro barrier model systems. J Lab Autom 20(2):107–126

Chapter 17

Utilizing Neonatal Tracheal Aspirate-Derived Bronchial Basal Stem Cells to Model Respiratory Syncytial Virus Infection

Caiqi Zhao, Gaurang M. Amonkar, Paul H. Lerou, and Xingbin Ai

Abstract

Respiratory syncytial virus (RSV) infection is a primary cause of bronchiolitis and hospitalization in infants and is associated with significant mortality and morbidity. To model RSV infection in infant bronchial epithelium, we developed a robust protocol for bronchial basal stem cell (BSC) derivation from tracheal aspirates (TAs) of intubated neonatal patients. TA BSCs can be expanded while retaining the ability to differentiate into all the major types of bronchial epithelial cells in air-liquid interface (ALI), making them an invaluable source to generate functional bronchial epithelium of infants. This protocol outlines the specific steps in utilizing TA BSCs to model RSV infection of infant bronchial epithelium in ALI culture.

Key words Respiratory syncytial virus (RSV), Infant, Tracheal aspirate (TA), Basal stem cell (BSC), Air-liquid interface (ALI), Airway epithelial cells (AEC)

1 Introduction

Respiratory syncytial virus (RSV) causes bronchiolitis and pneumonia that disproportionally affects children younger than 2 years of age, making RSV one of the most significant causes of hospitalization and mortality in infants [1]. RSV selectively infects ciliated cells in human bronchial epithelium [2]. The pathophysiology of infant RSV includes cytotoxicity, inflammation, and airway obstruction in both bronchi and bronchioles [3–6]. In addition to acute morbidity, RSV infection also predisposes young children to hyperreactive airway disease and asthma later in life [7–12].

To date, there is a lack of experimental models of RSV infection that recapitulate the pathophysiology of RSV disease in infants. A major obstacle is the narrow species tropism of RSV, rendering technical difficulties in modeling infant RSV disease using animal models [13]. The in vitro models of human cell lines, such as immortalized human HEp-2 and A549 cells, poorly represent

Bo Liang and Rachel Fearns (eds.), *Human Respiratory Syncytial Virus: Methods and Protocols*, Methods in Molecular Biology, vol. 2948, https://doi.org/10.1007/978-1-0716-4666-3_17,

human airway epithelium that is polarized and contains multiple cell types. Airway epithelial cells (AECs) generated in air-liquid interface (ALI) culture of human bronchial basal stem cells (BSCs) resemble the cellular complexity of human bronchial epithelium [14, 15] and thus provide a surrogate to model RSV infection in vitro. However, traditional approaches of BSC isolation from lung biopsy and airway brushing samples are not possible in infants. For this reason, prior studies of RSV infection employed BSCs derived from older children and adults [2, 16–18]. Although the results of these studies validate the ciliated cell specificity of RSV infection and RSV-induced cell sloughing and apoptosis in pediatric ALI cultures [2, 16, 17], how RSV interacts with infant bronchial epithelium to trigger severe diseases remains elusive.

To bypass the need for infant lung biopsy, we have developed a robust protocol of BSC derivation from tracheal aspirates (TAs). TAs can be collected during routine care of intubated patients and are generally considered medical waste [19–21]. Using TA BSCs isolated from newborns intubated for non-respiratory conditions, we generated differentiated AECs in ALI. We demonstrated that following RSV infection, neonatal AECs, compared to adult AECs, have increased viral load, inflammation, and epithelial damage [2, 22]. Here we provide a comprehensive protocol that describes the acquisition of neonatal TA BSCs, their differentiation into AECs in ALI, and the steps to establish RSV infection. Additional details of the primary culture of human TA BSCs are previously published [20]. This detailed protocol serves to facilitate basic and translational research of RSV infection in infants.

2 Materials

Prepare all mediums and reagents under a sterile workspace. Small Airway Epithelial Growth Medium (SAGM) and PneumaCult medium are two major mediums for the expansion and differentiation of neonatal TA BSCs. Prepare and store all reagents at 4 °C.

1. Opti-MEM medium: Opti-MEM supplemented with 2% fetal bovine serum (FBS), 1% GlutaMAX Supplement, and 1% antibiotics (*see* **Note 1**).
2. Methylcellulose medium: 4.5 g methylcellulose powder dissolved in 500 mL Opti-MEM medium (*see* **Note 2**).
3. Neutralization medium: RPMI 1640 medium supplemented with 10% FBS and 1% antibiotics.
4. SAGM with Rho/SMAD/mTOR inhibitors.

Component	Final concentration
SAGM	n/a
SAGM supplement mix	n/a
Y-27632 2HCl	5 μM
A 83–01	1 μM
CHIR 99021	0.4 μM
RAPAMYCIN 99%	5 nM
Normocin	100 μg/ml

5. PneumaCult medium with Rho/SMAD inhibitors (the first 7 days in ALI during 3-week differentiation).

Component	Final concentration
PneumaCult-ALI basal medium (450 mL)	n/a
PneumaCult-ALI 10 × supplement (50 mL)	n/a
PneumaCult-ALI supplement (5 × 1 mL)	n/a
Y-27632	5 μM
A 83-01	0.5 μM
Normocin	100 μg/mL

6. PneumaCult medium without inhibitors (the following 14 days in ALI during 3-week differentiation).

Component	Final concentration
PneumaCult-ALI basal medium (450 mL)	n/a
PneumaCult-ALI 10 × supplement (50 mL)	n/a
PneumaCult-ALI supplement (5 × 1 mL)	n/a
Normocin	100 μg/mL

3 Methods

3.1 Preparation of 804G-Conditioned Medium

The 804G-conditioned medium from the 804G rat bladder epithelial cell line is used to coat cell culture vessels to facilitate neonatal BSC attachment.

1. Remove the freezing vial of 804G cells (2 × 10^6 cells) from liquid nitrogen and thaw.

2. Transfer the cells to a 15 mL conical tube and add 5 mL RPMI 1640 medium with 10% FBS and 1% antibiotics.
3. Centrifuge the cells at 500 × g for 5 minutes (min) at room temperature.
4. Discard the supernatant and resuspend the cell pellet with 10–15 mL RPMI 1640 medium (10% FBS and 1% antibiotics) and plate into a T75 flask.
5. Place the flask in a 37 °C incubator with 5% CO_2 and change the medium every two days.
6. When the flask reaches 90% confluency and no contamination is present, add 5 mL 0.25% Trypsin to the T75 flask and incubate at 37 °C to dissociate the cells (5–10 min).
7. Transfer the cell suspension to a 15 mL conical tube and add 7 mL RPMI 1640 medium (10% FBS and 1% antibiotics).
8. Centrifuge the sample at 500 × *g* for 5 min at room temperature.
9. Discard the supernatant and resuspend the cell pellet with 25 mL RPMI 1640 medium (10% FBS and 1% antibiotics) and plate the cells into a T175 flask.
10. Place the flask in a 37 °C incubator with 5% CO_2 and change medium every two days.
11. When flask reaches 90% confluency, and no contamination is present, add 60 mL fresh RPMI 1640 medium (10% FBS and 1% antibiotics) to the T175 flask and change media every other day for 8 days.
12. Collect the supernatant every 2 days. This is the 804G-conditioned medium.
13. Sterilize with 0.22 μm filter and aliquot and store at −20 °C for future use.

3.2 Preparation of RSV Inoculum

1. Hep-2 cells were cultured in Opti-MEM supplemented with 2% fetal bovine serum (FBS) and 1% GlutaMAX Supplement. This is the complete medium for Hep-2 cells.
2. At 80% confluency, Hep-2 cells were infected with RSV A2 or A2-GFP strain in complete medium at a multiplicity of infection (MOI) of 0.1 for 1–2 hours (h) at 37 °C.
3. After infection, remove the RSV inoculum and add fresh complete medium.
4. Check the cells every day. When high cytopathic effects are observed at 3 ~ 5 days post-infection (dpi) in culture, scrape the cells into the medium.
5. Separate cell debris by centrifugation at 1500 × *g* for 10 min at 4 °C.

6. Collect the supernatant, aliquot, and store the supernatant at – 80 °C.
7. To measure RSV titer, 1.5×10^5 Hep-2 cells per well are seeded in a 24 well plate.
8. Incubate overnight at 37 °C with 5% CO_2. Cells should be 80–90% confluent (*see* **Note 3**).
9. Dilute virus stock in Opti-MEM at a series ratio of 1:10,000, 1: 100,000, 1:500,000, 1:1,000,000.
10. Remove the media from each well of Hep-2 cells and quickly replace with 0.5 mL of diluted virus.
11. Shake the plate to ensure the cells are covered and incubate at 37 °C for 2 h.
12. Remove inoculum and overlay each well with 2 mL 0.9% warm methylcellulose.
13. Incubate for 4 days at 37 °C with 5% CO_2 (*see* **Note 4**).
14. Remove as much methylcellulose as possible without disturbing the monolayer.
15. Fix cells by adding 2 mL of cold 80% methanol and incubate at 4 °C for 1 h.
16. Carefully wash the plates 3 times with ddH2O.
17. Perform the standard steps for incubation of blocking, primary antibody, and secondary antibody (*see* **Note 5**).
18. Wash and remove excess water, then add 200 μL of peroxidase solution substrate per well.
19. Wait for 5–10 min at room temperature or until plaques are easily visualized.
20. Remove substrate, count plaques, and calculate the virus titer.

3.3 Collection of Cells in Neonatal Tracheal Aspirate Samples

1. Add 5–10 mL 1 × PBS to dilute the TA samples (*see* **Note 6**). Mix well using pipette.
2. Lubricate a 70 μm cell strainer with 5 mL PBS to minimize cell attachment.
3. Put the 70 μm cell strainer on top of a 50 mL conical tube.
4. Filter the samples through the 70 μm cell strainer into the 50 mL conical tube.
5. Discard the cell strainer and centrifuge the 50 mL conical tube at 500 × *g* for 10 min at room temperature.
6. Gently discard the supernatant without disturbing the cell pellet (*see* **Note 7**).
7. Resuspend the cell pellet with 5 mL SAGM medium and transfer into a T25 flask pre-coated with the 804G-conditioned medium (*see* **Note 8**).

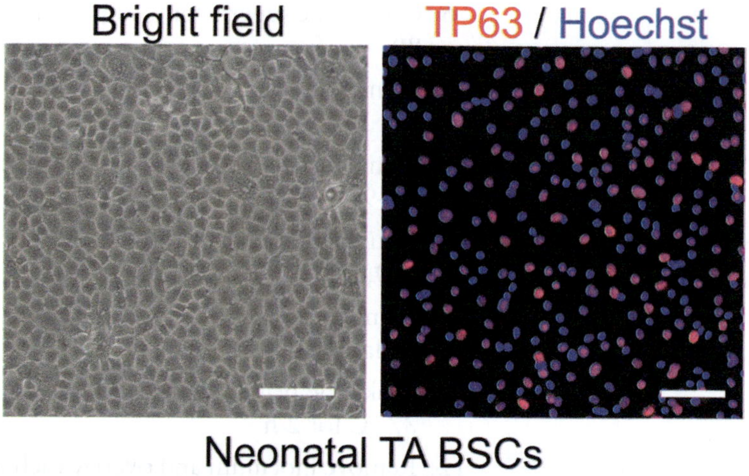

Fig. 1 In vitro culture of neonatal TA BSCs. Bright field image (left panel) and immunofluorescent TP63 staining (right panel) of neonatal TA BSCs. Scale bar, 50 μm

8. Culture the cells in the incubator at 37 °C with 5% CO_2.
9. Change the SAGM medium 1 day after isolation and every other day thereafter (*see* **Note 9**).
10. Cells with a cuboidal shape should appear after 1–2 weeks (Fig. 1). When the cells have reached 90% confluency without evidence of contamination, add 2 mL 0.25% Trypsin to the T25 flask and incubate at 37 °C for 5–10 min until the cells have detached. If cells remain adhering to the plate after 5 min, lightly tap the side of the flask to accelerate the dissociation.
11. Add 4 mL neutralization medium to the flask and transfer the cell suspension to a 15 mL conical tube.
12. Centrifuge the cell suspension at 500 × *g* for 5 min at room temperature.
13. Discard the supernatant and resuspend the cell pellet with 10 mL SAGM media.
14. Transfer the cell suspension into a T75 flask pre-coated with 804G-conditioned medium (*see* **Note 8**).
15. Culture the cells in the incubator at 37 °C with 5% CO_2 (*see* **Note 10**).
16. If needed, freeze neonatal TA BSCs from the T75 flask when 90% confluency is reached after checking the cell identity (Fig. 1) (*see* **Note 11**).

3.4 Differentiation of Neonatal TA BSCs in ALI Culture

1. When the cultured neonatal TA BSCs in the T75 flask have a proper cuboidal shape and have reached 90% confluency, add 5 mL 0.25% Trypsin to the flask to dissociate the cells as previously described.

2. Add 7 mL neutralization medium to the flask and transfer the cell suspension to a 15 mL conical tube.
3. Centrifuge the sample at 500 × *g* for 5 min at room temperature.
4. Resuspend the pellet with 1 mL SAGM medium and count the cells using a hemocytometer.
5. Calculate the volume of SAMG medium required to achieve a 1×10^6 cells/ml suspension and seed 200 μL medium (i.e., 2×10^5 cells) into the upper well of pre-coated 24-well inserts (*see* **Note 12**).
6. Add 700 μL SAGM medium into the basal side of the inserts.
7. Culture the cells in the incubator at 37 °C with 5% CO_2.
8. The next day after seeding, when the cells have reached complete confluence, remove SAGM medium from the basal side and upper well without touching the cells on the inserts.
9. Add 200 μL PneumaCult medium with inhibitors to the upper well and add 700 μL PneumaCult medium with inhibitors to the basal side (*see* **Note 13**).
10. The day after the addition of PneumaCult medium, remove the medium from the basal side and upper well, and only add fresh PneumaCult medium to the basal side. The day of removing medium from the upper well is day 0 of a 21-day ALI differentiation.
11. From now on, only change the PneumaCult medium of the basal side every other day (*see* **Note 13**).
12. On days 14 and 21, perform a PBS wash to remove excess mucus. Add 200-μL sterile PBS to the apical side and incubate at 37 °C for 20 min. After incubation, carefully remove the PBS from the insert without touching the cells on the insert.
13. On day 21 of differentiation, the neonatal ALI culture is fully differentiated and ready for RSV infection (Fig. 2) (*see* **Note 14**).

3.5 RSV Infection of Neonatal ALIs

1. Before performing RSV infection at day 21 of ALI differentiation, wash the apical side of the neonatal ALI culture 3 times with PBS at 37 °C for 20 min to remove mucus.
2. Calculate and dilute the RSV stock in PBS to make RSV inoculum at a concentration of 4×10^5 pfu per 200 μL (*see* **Note 15**).
3. Add 200 μL of inoculum into the apical well of the neonatal ALI culture.
4. Place the ALI in the incubator at 37 °C with 5% CO_2 for 1–2 h.
5. After inoculation, carefully remove the RSV inoculum from the apical side without touching the cells on the insert.

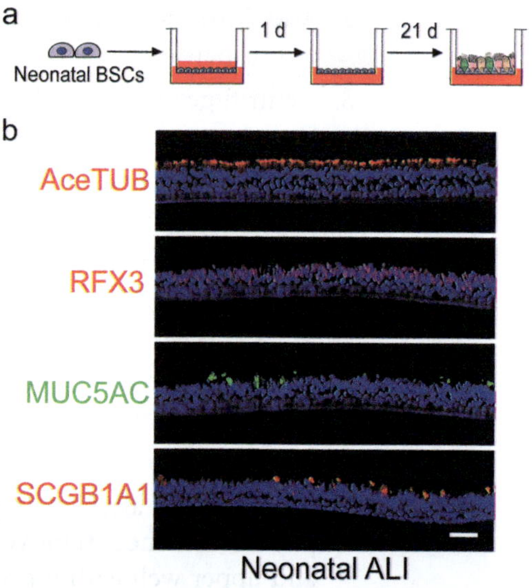

Fig. 2 Differentiation of neonatal TA BSCs using ALI culture. (**a**) Schematic of ALI differentiation of neonatal TA BSCs. (**b**) Representative antibody staining for markers of major epithelial cell types, including AceTUB (acetylated tubulin) and RFX3 for ciliated cells, MUC5AC for goblet cells, and SCGB1A1 for Club cells. Scale bar, 50 μm

6. Wash the apical side 3 times with PBS at 37 °C to remove unattached viruses.
7. Place the ALI in the incubator at 37 °C with 5% CO_2 for 2–4 days (*see* **Note 16**).
8. After infection, remove the basal medium and add 400 μL 4% PFA to the apical side and 750 μL 4% PFA to the basal side for cell fixation (*see* **Note 17**). Incubate for 15–20 min in the biosafety cabinet.

3.6 Sectioning and Immunofluorescence Staining of RSV Infected ALI

1. Using a scalpel and blunt forceps, carefully cut half of the membrane from the bottom of the insert transfer it into a 1.5 mL microcentrifuge tube.
2. Incubate the membrane in 10%, 15%, and 20% sucrose in 1× PBS on ice for 1 h each.
3. Incubate the membrane in 1:1 20% sucrose:OCT at 4 °C overnight.
4. Add 1:3 20% sucrose:OCT to halfway fill a cryomold on dry ice (*see* **Note 18**).
5. Once the solution turns opaque, place the membrane flat on top, noting its orientation (*see* **Note 19**).
6. Carefully add 1:3 20% sucrose:OCT to fill the rest of the cryomold, while preserving the orientation of the membrane, and incubate for 15 min or until solid.

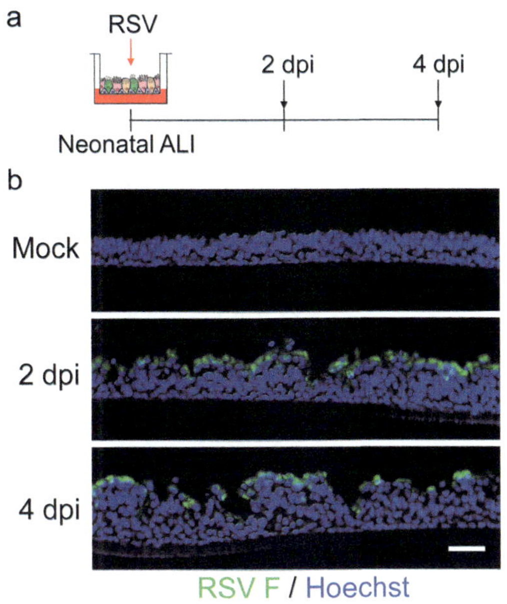

Fig. 3 RSV infection of neonatal ALI culture. (**a**) Schematic of RSV infection of neonatal ALI culture. (**b**) Representative antibody staining for RSV F protein. Scale bar, 50 μm

7. Perform cryosectioning on a microscope slide (*see* **Note 20**).
8. Wash the ALI sections with PBS to remove OCT.
9. Perform the standard steps for immunofluorescence staining: blocking, primary antibody, secondary antibody, and nuclear staining (*see* **Note 21**) (Fig. 3).
10. Once the staining is complete, add a drop of Fluoromount-G to ALI sections on the microscope slide.
11. Put the coverslip onto the microscope slide for imaging.
12. Allow the coverslip to dry completely before imaging.

4 Notes

1. FBS needs to be heat inactivated at 56 °C for 30 min.
2. To prepare methylcellulose, weigh the powder and put it into a glass bottle together with a magnetic stir bar and then autoclave. Shake to disperse any lumps and then add the Opti-MEM medium in the biosafety cabinet. Leave to stir in a cold room overnight.
3. If cells are too dense, the Hep-2 monolayer will tend to lift off during viral incubation, which will affect the accuracy of determining viral titer.

4. If the monolayer is overgrown and unhealthy, it is better to stop the incubation after 3 days; otherwise, the monolayer will become non-adherent. After 3–5 days, the plaques should be visible under the microscope.
5. Use 5% dried milk in 1× PBS for blocking buffer. The recommended time for the blocking step is no less than 1 h. For primary antibody, we recommend a mouse monoclonal antibody against RSV F protein (BIO-RAD, MCA490) for overnight. The secondary antibody (HRP) step is no less than 2 h. Careful washes with ddH_2O are essential between every step.
6. This protocol yields the best isolation of BSCs when the tracheal aspirates are freshly isolated. These samples can also be stored at 4 °C for up to 24 h after collection, but efficiency decreases over time. If the sample has an abundance of mucus, treat with 1 mL 0.1 mg/mL DTT in 1 × PBS. Incubate for 5–15 min at room temperature. Red blood cell lysis buffer may need to be used if the sample contains blood. The TA BSC isolation step should be sterile and performed in the biosafety cabinet.
7. If there is still a lot of mucus in the cell pellet, repeat the DTT treatment step. This treatment will help to increase the yield of TA BSCs recovery. EpCAM selection of airway BSCs can be performed to purify the cells.
8. Pre-coat the cell culture vessels with 804G-conditioned medium for at least 4 h at 37 °C before seeding neonatal BSCs. When possible, coat overnight for optimal cell adherence. Cell culture vessels can be coated with 804G-conditioned medium for up to 2 weeks. Alternatively, collagen solution (30 μg /ml) can be used for the coating step.
9. Monitor the flask for proper cuboidal cell shape and any possible microbial contamination every day. Check the flask under a 5×, 10×, and 20× microscope to visualize the cells and detect contamination.
10. The proliferative capacity and differentiation efficiency of neonatal TA BSCs likely decreases after every passage. It is best to perform experiments with TA BSCs prior to passage 5.
11. CryoStor CS10 (Biolife Solutions) is recommended as the freezing medium. The recommended cell number per freezing vial is 5×10^5.
12. We recommend the 6.5 mm Transwell® with 0.4 μm Pore Polyester Membrane Insert (Corning, Cat# 3470). Add 804G-conditioned medium to both the apical and basal well to pre-coat the inserts overnight.
13. For the first 7 days of a 21-day ALI differentiation, use PneumaCult medium with A 83-01 and Y-27632 inhibitors. For the remaining 14 days, use PneumaCult medium without A 83-01

and Y-27632 inhibitors. This is important for functional cilia analysis and RSV infection.

14. At day 21 of differentiation, cilia beating can be observed under bright field microscope to generally justify the differentiation status of neonatal ALI.
15. The infection dose should be adjusted and determined by the experimental goals. 4×10^5 pfu can be deemed as ~2 MOI for 24-well inserts. Under this infection condition, the epithelial damage is obvious at 2 dpi for neonatal ALIs.
16. RSV A2 strain with GFP expression can be employed to observe the progress of RSV infection. Usually, at 1 dpi, RSV-positive cells can be seen with fluorescent microscopy.
17. Similarly, RNA and protein samples can also be collected for analysis.
18. While loading the cryomold with 1:3 20% sucrose:OCT, it is important to avoid loading bubbles, as they can reduce the quality of sectioning.
19. The cryomold should be completely flat on the dry ice to prevent sample misorientation while the OCT solidifies.
20. We recommend that the thickness of the ALI section be between 8 and 12 μm, to optimize microscopy imaging quality.
21. Use 5% BSA and 0.1% triton in 1× PBS for the blocking buffer. The recommended times are: no less than 1 h for the blocking step, overnight for the primary antibody step, and no less than 2 h for the secondary antibody (fluorescently labeled) step.

Acknowledgments

This work was supported by an NIH grant to PHL (R21AI156597) and funds from the Department of Pediatrics at Massachusetts General Hospital for the Lung Cell Bank to XA. The content is solely the responsibility of the authors and does not necessarily represent the official views of the National Institutes of Health.

References

1. Hall CB, Weinberg GA, Iwane MK, Blumkin AK, Edwards KM, Staat MA, Auinger P, Griffin MR, Poehling KA, Erdman D, Grijalva CG, Zhu Y, Szilagyi P (2009) The burden of respiratory syncytial virus infection in young children. N Engl J Med 360(6):588–598. https://doi.org/10.1056/NEJMoa0804877
2. Zhang L, Peeples ME, Boucher RC, Collins PL, Pickles RJ (2002) Respiratory syncytial virus infection of human airway epithelial cells is polarized, specific to ciliated cells, and without obvious cytopathology. J Virol 76(11): 5654–5666. https://doi.org/10.1128/jvi.76.11.5654-5666.2002
3. Johnson JE, Gonzales RA, Olson SJ, Wright PF, Graham BS (2007) The histopathology of fatal untreated human respiratory syncytial virus infection. Mod Pathol 20(1):108–119.

https://doi.org/10.1038/modpathol.3800725

4. Welliver TP, Garofalo RP, Hosakote Y, Hintz KH, Avendano L, Sanchez K, Velozo L, Jafri H, Chavez-Bueno S, Ogra PL, McKinney L, Reed JL, Welliver RC Sr (2007) Severe human lower respiratory tract illness caused by respiratory syncytial virus and influenza virus is characterized by the absence of pulmonary cytotoxic lymphocyte responses. J Infect Dis 195(8):1126–1136. https://doi.org/10.1086/512615
5. Touzelet O, Power UF (2011) Cellular and molecular characteristics of RSV-induced disease in humans. In: Resch B (ed) Human respiratory syncytial virus infection. IntechOpen, Rijeka, pp 97–122. https://doi.org/10.5772/27650
6. Villenave R, Shields MD, Power UF (2013) Respiratory syncytial virus interaction with human airway epithelium. Trends Microbiol 21(5):238–244. https://doi.org/10.1016/j.tim.2013.02.004
7. Welliver RC (2003) Review of epidemiology and clinical risk factors for severe respiratory syncytial virus (RSV) infection. J Pediatr 143(5 Suppl):S112–S117. https://doi.org/10.1067/s0022-3476(03)00508-0
8. Sigurs N, Aljassim F, Kjellman B, Robinson PD, Sigurbergsson F, Bjarnason R, Gustafsson PM (2010) Asthma and allergy patterns over 18 years after severe RSV bronchiolitis in the first year of life. Thorax 65(12):1045–1052. https://doi.org/10.1136/thx.2009.121582
9. Vom Hove M, Prenzel F, Uhlig HH, Robel-Tillig E (2014) Pulmonary outcome in former preterm, very low birth weight children with bronchopulmonary dysplasia: a case-control follow-up at school age. J Pediatr 164:40–45. https://doi.org/10.1016/j.jpeds.2013.07.045
10. Maitre NL, Ballard RA, Ellenberg JH, Davis SD, Greenberg JM, Hamvas A, Pryhuber GS, Prematurity, Respiratory Outcomes P (2015) Respiratory consequences of prematurity: evolution of a diagnosis and development of a comprehensive approach. J Perinatol 35(5):313–321. https://doi.org/10.1038/jp.2015.19
11. Keller RL, Feng R, DeMauro SB, Ferkol T, Hardie W, Rogers EE, Stevens TP, Voynow JA, Bellamy SL, Shaw PA, Moore PE, Prematurity, Respiratory Outcomes P (2017) Bronchopulmonary dysplasia and perinatal characteristics predict 1-year respiratory outcomes in newborns born at extremely low gestational age: a prospective cohort study. J Pediatr 187(89–97):e89. https://doi.org/10.1016/j.jpeds.2017.04.026
12. Ptaschinski C, Lukacs NW (2019) Early life respiratory syncytial virus infection and asthmatic responses. Immunol Allergy Clin N Am 39(3):309–319. https://doi.org/10.1016/j.iac.2019.03.002
13. Pickles RJ (2013) Human airway epithelial cell cultures for modeling respiratory syncytial virus infection. Curr Top Microbiol Immunol 372:371–387. https://doi.org/10.1007/978-3-642-38919-1_19
14. Dvorak A, Tilley AE, Shaykhiev R, Wang R, Crystal RG (2011) Do airway epithelium air-liquid cultures represent the in vivo airway epithelium transcriptome? Am J Respir Cell Mol Biol 44(4):465–473. https://doi.org/10.1165/rcmb.2009-0453OC
15. Gray TE, Guzman K, Davis CW, Abdullah LH, Nettesheim P (1996) Mucociliary differentiation of serially passaged normal human tracheobronchial epithelial cells. Am J Respir Cell Mol Biol 14(1):104–112. https://doi.org/10.1165/ajrcmb.14.1.8534481
16. Villenave R, Thavagnanam S, Sarlang S, Parker J, Douglas I, Skibinski G, Heaney LG, McKaigue JP, Coyle PV, Shields MD, Power UF (2012) In vitro modeling of respiratory syncytial virus infection of pediatric bronchial epithelium, the primary target of infection in vivo. Proc Natl Acad Sci USA 109(13):5040–5045. https://doi.org/10.1073/pnas.1110203109
17. Guo-Parke H, Canning P, Douglas I, Villenave R, Heaney LG, Coyle PV, Lyons JD, Shields MD, Power UF (2013) Relative respiratory syncytial virus cytopathogenesis in upper and lower respiratory tract epithelium. Am J Respir Crit Care Med 188(7):842–851. https://doi.org/10.1164/rccm.201304-0750OC
18. Simpson J, Loh Z, Ullah MA, Lynch JP, Werder RB, Collinson N, Zhang V, Dondelinger Y, Bertrand MJM, Everard ML, Blyth CC, Hartel G, Van Oosterhout AJ, Gough PJ, Bertin J, Upham JW, Spann KM, Phipps S (2020) Respiratory syncytial virus infection promotes necroptosis and HMGB1 release by airway epithelial cells. Am J Respir Crit Care Med 201(11):1358–1371. https://doi.org/10.1164/rccm.201906-1149OC
19. Lu J, Zhu X, Shui JE, Xiong L, Gierahn T, Zhang C, Wood M, Hally S, Love JC, Li H, Crawford BC, Mou H, Lerou PH (2021) Rho/SMAD/mTOR triple inhibition enables long-term expansion of human neonatal tracheal aspirate-derived airway basal cell-like

cells. Pediatr Res 89(3):502–509. https://doi.org/10.1038/s41390-020-0925-3

20. Amonkar GM, Wagner R, Bankoti K, Shui JE, Ai X, Lerou PH (2022) Primary culture of tracheal aspirate-derived human airway basal stem cells. STAR Protoc 3(2):101390. https://doi.org/10.1016/j.xpro.2022.101390
21. Wagner R, Amonkar GM, Wang W, Shui JE, Bankoti K, Tse WH, High FA, Zalieckas JM, Buchmiller TL, Zani A, Keijzer R, Donahoe PK, Lerou PH, Ai X (2023) A tracheal aspirate-derived airway basal cell model reveals a proinflammatory epithelial defect in congenital diaphragmatic hernia. Am J Respir Crit Care Med 207:1214. https://doi.org/10.1164/rccm.202205-0953OC
22. Zhao C, Wang W, Bai Y, Amonkar G, Mou H, Olejnik J, Hume AJ, Muhlberger E, Fang Y, Que J, Fearns R, Ai X, Lerou PH (2023) Age-related STAT3 signaling regulates severity of respiratory syncytial viral infection in human bronchial epithelial cells. bioRxiv. 2023.2009.2020.558606. https://doi.org/10.1101/2023.09.20.558606

Chapter 18

Visualizing Respiratory Syncytial Virus Replication in Live Mice

Jean-François Eléouët, Marie Galloux, and Marie-Anne Rameix-Welti

Abstract

Modeling human respiratory syncytial virus (RSV) infection in vivo is an essential step in the search for novel vaccines, antiviral therapies, or preventive measures against RSV disease. The most commonly used experimental models of RSV infection are rodent models, in particular, inbred BALB/c mice and cotton rats (Bem et al., Am J Physiol Lung Cell Mol Physiol 301(2): L148–L156, 2011; Taylor, Vaccine 35(3): 469–480, 2017; Altamirano-Lagos, Front Microbiol 10: 873, 2019). However, depending on the RSV strain and viral titer, infection of mice with RSV can result in few or no symptoms, with viral load being estimated by RT-qPCR on lung tissue. By allowing the monitoring of the spread and intensity of the infection in the same animal over time, the use of recombinant RSV expressing the firefly luciferase represents a major advance in the study of RSV replication in the murine model.

Key words Mice, In vivo model, Visualizing, Replication, Luminescence, Luciferase, Preclinical model

1 Introduction

Laboratory mice (*Mus musculus*) are a convenient and cost-effective animal model for preclinical studies of RSV infection compared to other animals, such as cotton rats or non-human primates [1–3]. In addition, many molecular tools are available to study RSV infection in mice at the molecular and cellular levels. However, experimental infection with RSV in mice results in little or no clinical signs of disease and generally only mild lung pathology. Therefore, RSV replication is typically monitored by RT-qPCR using RNA extracted from homogenized lung tissue of infected animals, which is a laborious and time-consuming approach. Furthermore, mice must be sacrificed for each time point using this method, so the number of animals used is important, even if only for statistical data. As an alternative, we have developed a BALB/c mouse model based on a recombinant RSV virus expressing the firefly luciferase as

Bo Liang and Rachel Fearns (eds.), *Human Respiratory Syncytial Virus: Methods and Protocols*, Methods in Molecular Biology, vol. 2948, https://doi.org/10.1007/978-1-0716-4666-3_18,

a reporter [4]. The light emission is proportional to the amount of viral mRNA, which allows quantifying the viral replication and thus the potential effect of vaccines or antiviral drugs. Since RSV infection with this recombinant virus can be followed in the same mouse for the entire experiment duration, the number of animals can be drastically reduced. Moreover, this model reveals viral replication in the noses of the infected animals, which has not been studied before.

2 Materials

2.1 Cells and Media

To recover recombinant RSV, use the BSRT7/5 cells [5] for plasmid transfections. RSV (recombinant or WT) is grown on HEp-2 cells (ATCC number CCL-23). The cells are grown at 37 °C in 5% CO_2 with a humid atmosphere.

1. BSRT7/5 cells.
2. HEp-2 cells (ATCC number CCL-23).
3. Eagle's minimum essential medium.
4. Dulbecco's modified Eagle's medium.
5. MEM (10X), no glutamine.
6. 10% fetal calf serum (FCS).
7. 2 mM L-glutamine.
8. penicillin–streptomycin (10,000 U/mL).
9. Eagle's minimum essential medium supplemented with 2% FCS, 2 mM L-glutamine, and penicillin–streptomycin solution.
10. Avicel RC581 (FMC Biopolymer) (*see* **Note 6**).
11. Minimum essential medium (MEM).
12. Crystal violet solution: 8% in 20% ethanol, 2% Formaldehyde.

2.2 Mouse

House mice were kept in polypropylene cages in standard temperature- and humidity-controlled conditions with a 12-h light-dark rhythm and unlimited access to food, water, and enrichment (nests).

1. Mouse strain. Use BALB/c mice, usually around 8 weeks of age.
2. Mice bred in a pathogen-free biosafety laboratory 2 animal facilities.

2.3 Reverse Genetics

Use a pACNR1180 [6] or another similar low-copy plasmid containing a P15A origin of replication for cloning the RSV (Long strain ATCC VR-26) cDNA (*see* **Note 1**). The Firefly luc + NF gene

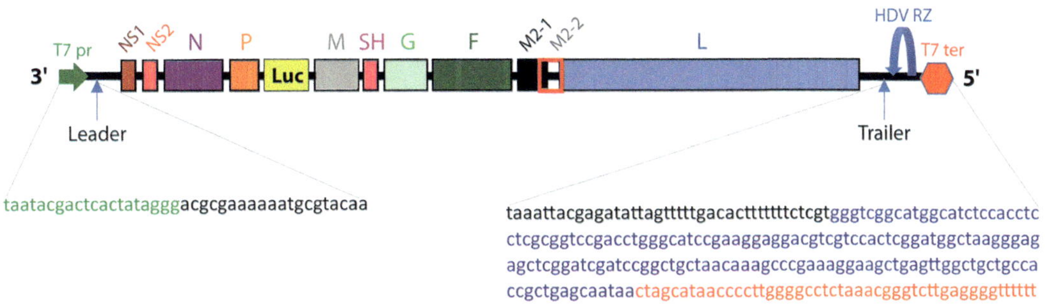

Fig. 1 Schematic of construction of the rRSV-Luc infectious clone (GenBank accession number: KF713491.1) derived from RSV Long strain (not to scale). Protein-encoding frames are shown as colored boxes, and leader and trailer regions are indicated by arrows. The firefly luciferase gene is inserted between the P and M genes, together with an upstream gene end (GE) and a downstream gene start (GS) signal derived from the N gene. Intergenic regions are shown as a black horizontal line. The T7 promoter is shown as a green arrow, the HDV ribozyme (RZ) is shown as a curved blue arrow, and the T7 transcription terminator (TT7) as a red box. The 5′ and 3′ ends of the viral genomic RNA are indicated. The cDNA sequences of junctions between the T7 promoter (green) and the anti-leader sequence (black) and between the HDV ribozyme (blue), followed by the T7 terminator (red) and the trailer sequence (black), are shown in the enlargements

(from pSP-luc + NF, Promega) is inserted between RSV P and M genes with an upstream "Gene Start" and a downstream "Gene End" signal derived from the RSV N gene (Fig. 1). The complete sequence of the RSV-Luc cDNA clone we use is available at GenBank (accession: KF713491.1) and can be engineered by gene synthesis. The RSV antigenomic cDNA sequence is under a T7 promoter and terminates with a hepatitis delta ribozyme just downstream of the RSV trailer sequence, followed by a T7 terminator. Use *E. coli* DH10B (Thermo Fisher Scientific) for plasmid amplification. Clone the N, P, L, and M2–1 original RSV (Long strain) genes in high-copy plasmids under a T7 RNA polymerase promoter and an IRES, and with a T7 terminator downstream. Alternatively, use synthetic codon-optimized RSV N, P, L, and M2-1 genes and use a pCDNA3 vector or equivalent for eukaryotic cell expression.

1. pACNR–rRSV-Luc plasmid DNA.
2. p-L, p-P, p-N, and p-M2-1 plasmid DNAs.
3. Lipofectamine2000 (Thermo Fisher Scientific).
4. Opti-MEM.
5. 10X RSV conservation buffer: 0.5 M Hepes pH 7.5, 1 M $MgSO_4$ (sterile).

2.4 Luminescence Quantification in Live Mice

1. The IVIS system (Xenogen Biosciences).
2. Living Image software (version 4.0, Caliper Life Sciences, Hopkinton, MA, USA).

2.5 Other Solutions

1. Luria Broth (LB).
2. Phosphate-buffered saline (PBS).
3. D-luciferin.
4. liquid nitrogen.
5. ketamine and xylazine.
6. isoflurane gas.

3 Methods

3.1 Cell Culture

Prepare media for a cell culture in a sterile environment for transfection and infection. Use DMEM with 2 mM L-glutamine supplemented with 10% fetal calf serum (FCS), 1000 units/mL penicillin, and 1 mg/mL streptomycin (or without antibiotics) and MEM with 2 mM L-glutamine supplemented with 0%, 2%, or 10% FCS, 1000 units/mL penicillin, and 1 mg/mL streptomycin, designated as "complete medium" in the following protocol.

Culture BSRT7/5 cells in complete DMEM and HEp-2 cells in complete MEM at 37 °C and 5% CO_2 in a sterile environment. Make stocks in complete medium supplemented with 10% dimethyl sulfoxide (DMSO) at 1 to 2 × 10^6 cells/mL. Conserve the cell stocks in liquid nitrogen.

Prepare a 10x RSV conservation solution (0.5 M HEPES and 1 M $MgSO_4$ [pH 7.5] in water) in a sterile environment.

3.2 Recovery of Recombinant Viruses

1. Plasmids p-N, p-P, p-L, and p-M2-1 containing the full-length RSV genome and N, P, L, and M2-1 genes must be grown at 28 °C in LB media using DH10B *E. coli* cells (*see* **Note 1**).
2. The day before transfection, make a suspension of the BSRT7/5 cell line at 5 × 10^5 cells/mL in complete medium. Distribute 2 mL of cell suspension per well in a 6-well plate. Prepare one well per virus that is going to be rescued and one additional well for a negative control. Incubate the plate at 37 °C and 5% CO_2. Check that the cells are at an 80–90% confluence the next day.
3. To rescue recombinant RSV-Luc from plasmids, transfect subconfluent BSR5T7/5 cells monolayers in 6 well plates. Mix 1.25 μg of vector pACNR–rRSV-Luc, 1 μg of p-N, 1 μg of p-P, 0.5 μg of p-L, and 0.25 μg of p-M2-1 DNAs to 250 μL of Opti-MEM, incubate for 5 min at room temperature (RT).
4. In parallel, mix 10 μL Lipofectamin2000 and 250 μL Opti-MEM and incubate for 5 min at RT.
5. Mix DNA and Lipofectamine2000 solutions and incubate for 30 min at RT.

6. Rinse cells with Opti-MEM and add 1.5 mL of Opti-MEM without antibiotics.
7. Add droplets of the DNA-Lipofectamine2000 mixture to the cells (500 μL).
8. Incubate at 37 °C, CO_2 5%.
9. At 72 h post-transfection, control the presence of syncytia by phase contrast microscopy and scrape the cells.
10. Vortex vigorously for 3 minutes.
11. Centrifuge at 400× *g* for 5 minutes and recover supernatants.
12. Amplify viruses on fresh HEp-2 cells at 37 °C in MEM 2% FCS at low multiplicity of infection (0.01 p.f.u. per cell) for 3 days, scrape the cells, vortex, and centrifuge for 5 minutes at 1000 g.
13. Make virus stocks in 1X conservation buffer by flash freezing them in liquid nitrogen (aliquots of 500 μL) and using appropriate freezing bulbs. Alternatively, immerse tubes in −80 °C ethanol precooled with dry ice (*see* **Note 7**).
14. Store at −80 °C (*see* **Note 2**).

3.3 Virus Titration by a Plaque Assay

1. Prepare a sterile microcrystalline cellulose suspension (2.4% [w/v] in water). Dilute 2.4 g of microcrystalline cellulose Avicel 581 in 100 mL of distilled water and autoclave at 121 °C for 20 min. Store at room temperature (the suspension is stable for 1 year). After opening the solution in a sterile environment, store it at 4 °C for 6 months. Always mix the suspension before use (by hand shaking or vortexing) to make sure it is homogeneous.
2. Plate HEp-2 cells in 12 well plates the day before.
3. The next day, prepare the 2x MEM in a sterile environment. Dilute commercial MEM 10x with sterile water and add L-glutamine, 1000 units/mL penicillin, and 1 mg/mL streptomycin. Shake the dilution vigorously and store it at 4 °C.
4. Prepare six tubes containing 900 μL of MEM without FCS per virus to be titrated (the titration tubes). Thaw the virus aliquots, vortex them vigorously for 5 s, and transfer 100 μL to the first titration tube.
5. Perform a tenfold dilution 6x, as follows. Add 100 μL of virus to 900 μL of medium in the first tube, put the cap on the tube, and mix its contents by vortexing for a few seconds. Change the tip on the pipette, add 100 μL of the first dilution to 900 μL of medium in the second tube, put the cap on the tube, and vortex. Repeat the procedure until the sixth tube (*see* **Note 8**).
6. Write the virus name and the fold dilutions on the HEp-2 12-well plates. Add a mark to match the plate and its cover because they may be separated during staining (**step 4.9**).

Remove the medium from the plates and distribute 400 μL of one dilution per well. Incubate the plates at 37 °C for 2 h, for virus adsorption (*see* **Note 9**).

7. Prepare the microcrystalline cellulose overlay during the virus adsorption (extemporaneous preparation). To obtain 100 mL of overlay, mix 10 mL of 2.4% microcrystalline cellulose suspension, 10 mL of 2x MEM, and 80 mL of MEM with 2% FCS.
8. Add the microcrystalline cellulose suspension and the MEM and mix vigorously.
9. At the end of the 2 h incubation, add 2–3 mL of overlay to each well of the 12-well plates without removing the inoculum. Be careful to avoid the contamination of the adjacent wells with high viral titer inocula. Incubate the plate at 37 °C and 5% CO_2 for 6 days. Do not move the plate and do not move the incubator during incubation.
10. After 6 days at 37 °C and 5% CO_2, remove the cellulose medium and reveal plaques by adding 0.5% crystal violet/ 20% ethanol solution staining to the cell layers for 30 min. Wash 3 times with tap water, drain them invertedly for a couple of minutes, and count the number of p.f.u. per well by the naked eye.
11. Calculate virus titers. Count the plaques in the wells of the dry plates, which are visible to the naked eye. Check that the number of plaques of the different dilutions is coherent (factor 10 between each dilution). Choose the well on which the plaques are the easiest to count. Assess the number of plaques versus the inoculum volume and the dilution.

3.4 Mouse Infections

1. Anesthetize mice with a mixture of ketamine and xylazine (1 and 0.2 mg per mouse, respectively, I.P. route) (*see* **Note 3**).
2. Infect intranasally with 50 μL of PBS containing 6×10^4 p.f.u. of rRSV-Luc per mouse.
3. Control mice are inoculated with PBS only.

3.5 In Vivo Luminescence Measurements

1. Anaesthetize mice with a mixture of ketamine and xylazine (1 and 0.2 mg per mouse, respectively, I.P. route), or daily using isofluorane gas anesthesia (*see* **Notes 3**, **4**, and **5**).
2. Inoculate 50 μL of PBS containing D-luciferin using $0.75 \text{ mg} \times \text{kg}^{-1}$ by intranasal route.
3. Measure luminescence 5 min after intranasal injection of luciferin (*see* **Note 4**).
4. Measure photon emission of mock-infected and rRSV-Luc-infected mice using the IVIS 200 imaging system (Fig. 2). Use the Living Image software to measure the luciferase activities. Bioluminescence images are acquired for 1 min with

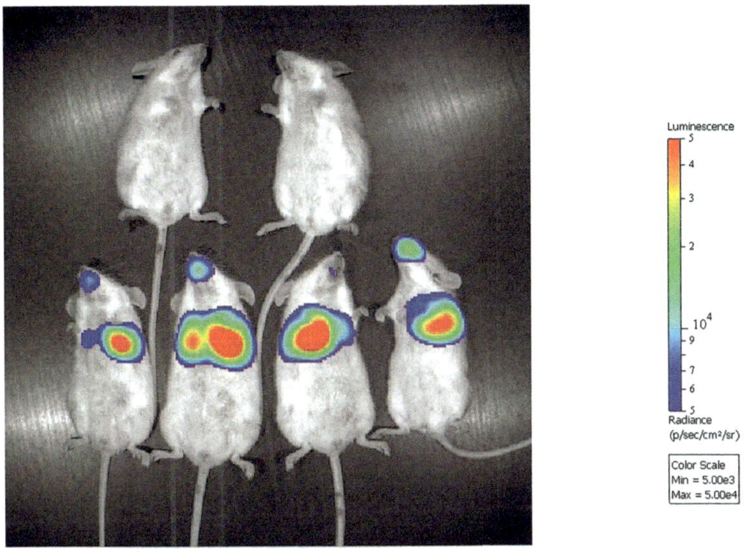

Fig. 2 RSV-Luc replication in whole BALB/c mice visualized by luminescence using the IVIS system. Adult BALB/c mice were intranasally infected with rRSV-Luc. At 4 days post-infection (d.p.i.), mice received luciferin intranasally (1.5 mg under 50 μL), and luciferase activity was visualized in the IVIS-200 imaging system. The two mice at the top of the image were mock-infected. The scale on the right indicates the average radiance: the sum of the photons per second from each pixel inside the region of interest/number of pixels (p/s/cm2/sr). A red signal indicates a high radiance and, consequently, a high luciferase activity, whereas a blue signal indicates a low radiance and a low replication

f/stop 1/4 1 and binning 1/4 8. A digital false-color photon emission image of the mice is generated, and photons are counted within a constant region of interest corresponding to the surface of either the chest encompassing the whole airway area or, specifically the nose or lungs of each animal. Photon emissions are measured as radiance in $ps^{1}cm^{2}sr^{1}$. The measure can be done every 2 days when mice are anesthetized with a mixture of ketamine and xylazine and daily with isoflurane.

4 Notes

1. The RSV genomic cDNA and the L gene can be toxic for bacteria at 37 °C; thus, they must be grown at 25–28 °C for plasmid amplification in *E. coli* DH10B cells.
2. The virus titers can drop after one year of storage at −80 °C.
3. Animal experiments must be approved by the local ethics committee and performed according to the local animal care rules.
4. Only white mice like the BALB/c stain can be used for luminescence measurement on live animals.

5. The peak of infection by RSV-Luc on Balb/c adult mice is 4 days post-inoculation.
6. Avicel is the foreign trade name, composed by mixing the product with sodium carboxymethyl cellulose (CMC) in different proportions. For example, Avicel -RC-581 contains 8.5% (±1)CMC.
7. Label tubes with alcohol-resistant tags.
8. It is very important to change the tip for each dilution.
9. Change the pipette tip between each inoculum or proceed from lowest to highest concentration with the same tip. Inoculate a limited series of plates (1–2) simultaneously to avoid the cells drying.

References

1. Bem RA, Domachowske JB, Rosenberg HF (2011) Animal models of human respiratory syncytial virus disease. Am J Physiol Lung Cell Mol Physiol 301(2):L148–L156. https://doi.org/10.1152/ajplung.00065.2011
2. Taylor G (2017) Animal models of respiratory syncytial virus infection. Vaccine 35(3): 469–480. https://doi.org/10.1016/j.vaccine.2016.11.054
3. Altamirano-Lagos MJ, Diaz FE, Mansilla MA, Rivera-Perez D, Soto D, McGill JL, Vasquez AE, Kalergis AM (2019) Current animal models for understanding the pathology caused by the respiratory syncytial virus. Front Microbiol 10: 873. https://doi.org/10.3389/fmicb.2019.00873
4. Rameix-Welti MA, Le Goffic R, Herve PL, Sourimant J, Remot A, Riffault S, Yu Q, Galloux M, Gault E, Eleouet JF (2014) Visualizing the replication of respiratory syncytial virus in cells and in living mice. Nat Commun 5:5104. https://doi.org/10.1038/ncomms6104
5. Buchholz UJ, Finke S, Conzelmann KK (1999) Generation of bovine respiratory syncytial virus (BRSV) from cDNA: BRSV NS2 is not essential for virus replication in tissue culture, and the human RSV leader region acts as a functional BRSV genome promoter. J Virol 73(1): 251–259
6. Ruggli N, Tratschin JD, Mittelholzer C, Hofmann MA (1996) Nucleotide sequence of classical swine fever virus strain Alfort/187 and transcription of infectious RNA from stably cloned full-length cDNA. J Virol 70(6): 3478–3487. https://doi.org/10.1128/JVI.70.6.3478-3487.1996

Chapter 19

Cotton Rat Model of Respiratory Syncytial Virus Infection

Jonathan Miller and Stefan Niewiesk

Abstract

The hispid cotton rat (*Sigmodon hispidus*) has demonstrated excellent translational and predictive value in the study of respiratory syncytial virus (RSV). While the cotton rat model has many advantages over the mouse model, unfamiliarity with the species can serve as a deterrent to its use. Here, we review the most common protocols for studying RSV in the cotton rat model, including housing and handling, anesthesia and intranasal infection, drug administration, blood draws, and postmortem tissue collection. Using these basic protocols, investigators can pursue downstream applications similar to those available in other animal models, limited only by available species-specific or cross-reactive reagents.

Key words Cotton rat, Animal model, Respiratory syncytial virus, *Sigmodon hispidus*

1 Introduction

Several animal models have been used to study respiratory syncytial virus (RSV) pathogenesis and to develop interventions, including cotton rats (*Sigmodon hispidus*), Syrian hamsters, mice, ferrets, guinea pigs, sheep, and nonhuman primates [1]. Of these, cotton rats have emerged as the preferred model for testing vaccines and therapeutics due to the predictive value of findings for humans in this animal model. They have been the primary model used for the development of prophylactic antibody formulations such as palivizumab (Synagis) and Nirsevimab (Beyfortus) [2, 3]. Cotton rats accurately mirror vaccine-enhanced disease following immunization with formalin-inactivated RSV and subsequent challenge, and their use in safety and efficacy trials is now standard in the vaccinology industry [4].

Experimentally, the cotton rat model has several advantages, leading to its widespread use in studies of RSV infection. Cotton rats are as much as 1000-fold more permissive to RSV replication than common inbred mouse strains [5]. In addition to lab-adapted strains of RSV, clinical isolates and recombinant viruses (including those with attenuating mutations in the NS, SH, L, F, and/or G

Bo Liang and Rachel Fearns (eds.), *Human Respiratory Syncytial Virus: Methods and Protocols*, Methods in Molecular Biology, vol. 2948, https://doi.org/10.1007/978-1-0716-4666-3_19,

proteins) grow well in cotton rats [6–9]. Cotton rats express an RSV receptor, CX3CR1, on the respiratory epithelium [10] and RSV replication occurs at a similar magnitude in both the upper and lower respiratory tracts. Cotton rats mirror the airway anatomy of humans in their distribution of ciliated respiratory epithelium through the respiratory bronchioles. By contrast, mice have short or absent respiratory bronchioles and significantly less ciliated columnar epithelium in their pulmonary parenchyma as a result [11]. Consequentially, in both cotton rats and humans, RSV targets ciliated epithelial cells, whereas in the mouse, it infects pneumocytes [12, 13]. Cotton rats have fully functional Mx genes encoding the Mx1 and Mx2 proteins that contribute to innate antiviral defense, as seen in humans [14]. This contrasts with laboratory mice with absent or defective cytoplasmic Mx proteins.

While cotton rats have excellent translational value for RSV studies, several limitations to the model should be considered before and during experimental design. RSV replicates to high titers in the upper and lower airways in cotton rats but does not induce significant clinical disease. The list of species-specific reagents for cotton rats is growing (*see* **Note 1**) but lags significantly behind mice. As such, there are some applications for the model that are not feasible at this time. Finally, cotton rats are unfamiliar animals for most investigators. Initiating studies with any new animal model can be challenging, and the relative paucity of instructional materials regarding cotton rat use may feel like an obstacle. This series of protocols relating to common methods using cotton rats in RSV research seeks to alleviate this challenge. In addition, several reviews detailing the cotton rat model and its use in infectious disease research are available [4, 5, 15–17].

2 Materials

2.1 Housing and Handling

1. Inbred hispid cotton rats. Sources include Inotiv (formerly Envigo), Ace Animals, and Sigmovir.
2. Individually vented, polycarbonate microisolation cages sized for rats (approximately 900 cm^2; *see* **Note 2**).
3. Dry pelleted rodent feed, fed ad libitum (*see* **Note 3**).
4. Hard plastic environmental stimuli (e.g., petite Nylabones or short sections of polyvinyl chloride pipe) and/or hazel wood sticks.
5. Chip-wood or corncob bedding.
6. A custom-made handling device is constructed by welding a pair of barbecue tongs to two half-cylinders of wire cage lid (*see* Fig. 1 for schematic). We have had success working with local workshops or university physics labs to make and reproduce

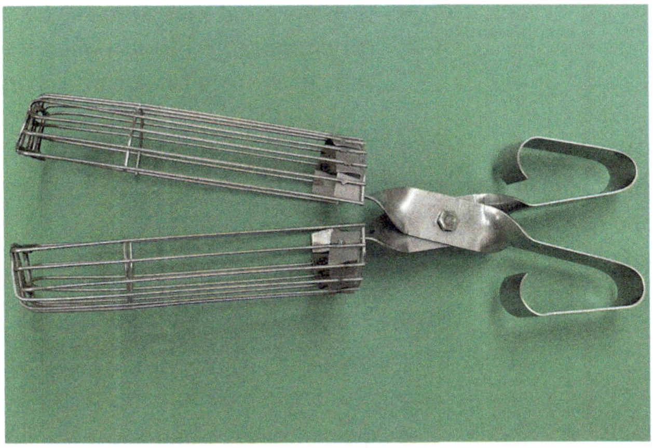

Fig. 1 Cotton rat handling device. Two half cylinders measuring 17 cm in length and 8 cm in diameter are constructed from metal caging. The half cylinders are welded to a pair of barbecue tongs [30]

this device. For an alternative to handling cotton rats use leather garden gloves, *see* **Note 4**.

2.2 Anesthesia and Intranasal RSV Infection

1. Liquid isoflurane.
2. Isoflurane vaporizer system with vacuum exhaust system or charcoal scavenging system.
3. Anesthesia induction chamber.
4. Biosafety cabinet station was set up with sterile pads, pipettor, and tips, and RSV diluted in PBS to the desired concentration for infection (*see* **Note 5**).

2.3 Drug Administration

1. Liquid isoflurane.
2. Isoflurane vaporizer system with vacuum exhaust system or charcoal scavenging system.
3. Anesthesia induction chamber.
4. Biosafety cabinet station (for RSV-infected cotton rats) or other clean workstations, set up with sterile pads.
5. Pipettor and tips (intranasal administration), syringe and 22-gauge or smaller needle (subcutaneous administration), syringe and 23-gauge or smaller needle (intramuscular or intraperitoneal administration), or 16–18-gauge plastic gavage tubes (oral gavage).

2.4 Retroorbital Blood Collection

1. Liquid isoflurane.
2. Isoflurane vaporizer system with vacuum exhaust system or charcoal scavenging system.
3. Anesthesia induction chamber.

4. Biosafety cabinet station set up with sterile pads, capillary tubes, suitable blood collection microtainers (e.g., yellow-top activator tubes for serum separation), and an empty syringe.

2.5 Collection and Homogenization of Lung and Nasal Turbinate for RSV Titration

1. Sterile surgical tools for dissection. Include a separate pair of Metzenbaum forceps for collecting nasal samples.
2. Sterile razor blades.
3. Spray bottle with 70% ethanol.
4. Pre-weighed and labeled microcentrifuge tubes for tissue samples.
5. Dulbecco's Modified Eagle Medium (DMEM).
6. Sterile glass Dounce homogenizers.
7. Sterile mortars and pestles.
8. Sterile sand.
9. Labeled cryovials.
10. Dry ice.

2.6 Collection of Lung and Nasal Tissue for Histological Analysis

1. Sterile surgical tools for dissection.
2. Spray bottle with 70% ethanol.
3. 3 mL syringe filled with 10% neutral buffered formalin (*see* **Note 6**) attached to a 22-gauge needle.
4. Suture thread (material and thickness are not critical).
5. Container for each tissue per animal filled with 10 mL of 10% neutral buffered formalin.
6. 0.35 M EDTA solution in distilled water (*see* **Note 7**).
7. Cutting board.
8. Histology-trimming blade.
9. Histology tissue cassettes.

3 Methods

Follow all institutional environmental health and safety guidelines when working with isoflurane or other vaporized anesthetic to minimize exposure to waste anesthetic gas.

3.1 Housing and Handling

Cotton rats are diurnal in laboratory settings. Facilities should be kept at 22 ± 2 °C, 20–70% relative humidity, and a 12:12 h dark: light cycle. Biosafety protocols for RSV infection do not differ from other immunocompetent laboratory rodents. Cotton rats differ from mice in that the federal requirements in the United States for their transportation, sale, and handling are covered by the US

Department of Agriculture Animal Welfare Act (9 CFR AWA). For investigators seeking to maintain a breeding program, *see* **Note 8**.

1. Cotton rats can be maintained in same-sex groups of up to 3 per cage in standard polycarbonate microisolation cages sized for rats (approximately 2065 cm^2 floor area; *see* **Note 2**). For suggestions regarding housing to minimize fighting, *see* **Note 9**.
2. Allow ad libitum access to standard dry pelleted rodent chow and water. Importantly for bottle-based systems, cotton rats tend to drink more than rats (>12 mL water per 100 grams of body weight per day).
3. Include hard objects in cages as environmental stimuli and to wear down cotton rats' hypselodont (continuously growing throughout life) incisors. Options include petite nylon bones (manufacturers include Bio-Serv and Nylabone) and pieces of polyvinyl chloride pipe.
4. Use chip-wood or corncob bedding for both breeding and maintenance.
5. Cotton rats bite readily as a defensive mechanism. Using a custom-made handling device (Fig. 1) is strongly recommended to reduce stress during handling and minimize risk of injury to the handler. Using such a device, the hand of the handler remains far enough away to not alarm cotton rats. *See* **Note 4** for an alternative handling strategy if construction of such a device is not feasible. Never grab or hold a cotton rat directly by the tail, since they attempt to free themselves in a twisting motion, and the tail will deglove.

3.2 Anesthesia and Intranasal RSV Infection

1. Set up a sterile infection station in a biosafety cabinet prior to inducing anesthesia. Dilute the stock of RSV in PBS to the desired concentration (*see* **Note 5**). Minimizing time under anesthesia reduces the risk of hypothermia and speeds up animal recovery.
2. Set up a sterile pad within the biosafety cabinet. Placing the cotton rats on a padded surface decreases thermal transfer compared to metal and reduces the risk of hypothermia.
3. Draw up 100 μL of RSV suspension with a pipette. *See* **Note 10** regarding volume considerations for intranasal inoculation.
4. Connect an anesthesia vaporizer system with liquid isoflurane to an anesthesia induction box.
5. Set the isoflurane concentration to 4–5% and turn on oxygen flow (2 liters per minute). Allow the isoflurane concentration to increase in the induction chamber for 1 min before adding cotton rats.

6. Gently drop cotton rats into the anesthesia induction chamber. If the cotton rat was captured using a handling device, open the arms in the induction chamber and gently shake downward. Cotton rats usually fall recumbent after a few seconds in the induction chamber, but be prepared to quickly close the chamber to reduce the risk of animal escape and minimize exposure of the researcher to isoflurane.
7. Cotton rats are typically in a light plane of anesthesia within 1 min. To check if a cotton rat is completely anesthetized, open the induction chamber and lift by the scruff. If it is completely limp, proceed to **step 8**. If it arches its neck and/or shoulders dorsally, place the cotton rat back in the induction chamber until it is completely anesthetized. Do not lift a cotton rat by the tail, even when anesthetized, as they are more susceptible to degloving than other rodent species. As a rule of thumb, a cotton rat will stay anesthetized for the same amount of time it spent in the anesthesia chamber.
8. Transfer the anesthetized cotton rat to the biosafety cabinet and hold it in your non-dominant hand. Hold the cotton rat in dorsal recumbency with the lower back sitting on the sterile pad or your palm and the upper back supported by your pinky, ring finger, and middle finger (Fig. 2). Support the back of the

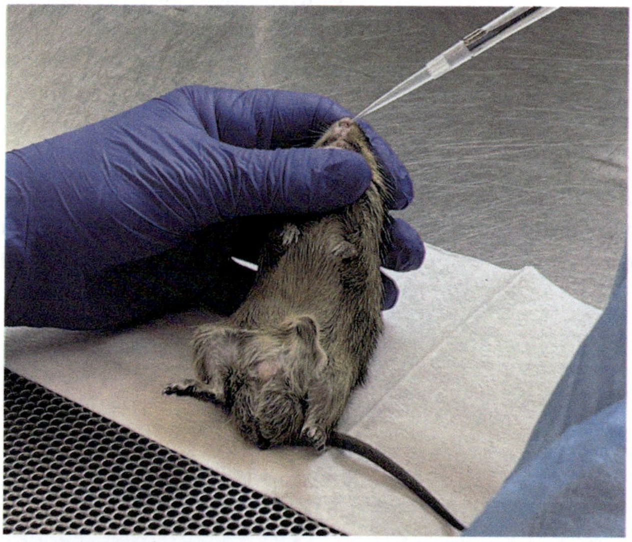

Fig. 2 Positioning of anesthetized cotton rats for intranasal inoculation. The lower back rests on a sterile pad while the handler supports the upper back with the pinky, middle finger, and ring finger of the non-dominant hand. The handler places their index finger on the back of the head while gently holding the mouth closed with their thumb, being careful not to press on the neck. The elbow of the dominant hand is rested on the biosafety cabinet surface for stabilization while the inoculum is slowly dispensed directly onto the nares

head with your index finger so that the nose is oriented vertically. Gently press upward on the mandible with your thumb to hold the mouth closed. It is important not to apply significant pressure to the neck, as the trachea will be occluded.

9. With your dominant hand, slowly dispense the pre-drawn RSV suspension directly onto the nasal planum with the pipette. Pause dispensing periodically when a bead has formed on the nose, as it can take several seconds for the cotton rat to inhale. Once the bead of liquid has been inhaled, resume slowly dispensing.
10. Transfer the cotton rat back to its cage and lay it gently on its back or side.
11. Inhalant anesthetic is the preferred method of anesthesia for infection, injections, and blood drawing. For injectable anesthesia, *see* **Note 11**.

3.3 Drug Administration

1. The standard routes of administration of experimental substances in cotton rats are intranasal, subcutaneous, intraperitoneal, intramuscular, and oral applications.
2. Intranasal administration is a reliable route. Follow Subheading 3.2, **steps 4–8**, for anesthesia and intranasal inoculation. *See* **Note 10** regarding volume considerations—these are important when administering multiple boluses within a 24-h timespan.
3. Subcutaneous, intraperitoneal, and intramuscular administration can be performed as in mice or rats. We typically use a 22-gauge needle or smaller for subcutaneous injections and a 23-gauge needle or smaller for other routes. Restrict injection volumes to 10 mL/kg body weight for intraperitoneal injection, 0.2 mL/site for intramuscular injection, and 5 mL/kg for subcutaneous injection.
4. Oral administration can be performed by gavage in anesthetized cotton rats, following institutional protocols for gavage in rats. We recommend using 16–18-gauge, rounded-tip, flexible plastic gavage tubes. After insertion of the gavage tube into the stomach, the placement may be checked by palpating the rounded tip through the abdominal wall.
5. Intravenous administration of substances in cotton rats is not feasible because the tail is very fragile, and the tail vein is not visible through the dark skin. Intracardiac injection has been described [18] but should only be attempted by trained individuals if necessary.

3.4 Retroorbital Blood Collection

1. The retroorbital plexus is the recommended blood sampling site in cotton rats due to ease of access and the relatively large sample volume that can be obtained. In trained hands, adverse effects are rare. However, in unskilled hands, retroorbital bleeding has greater potential for clinical complications than other routes. This guide is intended as an introduction to the workflow of retroorbital bleeding with some helpful considerations. Any inexperienced personnel should train with an experienced veterinarian or laboratory animal technician prior to using this technique. *See* **Note 12** regarding alternative blood sampling sites.
2. Prior to inducing anesthesia, lay out a workstation in a biosafety cabinet with a sterile pad, capillary tubes, blood collection tubes, and an empty syringe.
3. Retroorbital bleeding must be performed under general anesthesia. Follow the isoflurane anesthesia protocol outlined in Subheading 3.2, **steps 4–7**.
4. Transfer the cotton rat to the sterile pad and place it in lateral recumbency.
5. Insert a capillary tube at the medial canthus at a 45-degree angle oriented toward the back of the orbit. Once resistance is met, pull back slightly until blood flows up the tube. In order to encourage blood flow, a gentle circular twisting motion may be used.
6. Once a suitable blood volume (*see* **Note 13**) has been obtained, transfer the blood to a collection tube. Set the tip of the capillary tube into the collection tube and slowly eject 1 mL of air from an empty syringe into the clean end of the capillary tube.
7. Hold the eyelids of the sampled eye closed with a pad and apply light pressure to help initiate hemostasis. After 20 s, or once the cotton rat starts arching its neck or shoulders dorsally, transfer the cotton rat back to its cage.
8. Monitor cotton rats following the procedure to ensure adequate hemostasis.
9. A minimum interval of 2 weeks is required between repeated samplings of the same retroorbital plexus to allow the tissue to heal and reestablish normal blood flow. Alternating orbits can allow an investigator to collect blood by this route weekly, but repeated sampling still carries increased risk of prolonged hemorrhage and hematoma formation.

3.5 Collection and Homogenization of Lung and Nasal Turbinates for RSV Titration

1. Any standard method of euthanasia for laboratory animals can be performed in cotton rats. We routinely euthanize by CO_2 inhalation; briefly, cotton rats are placed in a transparent rodent euthanasia chamber with a CO_2 flow rate of 50% of the chamber volume per minute. Following recumbency and cessation of agonal breathing, confirm death by touching the cornea and observing the lack of a palpebral reflex. For all methods of euthanasia, follow the laboratory animal guidelines in the AVMA Guidelines for the Euthanasia of Animals: 2020 Edition.
2. Perform all dissections and sample collections on a sterile surface. Wet the ventral midline and head with 70% ethanol to minimize the risk of contamination of samples with hair.
3. Open the abdomen of the euthanized cotton rat and cut the renal artery to minimize blood within the lung.
4. Open the thorax and collect a standardized sample of lung tissue. We prefer to collect the left lung lobe for viral titration. Place the lung sample in a pre-weighed and labeled microcentrifuge tube on ice.
5. To collect the nasal turbinates, reflect the skin away from the skull and remove the mandible from the maxilla. It is helpful to perform this step while the head is still attached to the body.
6. Decapitate the cotton rat. Bisect the skull with a mid-sagittal cut using a sharp, sterile razor blade.
7. Collect the nasal turbinates and mucosa by scraping each bisected nasal cavity with a pair of Metzenbaum forceps. Place the sample in a pre-weighed and labeled microcentrifuge tube on ice.
8. Weigh lung and nasal turbinate samples.
9. For lung samples, add 2 mL of chilled DMEM into a sterile 15 mL glass Dounce homogenizer on ice. Transfer a lung sample to the Dounce homogenizer and grind the tissue until no visible lung pieces remain. With a sterile serological pipette, transfer the lung homogenate to a labeled cryovial. Snap freeze the homogenate on dry ice and store at −80 °C (*see* **Note 14**) until ready to titer. For an alternative to manual lung homogenization, *see* **Note 15**.
10. For nasal turbinate samples, add 3 mL of chilled MEM into a sterile mortar on ice. Add a pinch of the sterile stand using a metal spatula. Transfer a nasal turbinate sample to the mortar using sterile forceps. Grind the nasal turbinate until no visible tissue pieces remain. With a sterile serological pipet, transfer 2 mL of the homogenate to a labeled cryovial, minimizing aspiration of the sand. Snap freeze the homogenate on dry ice and store at −80 °C (*see* **Note 14**) until ready to titer.

3.6 Collection of Lung and Nasal Tissue for Histological Analysis

1. Following Subheading 3.4, **steps 1–3**, euthanize cotton rats, wet the ventral abdomen and thorax with 70% ethanol, open the abdomen, and cut the renal artery.
2. Reflect the skin from the ventral thorax, transect the diaphragm, and fully remove the ventral rib cage to visualize the thoracic cavity.
3. Hold the trachea near the first bifurcation of primary bronchi with a pair of curved or Metzenbaum forceps. Transect the trachea and esophagus in the mid-cervical region.
4. With forceps, maintain a loose grip on the trachea to not occlude the lumen. Hold the trachea vertically to prevent back-flow of formalin. Using a 3 mL syringe filled with 10% neutral buffered formalin (*see* **Note 14**) attached to a 22-gauge needle, slowly eject formalin into the trachea at a rate of <1 mL per 5 s. Once all lobes are inflated and slightly pale, stop ejecting formalin. The accessory lobe is typically the last to inflate, and as such can serve as a queue to stop.
5. Withdraw the needle. Tie off the trachea near the tracheal bifurcation with 5 inches of suture to ensure that the lungs stay inflated during fixation.
6. Holding the trachea and reflecting ventrally and caudally, gently pull the pluck (lungs, heart, and mediastinum) away from the thoracic body wall. Cut or bluntly dissect any connective tissue attachments to the body wall. Transect the esophagus, aorta, and caudal vena cava at the level of the diaphragm to free the pluck.
7. Remove the heart from the pluck. Place the lungs in a labeled, sealable container with 10 mL of 10% neutral buffered formalin to maintain a formalin:tissue volume ratio of 10:1 or greater.
8. Decapitate the head and remove the skin. Remove the mandible. Place the skull in a labeled, sealable container with 10 mL of 10% neutral buffered formalin to maintain a formalin:tissue volume ratio of 10:1 or greater.
9. After fixing the head for 24 h in formalin at room temperature, transfer into ≥20 mL of 0.35 M EDTA solution in distilled water (*see* **Note 7**). Change the decalcification solution daily and keep at room temperature until the skull is decalcified (14 to 21 days; required time increases with age).
10. Trim a standardized section of each lung lobe for histologic processing after ≥24 h of fixation in formalin at room temperature (*see* **Notes 16** and **17**). In a fume hood, remove the lungs from formalin and place them on a cutting board. Using a trimming blade, cut sections that are 3–5 mm thick. From the left lobe, right caudal lobe, and right middle lobes, trim a longitudinal section incorporating the lobar bronchus. From

the right cranial lobe and accessory lobe, trim transverse sections (cross-sections). For helpful diagrams, utilize the Revised guides for organ sampling and trimming in mice and rats [19] protocol for lung trimming in inhalational studies in rats. Place trimmed sections face down in a histology cassette in 10% neutral buffered formalin and submit to your histology laboratory for processing.

11. Decalcification of the skull is complete when a sharp histology-trimming blade can easily cut through the tissue. Encountering hard resistance with the blade indicates that the tissue is still mineralized and should be further decalcified. In a fume hood, remove the decalcified skull from formalin and place it on a cutting board. Cut 3–5 mm thick, standardized, transverse sections of the rostral skull through the nasal cavity at the 4 following landmarks: immediately caudal to the incisors; at the level of the incisive papilla; halfway between the incisive papilla and the first molar teeth; and at the level of the first molar teeth. For helpful diagrams, utilize the Revised guides for organ sampling and trimming in mice and rats [19] protocol for nasal cavity trimming in rats. Place trimmed sections face down in a histology cassette in 10% neutral buffered formalin and submit to your histology laboratory for processing.
12. Histochemical stains rely on chemical constituents of tissue for differential staining and typically have minimal inter-species variability. As such, most histochemical stains can be used for cotton rat tissue as for any other laboratory animal species. Examples with application for RSV research include routine hematoxylin and eosin, periodic acid-Schiff or Alcian blue for mucus production quantification [20], Giemsa to highlight mast cells, and Luna to highlight eosinophils [20, 21].

4 Notes

1. The list of possible downstream assays and experiments using RSV-infected cotton rats is growing as additional reagents for the species are identified. R&D Systems (Minneapolis, MN) has numerous recombinant cotton rat cytokines and antibodies against cotton rat proteins. Recently, the number of newly characterized cross-reactive antibodies has outpaced cotton rat-specific antibodies. As such, investigators should review the literature for instances of cross-reactive antibodies when considering flow cytometry, immunohistochemistry, or Western blot assays. Sequences for many cotton rat genes are partially or completely characterized, making primer design relatively straightforward. A transcriptome for the cotton rat

has been published [25] and genome information is available (taxonomy ID: 42415) [26, 27]. Temporary transgenesis of cotton rat lungs is accomplished through the application of adenovirus-associated vector type 5, with expression of foreign protein in cotton rat airway epithelium for at least 4 weeks [28]. Temporary *in vivo* reduction in the expression of proteins in the lung can be performed through intranasal delivery of peptide-conjugated morpholino oligomers [10]. Other assays, such as routine histopathology and serum chemistry panels, do not require species-specific reagents, but interpretation should be made in light of species variations and reference intervals (as an example, cotton rats have a high baseline percentage of eosinophils in peripheral blood and lung) [21]. Pulmonary function analysis using arterial blood gases and lung mechanical properties as metrics can be performed using the flexiVent® system (SCIREQ, Montreal, Quebec) [20].

2. Cotton rats should be housed in polycarbonate cages sized for rats. We keep 3 cotton rats per cage or one breeding pair with its litter before weaning at 21 days.
3. Cotton rat nutritional requirements are similar to rats, and standard rodent chow is adequate for nutrition during both maintenance and reproduction [22, 23].
4. Protective gloves, such as leather garden gloves, can be used instead of a custom-made device for handling cotton rats. Grab and hold animals by the scruff or torso, never the tail. Moving animals for cage changing or anesthesia can be performed inside a tall plastic tub or barrel as a secondary container to prevent escape.
5. Intranasal inoculation with 10^5 $TCID_{50}$ is widely accepted as the standard dose for vaccine challenge or testing antiviral efficacy. We and others [4] have found that increasing the dose of RSV in challenge studies can promote breakthrough infections to stratify vaccine candidates. Higher doses of RSV also induce increased pulmonary histopathologic lesions, but doses of 10^6 $TCID_{50}$ and higher may not represent realistic natural conditions.
6. Fresh 4% formaldehyde (paraformaldehyde) can be used interchangeably with 10% neutral buffered formalin for fixation and lung inflation.
7. There are numerous alternative decalcification solutions that have been used in laboratory settings [29]. We highly recommend avoiding even brief decalcification using strong acids such as nitric acid or hydrochloric acid, as overexposure of tissue to these acids impairs differential staining. In our experience, 0.35 M EDTA solution results in more rapid decalcification of cotton rat skulls compared to other mild solutions, such

as 10% formic acid. Decalcification on a slow shaker can speed up the process. Decalcification should be performed at room temperature; decalcifying fixed tissues at 4 °C is unnecessary, and the process is significantly slowed at lower temperatures. Many commercially available mild decalcification/fixation solutions are available.

8. Maintaining a breeding program is not always necessary, as animals can be procured from sources described in Subheading 2.1 **item 1**. One advantage of breeding cotton rats in-house is a more reliable access to juvenile animals, an age group that may be of interest to some investigators. Additionally, cotton rats raised in a single facility tend to be significantly less stressed and easier to handle. We separate males and females at the time of weaning and set up monogamous breeding pairs one week later or before 8 weeks of age at the latest for best results. Inter-animal aggression seems to be reduced when introducing the female into the cage of the male. Guidelines for food, water, bedding, and environmental stimuli for breeding cages are the same as for maintenance cages. Additionally, breeding cages should be set up with shreddable cotton or other nesting material. Gestation is approximately 27 days, and litters of 2–9 animals are typical. Pups can be weaned as early as 2 weeks of age, but we recommend waiting until 3 weeks. Cotton rats reach sexual maturity by 6–8 weeks of age. Breeding pairs should not be separated between litters, as re-introducing pairs increases the risk of fighting.
9. Minimizing aggression and fighting is important since cotton rats can cause fatal skin and tail wounds to one another. Males are much more prone to fighting than females. Only littermates should be housed together, and cage-mates should be placed together before 7 weeks of age [22]. We prefer to establish cage-mates at the time of weaning, or approximately 3 weeks of age. If a cotton rat presents with any fight wounds, all animals from that cage should be separated into their own cages as the aggressor will continue to fight with its cage mates, leading to severe wounds and often death.
10. Adult cotton rats tolerate 100 μL intranasal boluses well, which is the volume we typically use for RSV infection. The minimum volume for infecting or administering drugs intranasally is 50 μL to reach the lower airways reliably. The maximum volume for intranasal administration is 200 μL; going over this volume, including with multiple boluses over the course of 24 h, risks flooding the alveoli.
11. Options for injectable anesthesia in cotton rats include a mixture of ketamine (25 mg/kg) and acepromazine (2.5 mg/kg), with or without xylazine (7.5 mg/kg).

12. Other sites for blood sampling in cotton rats that have been described include a subzygomatic venous sinus and the sublingual vein [24]. The tail vein is not recommended, as it is difficult to visualize, and the tail is fragile. If collecting blood at a terminal time point at the time of euthanasia, collection from the right cardiac ventricle is reliable.
13. The amount of blood collected by retroorbital bleeding should be limited to the minimum required volume, which varies by application. The maximum volume that should be collected at once should be restricted to 7.5% of the estimated blood volume of the cotton rat, and this should be limited to once per week. Following estimates for rats, the total blood volume of a cotton rat is approximately 64 mL/kg body weight.
14. RSV in homogenates of lung and nasal turbinate is stable at − 80 °C and does not need to be titered the same day as tissue collection and homogenization. Thaw homogenates on ice when ready to measure RSV titer.
15. Lung can be homogenized with a Precellys® (Bertin Technologies, Paris) tissue homogenizer for RSV titration if used in conjunction with a Cryolys® Evolution (Bertin) cryogenic cooling system. The tissue must be maintained at 4 °C during the homogenization. We shake the left lung in 2 mL Precellys® tubes with 1.4 mm ceramic beads and 1 mL MEM at 6000 RPM, 3 times for 30 s each with 10 s breaks. Using these settings, we do not see a drop in RSV titer compared to hand homogenization.
16. Many histology laboratories can and prefer to trim tissue sections if an investigator's team is not familiar or comfortable with trimming.
17. Prolonged exposure of tissue to formalin, particularly if not properly buffered, can result in the formation of acid formaldehyde hematin pigment in tissues. While this pigment can be readily identified as an artifact in histopathological evaluation, it should be avoided.

References

1. Taylor G (2017) Animal models of respiratory syncytial virus infection. Vaccine 35(3): 469–480. https://doi.org/10.1016/j.vaccine.2016.11.054
2. Zhu Q, McLellan JS, Kallewaard NL et al (2017) A highly potent extended half-life antibody as a potential RSV vaccine surrogate for all infants. Sci Transl Med 9(388):eaaj1928. https://doi.org/10.1126/scitranslmed.aaj1928
3. Malley R, DeVincenzo J, Ramilo O et al (1998) Reduction of respiratory syncytial virus (RSV) in tracheal aspirates in intubated infants by use of humanized monoclonal antibody to RSV F protein. J Infect Dis 178(6): 1555–1561. https://doi.org/10.1086/314523
4. Boukhvalova MS, Yim KC, Blanco J (2018) Cotton rat model for testing vaccines and antivirals against respiratory syncytial virus. Antivir

Chem Chemother 26:2040206618770518. https://doi.org/10.1177/2040206618770518
5. Niewiesk S, Prince G (2002) Diversifying animal models: the use of hispid cotton rats (*Sigmodon hispidus*) in infectious diseases. Lab Anim 36(4):357–372. https://doi.org/10.1258/002367702320389026
6. Tiffany J, Rongzhang W, Olivia H et al (2021) A novel live attenuated respiratory syncytial virus vaccine candidate with mutations in the L protein SAM binding site and the G protein cleavage site is protective in cotton rats and a rhesus macaque. J Virol 95(3). https://doi.org/10.1128/jvi.01568-20
7. Stobart CC, Rostad CA, Ke Z et al (2016) A live RSV vaccine with engineered thermostability is immunogenic in cotton rats despite high attenuation. Nat Commun 7:13916. https://doi.org/10.1038/ncomms13916
8. Whitehead SS, Hill MG, Firestone CY et al (1999) Replacement of the F and G proteins of respiratory syncytial virus (RSV) subgroup A with those of subgroup B generates chimeric live attenuated RSV subgroup B vaccine candidates. J Virol 73(12):9773–9780. https://doi.org/10.1128/JVI.73.12.9773-9780.1999
9. Rostad Christina A, Stobart Christopher C, Gilbert Brian E et al (2016) A recombinant respiratory syncytial virus vaccine candidate attenuated by a low-fusion F protein is immunogenic and protective against challenge in cotton rats. J Virol 90(16):7508–7518. https://doi.org/10.1128/jvi.00012-16
10. Gia G, Johnson Sara M, Heather C et al (2021) CX3CR1 is a receptor for human respiratory syncytial virus in cotton rats. J Virol 95(16). https://doi.org/10.1128/jvi.00010-21
11. Martinez-Sobrido L, Negin G, Fernandez-Sesma A et al (2006) Protection against respiratory syncytial virus by a recombinant Newcastle disease virus vector. J Virol 80(3):1130–1139. https://doi.org/10.1128/jvi.80.3.1130-1139.2006
12. Johnson JE, Gonzales RA, Olson SJ et al (2007) The histopathology of fatal untreated human respiratory syncytial virus infection. Mod Pathol 20(1):108–119. https://doi.org/10.1038/modpathol.3800725
13. Harkema JR, Nikula KJ, Haschek WM (2013) Chapter 51 – Respiratory system. In: Haschek WM, Rousseaux CG, Wallig MA (eds) Haschek and Rousseaux's handbook of toxicologic pathology, 3rd edn. Academic Press, Boston, pp 1935–2003
14. Pletneva LM, Haller O, Porter DD et al (2006) Interferon-inducible Mx gene expression in cotton rats: cloning, characterization, and expression during influenza viral infection. J Interf Cytokine Res 26(12):914–921. https://doi.org/10.1089/jir.2006.26.914
15. Curlee JF, Cooper DM (2012) Chapter 49 – Cotton rat. In: Suckow MA, Stevens KA, Wilson RP (eds) The laboratory rabbit, guinea pig, hamster, and other rodents. Academic Press, Boston, pp 1105–1113
16. Boukhvalova MS, Prince GA, Blanco JCG (2009) The cotton rat model of respiratory viral infections. Biologicals 37(3):152–159. https://doi.org/10.1016/j.biologicals.2009.02.017
17. Green MG, Huey D, Niewiesk S (2013) The cotton rat (*Sigmodon hispidus*) as an animal model for respiratory tract infections with human pathogens. Lab Anim 42(5):170–176. https://doi.org/10.1038/laban.188
18. Rojas-Martinez A, Wyde PR, Montgomery CA et al (1998) Distribution, persistency, toxicity, and lack of replication of an E1A-deficient adenoviral vector after intracardiac delivery in the cotton rat. Cancer Gene Ther 5(6):365–370
19. Kittel B, Ruehl-Fehlert C, Morawietz G et al (2004) Revised guides for organ sampling and trimming in rats and mice – Part 2. A joint publication of the RITA and NACAD groups. Exp Toxicol Pathol 55(6):413–431. https://doi.org/10.1078/0940-2993-00349
20. Martinez ME, Harder OE, Rosas LE et al (2020) Pulmonary function analysis in cotton rats after respiratory syncytial virus infection. PLoS One 15(8):e0237404. https://doi.org/10.1371/journal.pone.0237404
21. Green MG, Petroff N, La Perle K, M.D. et al (2018) Characterization of cotton rat (*Sigmodon hispidus*) eosinophils, including their response to respiratory syncytial virus infection. Comp Med 68(1):31–40
22. Faith RE, Montgomery CA, Durfee WJ et al (1997) The cotton rat in biomedical research. Lab Anim Sci 47(4):337–345
23. Randolph JC, Cameron GN, McClure PA (1995) Nutritional requirements for reproduction in the hispid cotton rat, *Sigmodon hispidus*. J Mammal 76(4):1113–1126. https://doi.org/10.2307/1382603
24. Ayers JD, Rota PA, Collins ML et al (2012) Alternatives to retroorbital blood collection in hispid cotton rats (*Sigmodon hispidus*). J Am Assoc Lab Anim Sci 51(2):239–245
25. Rajagopala SV, Singh H, Patel MC et al (2018) Cotton rat lung transcriptome reveals host immune response to respiratory syncytial virus infection. Sci Rep 8(1):11318. https://doi.org/10.1038/s41598-018-29374-x

26. NCBI Taxonomy browser (*Sigmodon hispidus*). National Center for Biotechnology Information. Available via https://www.ncbi.nlm.nih.gov/Taxonomy/Browser/wwwtax.cgi?mode=Info&id=42415
27. Schoch CL, Ciufo S, Domrachev M et al (2020) NCBI taxonomy: a comprehensive update on curation, resources and tools. Database (Oxford) 2020. https://doi.org/10.1093/database/baaa062
28. Martinez ME, Capella Gonzalez C, Huey D et al (2021) Immunogenicity and inflammatory properties of respiratory syncytial virus attachment G protein in cotton rats. PLoS One 16(2):e0246770. https://doi.org/10.1371/journal.pone.0246770
29. Liu H, Zhu R, Liu C et al (2017) Evaluation of decalcification techniques for rat femurs using HE and immunohistochemical staining. Biomed Res Int 2017:9050754. https://doi.org/10.1155/2017/9050754
30. Niewiesk S, Voelp F, ter Meulen V (1997) A maintenance and handling device for cotton rats (*Sigmodon hispidus*). Lab Anim (26):32–33

Chapter 20

Lamb Model of RSV Infection

Sarhad Alnajjar, Jack Gallup, and Mark Ackermann

Abstract

Respiratory syncytial virus (RSV) is one of the leading causes of lower respiratory tract infections in infants and older adults. There are several animal models to study RSV infection and test the efficacy of antiviral compounds and vaccines. The lamb model has some advantageous features that contribute to it being one of the best animal models of RSV due to the similarity in size, biology, immune responses, and cellular/structural composition of the respiratory system to infants, the susceptibility to human RSV, and the possibility to investigate multiple approaches and experimental designs. Thus, the lamb model is considered and used as a preclinical model to evaluate the effectiveness and characterize the dosing and mechanisms of newly developed antiviral compounds before they move to human clinical trials. In this chapter, the detailed protocol for the lamb model of RSV infection is explained, including all the assays used to evaluate the severity of infection.

Key words Respiratory syncytial virus (RSV), Lamb model of RSV infection

1 Introduction

Respiratory syncytial virus (RSV) is one of the leading causes of RSV infection in children worldwide, with an estimated 3.4 million hospital admissions due to RSV-associated acute lower respiratory tract infections with significant mortality [1, 2]. In addition to children, RSV causes significant morbidity and mortality in older adults and immune-compromised individuals [3]. An effective RSV vaccine has recently been approved in the US for individuals ≥60 years of age, but it is still limitedly approved or not yet approved in other countries [4, 5]. However, there is still a lack of an approved RSV treatment, with some having been advanced into clinical development. Only two monoclonal antibody therapies have been approved for prophylactic use, and they are limited to

Supplementary Information The online version contains supplementary material available at https://doi.org/10.1007/978-1-0716-4666-3_20.

Bo Liang and Rachel Fearns (eds.), *Human Respiratory Syncytial Virus: Methods and Protocols*, Methods in Molecular Biology, vol. 2948, https://doi.org/10.1007/978-1-0716-4666-3_20,

high-risk individuals [6]. Animal models play important roles in the development of new therapies.

There are multiple animal models of RSV infection. However, each has its benefits and limitations [7, 8]. This chapter discusses the neonatal lamb model, a well-established preclinical RSV infection model. The most significant advantage of the lamb model is that lambs are susceptible to human RSV, and the same viral strain used in human clinical trials is used in lambs, i.e., suggesting the transferability of data to human infection. Other key criteria include similarity in respiratory anatomical and histological structure to humans and similarity in immune response to infection [9, 10]. Lambs develop Acute Lower Respiratory tract Infection (ALRTI) after nebulization with RSV, which is similar to infection developed in severely RSV-ill individuals, including severe bronchiolitis and pneumonia [11]. The similarity in immune response, ability to use colostrum-deprived lambs, development of antibodies to infection and vaccination, and transfer of maternal immunity makes lambs a good model for vaccine studies [12, 13]. Similar body size and lung size to humans make lamb lungs accessible for evaluation and procedures, provide the capability to use different routes of treatment administration, and provide an adequate sample size for testing. Different therapeutic treatment routes have been investigated in lambs, including intramuscular injections, inhalation, and oral administration [14, 15]. The limitations with the lamb model are minimal, including the difficulty of housing a large number of lambs at a BSL2 level, the cost of the project, and the availability of reagents specific to sheep. However, these have been overcome throughout the years of experience establishing the model.

This chapter will explain the methodology of the lamb model in detail by reviewing the multiple assays used in each study. The protocols will focus on laboratory assays specific to the lamb model. Some assays, such as the hematoxyline-eosin stain and RNAscope assay, will not be explained since detailed protocols are provided by the kit manufacturers or it is a standard procedure.

2 Materials

Prepare material in advance unless it needs to be used immediately. All solutions are prepared using ultrapure water. Prepare and store solution at room temperature unless indicated otherwise. pH is critical for some of the solutions. Thus, re-check the pH before use.

2.1 Animals

Neonatal lambs (Suffolk, Polypay, Dorsett cross) aged 1–3 days with 2–7 kg body weights are obtained for this experiment. Lambs should be deprived of colostrum and fed a colostrum-free milk replacer (Land O'Lakes Ultra Fresh Lamb Milk Replacer) (*see* **Note 1**).

2.2 M37 RSV Viral Stock Preparation

1. HEp-2 starter cultures in 75 cm^2 and 300 cm^2 flasks with 70–80% confluency.
2. Complete DMEM: supplement DMEM with 10% heat-inactivated FBS and 50 μg/mL kanamycin sulfate.
3. 60% sucrose solution: dissolve 60 gm sucrose in 100 mL PBS. Sterilize by filtration and store at 4 °C.
4. RSV viral inoculum.

2.3 Animal Monitoring for the Appearance of Clinical Signs and the Clinical Score

1. Stethoscope.
2. One UCT-2112 (Unified Information devices, USA) for each lamb.

2.4 Blood Sampling for PK Analysis

1. 3 mL blood collection tubes containing K_2EDTA anticoagulant.
2. 2 mL cryovials.

2.5 Lung Collection and Processing

The following materials need to be prepared 1–2 days before the necropsy.

1. Liquid nitrogen.
2. 10% neutral-buffered formalin (NBF).
3. BALF/lung retrieval solution: mixing 42.5% Iscove's modified Dulbecco's medium, 7.5% glycerol, 1% heat-inactivated FBS, 49% DMEM, and 5 μg/mL kanamycin sulfate. Store at 4 °C.
4. 1 mL Eppendorf vials filled with 1 mL TRIzol for each lamb. Store on ice before necropsy.
5. Normal 0.9% saline.

2.6 Infectious Focus Forming Units (IFFU)

1. 850 μL-capacity 0.45 μm Costar SPIN-X filters (Costar Cat No: 8162).
2. Seeding 12 well plates: 12-well culture plates seeded with HEp-2 cells grown to 70% confluency to be ready at the time of infection.
3. Complete DMEM: DMEM media supplemented with 10% heat-inactivated fetal bovine serum (FBS) and 50 μg/mL kanamycin sulfate. Sterilized by filtration and stored at 4 °C till use.
4. Acetone/ Methanol fixation solution: Add 60 mL acetone to 40 mL methanol to make a 60%/40% acetone/methanol solution. Keep it cold by storing it at −20 °C.

5. TBS-T solution: add 500 μL tween 20 for each 1 L TRIS buffered saline.
6. BSA solution: make 3% BSA solution by adding 3 gm Bovine Serum Albumin for each 100 mL TBS-T.
7. Primary antibody: dilute goat anti-RSV antibody 1:800 in BSA solution. Make immediately before use.
8. Secondary antibody: Alexa Fluor® 488 F(ab')2 fragment of rabbit anti-goat IgG 1:800 in BSA solution. Make immediately before use.

2.7 Quantitative Reverse Transcription Polymerase Chain Reaction (RT-qPCR)

1. TRIzol solution.
2. Chloroform.
3. Isopropanol: cold (−20 °C) isopropanol.
4. 75% ethanol: add 25 mL of nuclease-free water to 75 mL ethanol. Stored and used cold at 4 °C or on ice.
5. 200 μL thin-walled tubes.
6. 1.5 mL Eppendorf tubes.
7. TURBO DNA-free™ Kit.
8. RNAseOUT.
9. Probs: RSV custom probe forward primer: 5′-GCTCTTAG CAAAGTCAAGTTGAACGA; reverse primer: 5′-TGCTCCG TTGGATGGTGTATT ; hydrolysis probe: 5′-6FAM-ACACTCAACAAAGATCAACTTCTGTCATCCAGC - TAMRA.
10. qScript™ One-Step Fast qRT-PCR Kit, ROX reaction mix (Quantabio Cat No: 95080-500).
11. 96 well qPCR plates.

2.8 Immunohistochemistry (IHC) of Lung Sections

1. Bio SB TintoDetector Cap Gap Plus Slide.
2. TintoDetector slide holder.
3. Xylene.
4. Ethanol: For rehydration and dehydration, a gradual concentration of 100%, 95%, and 70% ethanol is needed.
5. 3% Peroxide in Methanol: add 60 mL of a 30% peroxide solution to 540 mL of methanol to make 600 mL of 3% peroxide in methanol.
6. pH 9. TRIS-base antigen retrieval solution: make 1 L of antigen retrieval solution by adding 1.21 g TRIZMA Base, 0.37224 g EDTA (or 2 mL of 0.5 M EDTA) to 1 L water. pH to 9.0. Add 500 μL Tween 20.
7. TintoRetriever pressure cooker.

8. TBS-T solution: add 500 μL tween 20 for each 1 L TRIS buffered saline.
9. 3% BSA solution: Make a 3% BSA solution by adding 3 gm of Bovine Serum Albumin for each 100 mL TBS-T.
10. 30% (normal swine serum) NSS solution: add 2 mL normal swine serum to 8 mL TBS-T.
11. Antibody diluent: add 10 mL NSS to 90 mL 3% BSA solution.
12. Primary antibody: dilute goat anti-RSV antibody 1:800 in antibody diluent. Make immediately before use.
13. Secondary antibody: biotinylated Rabbit anti-Goat antibody 1: 800 dilution in antibody diluent. Make immediately before use.
14. 3% peroxide in TBS-T: to make 18 mL of 3% peroxide, add 1.8 mL 60% peroxide to 16.2 mL of TBS-T. Make fresh before use.
15. SS-HRP: to make 18 mL of 1:200 Streptavidin-horse reddish peroxidase, add 90 μL SS-HRP to 17.91 mL TBS-T.
16. Nova Red working solution according to the manufacturer. Make fresh before use.

3 Methods

As stated before, the method section will explain in detail the protocol of some of the laboratory assays. The manufacturer provides a detailed protocol for other laboratory assays, such as RNAscope.

3.1 Animals

1. House the colostrum-deprived neonatal lambs in a BSL2 animal housing facility.
2. Treat lamb with prophylactic antibacterial such as Naxcel injection and oral neomycin sulfate.

3.2 Preparation of M37 RSV Viral Stock

The RSV strain we use in our lab is RSV M37, purchased from Meridian BioSciences (Memphis, TN, USA). This strain is a wild Type A RSV isolated from the respiratory secretions of an infant hospitalized for bronchiolitis [16, 17] (*see* **Note 2**).

1. Prepare starter cultures of HEp-2 cells in 75 cm^2 cell culture flasks in complete DMEM.
2. Add 200 μL of the original RSV M37 to 1 mL DMEM, then add the 1.2 mL diluted virus (4×10^6 IFFU/mL) to the 80–90% confluent HEp-2 cells in 75 cm^2.
3. Incubates at 37 °C, 5% CO_2 for 1–2 h with gentle manual rocking of the flasks every 10 min.

4. Add 11 mL of complete DMEM to each flask and incubate at 37 °C, 5% CO_2 for 2–3 days.
5. Observe the cells daily for cytopathic effects (CPE).
6. Freeze the flask/cells at 50% dead cells or more than 80% CPE and keep frozen until next step (*see* **Note 3**).
7. Prepare several HEp-2 cells in 300 cm^2 cell culture flasks and start the process when the growth reaches 70–80% confluency.
8. Thaw the RSV-infected 75 cm^2 flasks (from **step 6**) at room temperature and pool all cells and media into one 50 mL tube. Then, vortex for 10 s at high speed and centrifuge at 2500 rpm (~1250× *g*) for 10 min at 4 °C.
9. Collect the supernatant in one tube and add 4 mL for each 300 cm^2 flask.
10. Incubates at 37 °C, 5% CO_2 for 1–2 h with gentle manual rocking of the flasks every 10 min.
11. Add 25–30 mL of complete DMEM to each flask and incubate at 37 °C, 5% CO_2 for 2–3 days.
12. Observe the cells daily for cytopathic effects (CPE).
13. Harvest the cells at 50% dead cells or more than 80% CPE with sterile tissue scraper, and pool the collected cells and media into 50 mL conical tube(s) (*see* **Note 3**).
14. Vortex the tubes for 10 s at high speed and centrifuge at 2500 rpm (~1250 g) for 10 min at 4 °C.
15. Pool the resulting supernatant into a sterile 500 mL bottle and save it aside.
16. Pool all the associated pellets into one tube and resuspend in a 5–10 mL of the saved pooled supernatant.
17. Sonicate the pellet in the 50 mL tube using a tip-style sonicator for 5 × 2 s pulses with 5 s in between. *Sonication is a highly aerosolizing process. Thus, this is done strictly inside the biological safety cabinet, allowing air to settle and disinfect before moving to the next step.*
18. Centrifuge the tube at 2500 rpm (1250 g) for 10 min at 4 °C and add the supernatant to the initially saved supernate in the 500 mL bottle (**step 15**).
19. Mix the solution well and add 20 mL of 60% sucrose in PBS to every 40 mL of the viral suspension to obtain a 20% final sucrose concentration [18].
20. aliquot the final virus suspension into 50 mL tubes and small portion in several 1 mL tubes to be used for plaque assay.
21. Save the vial stocks at −80 °C until further use.

3.3 RSV Infection

1. Animal work must be approved by your organization's Institutional Animal Care and Use Committee before the project is conducted to ensure the human use and treatment of laboratory animals and avoid/reduce unnecessary distress.
2. Divide the lambs randomly into different groups according to the experimental design.
3. Put at least 6 lambs in each group.
4. have at least one RSV-infected control group, one RSV-free control group, and several treatment groups as research needs.
5. Separate RSV-infected lambs from non-infected control to avoid cross-contamination.
6. Separate the RSV-infected groups from each other by placing them in different cages.
7. Avoid having less than 3 lambs in a cage. Sheep are social animals, and separating them will create a stressful environment (consult with Laboratory animal veterinarian before work).
8. Nebulize 6 mL of 1.27×10^7 Infectious Forming Unit (IFFU)/mL for each lamb using either PARI LC Sprint™ nebulizers or Aerogen mesh nebulizers over 15–30 min [18] (*see* **Note 4**).
9. Nebulized the No-RSV control negative lambs with 6 mL mock media (cell-conditioned media with no virus).
10. Monitor the development of infection and appearance of disease in the 5 days post-RSV inoculation (*see* **Note 5**).
11. Humanely euthanize lambs at day 6 post-infection since it is the time of peak viral infection and damage to the lung.

3.4 Animal Monitoring for the Appearance of Clinical Signs and the Clinical Score

1. Check the lambs twice daily for health and wellbeing throughout their stay at the housing facility.
2. Monitor feeding, body weight, temperature, heart rate, and respiratory rate.
3. Tube feed/ train the lamb that has difficulty suckling individually until they are comfortable with the bucket feeder.
4. Upon arrival, implant the UCT-2112 subcutaneously in the inguinal region and check the temperature three times daily. The UCT-2112 can be implanted intramuscularly or intraperitoneally, but subcutaneous implantation is preferred due to the lamb's fragile and immune-deprived status.
5. Record general activity such as feeding, interaction with other lambs and individuals, recumbency, weaknesses, and stool consistency.

6. Evaluate and examine the lambs for the development of respiratory-associated signs such as respiratory rate, wheezing, expiratory effort, and cough. Wheezing and expiratory effort were evaluated by auscultation [16].

3.5 Blood Sampling for PK Analysis

1. Collect blood at different time points according to the experimental design.
2. Collect 3 mL blood from the jugular vein in K_2EDTA blood tubes.
3. Keep collected blood samples at 4 °C till centrifugation.
4. Centrifuge the blood at 1600 g for 10 min at 4 °C to obtain the plasma.
5. Aliquot the plasma to 2 mL cryovials and store at −80 °C.

3.6 Lung Collection and Processing

This process is performed at necropsy immediately after humane euthanizing of the lambs.

1. Open the thorax and examine the lung in situ, photograph the lung, and evaluate the RSV gross lesion (*see* **Note 6**).
2. Excise the right middle and right caudal lung lobes for the BALF sample collection procedure (Fig. 1).
3. Put a random piece of lung tissue ~1 gm in a 50 mL tube. Add 1/10 weight to the volume of lung retrieval solution and keep on ice. This will be used for Lung IFFU if needed.
4. Separate the remaining Left and right lung lobes (i.e., 4 lobes—Right Cranial, Left Cranial, Left Middle, and Left Caudal).
5. Take two samples from each of these lobes, place them in tissue cassettes, and put them in 10% neutral-buffered formalin (NBF) for histological and immunohistochemical analysis.
6. Place the remaining lung tissue in cryovials and snap freeze in liquid nitrogen. These samples will be used for lung pharmacokinetics, urea concentration, lung qPCR, and any other assay.

3.7 BALF Collection

BALF samples from each animal are collected immediately after euthanasia on Day 6 p.i. from the right middle and right caudal lobes [19].

1. Use a pipette gun and a 10 mL serological pipette filled with 5 mL of BALF retrieval solution.
2. Instill the excised Right Caudal lobe with the 5 mL solution through the main bronchi and retrieve it several times.
3. Retrieve the solution from the bronchi, add it to a 1.5 mL Eppendorf tube, and keep it on ice. There will be ~1–2 mL of BALF retrieved since some of the solutions will be lost within the lung lobe.

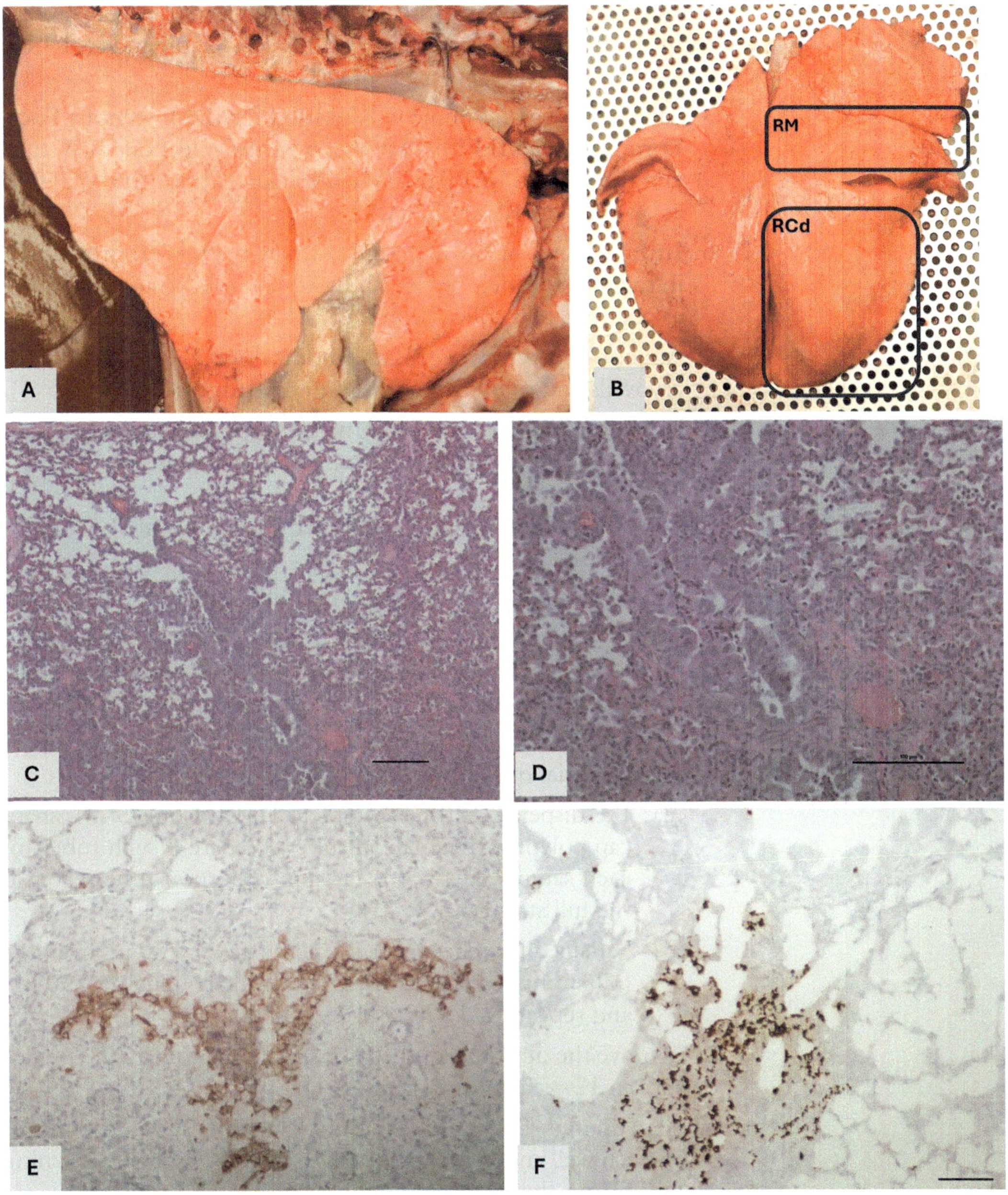

Fig. 1 Photo representation of (**a**) gross lesion associated with RSV infection in lambs, (**b**) same lung full view highlighting the Right Medial (RM) and Right Caudal (RCd) lobe that was used to harvest 1 mL BALF for compound concentration and 1–2 mL BALF in DMEM for viral IFFU assay; (**c&d**) represent microscopic lesion associated with RSV infection; (**e**) RSV antigen by IHC; (**f**) RSV mRNA by RNAscope

4. Immediately obtain 100 μL of the collected BALF using a microtiter pipette, add it to the 1 mL TRIzol, and store it at −80 °C.
5. The 1 mL BALF sample will be used for the Infectious Focus Forming Unit (IFFU) assay to assess the infectious RSV titer, and the 1.1 mL BALF in TRIzol will be used for the qRT-PCR assay to assess RSV mRNA.
6. Instill 1–3 mL of Normal Saline using a 1 mL microtiter pipette in the main bronchi of the Right Middle lobe, retrieve as much as possible, and collect it in a 15 mL tube.
7. Aliquot the retrieved Right Middle lobe BALF into 1 mL cryovials and snap freeze with liquid nitrogen.
8. The Right Middle lobe BALF will be used for pharmacokinetic analysis and blood urea nitrogen concentrations.

3.8 Infectious Focus Forming Unit (IFFU)

1. Obtain the Right Caudal lobe BALF Eppendorf tubes collected at necropsy.
2. Spin down the samples for 5 min at 3000 g to pellet large debris.
3. Collect approximately 800–850 μL of each supernatant and then spin through 850 μL capacity 0.45 μm Costar SPIN-X filter for 5 min.
4. Prepare serial dilution of the sample as follows:
   (a) Pre-dispense 720 μL complete DMEM into 5 mL snap-cap polypropylene falcon tubes. Prepare 3 such tubes for each sample, as the Costar spin collection tube serves as the full strength.
   (b) Serially dilute the samples 1:10 by moving 80 μL from the full-strength BALF to the 5 mL tube containing 720 μL, and repeat for 1:100 and 1:1000 dilution.
5. Remove the media from the HEp-2 cells grown to 70% confluence in 12-well culture plates.
6. Plate diluted samples in triplicate by adding 200 μL of sample for each well and incubate for 1–2 h at 37 °C, 5% CO_2, turning and mixing the plate every 15 min.
7. Add 1 mL complete DMEM and incubate for 48 h at 37 °C, 5% CO_2.
8. Remove the media from the wells and add 0.5 mL cold acetone/methanol (60%/40%) for 1 min.
9. Remove the fixative and let the plates dry.
10. Store the dried plates at −20 °C or proceed to the next step (staining).

11. Rehydrate the plate by adding ~1 mL TBS-T to each well and rotating on a plate rotator for 5 min.
12. Dump the TBS-T, add 1 mL 3% BSA, and rotate for 30 min.
13. Dump the blocking solution, add 325 μL primary antibody solution, and incubate overnight in a 4 °C cold room or rotator in a fridge.
14. Dump the primary antibody and wash the plates 3 times with TBS-T by adding 1 mL TBS-T, rotate for 5 min, then dump it and repeat 3 times.
15. After the third wash, add 325 μL secondary antibody solution and incubate for 30 min at room temperature.
16. Dump the secondary antibody solution and wash the plates 3 times with TBS-T.
17. Finally, add 1 mL TBS-T to each well and store the plates at 4 °C.
18. Examine the plates with a fluorescent inverted microscope using the FITX/GFP filter.
19. Count the total number of IFFUs (which is defined here as 3 or more fluorescing cells) for each well.
20. Obtain IFFU/mL calculations by multiplying the average number resulting from triplicate well counts by the initial BALF sample dilution factor and multiplying that value by 5 to obtain counts/mL since 1000 μL/200 μL (the actual sample applied for each well) = 5.

3.9 Quantitative Reverse Transcription Polymerase Chain Reaction (RT-qPCR)

3.9.1 BALF RNA Isolation (Using TRIzol)

1. Obtain the 1.1 mL Right Caudal lobe BALF in TRIzol from − 80 °C and thaw at room temperature.
2. Vortex each for 5 s, then let them stand for 5 min.
3. Using a 1 mL pipet, add 200 μL chloroform to each tube, vortex 5 s, and incubate for 3 min. at r.t., then microfuge for 15 min at 12,000 *g* at 4 °C.
4. In one motion, carefully remove and save exactly 550 μL of each aqueous (top) layer into new 1.5 mL tubes (entirely avoid the white gDNA-protein interface layer as much as absolutely possible during transfer).
5. Add 500 μL cold (−20 °C) isopropanol to each water layer, vortex briefly, and let sit for 10 min at r.t., then microfuge for 10 min at 12,000 g at 4 °C; pellets **ARE NOT** visible, but they are there!
6. Dump off isopropanol gently, and add 1 mL 75% nuclease-free ethanol to each vial, vortex briefly, then spin 5 min at 16,100× *g* at 4 °C.

7. Dump off isopropanol and add 1 mL of 75% nuclease-free ethanol to each vial. Vortex briefly, then spin for 5 min at 16,100× *g* at 4 °C.
8. Gently dump off ethanol from each and lay tubes (opened) for about 35 min to evaporate most of the ethanol away.
9. Add 100 μL of nuclease-free water to each dried pellet and vortex each tube for 5–10 s to help resolubilize the RNA pellet. To help solubilize the RNA, spin tubes down briefly, heat at 60 °C for about 1.5 min, vortex each for 10 s, spin down briefly, then store at 4 °C until the next procedure or continue with the procedure of making 1:10 dilutions (*see* **Note** 7).
10. 80 μL of each 100 μL RNA isolate is diluted 1:10 with 710 μL nuclease-free water and 10 μL RNaseOUT (to get 800 μL samples each). The samples are analyzed for RSV only.
11. For RT-qPCR: All BALF RNA samples are run at full strength (the 1:10 diluted samples). A serial ninefold dilution standard curve using a Stock I of the 1:10 BALF RNA isolates from selected known infected samples was used to compute RT-qPCR reaction efficiencies for each target (*see* **Note 8**).
12. qScript™ One-Step Fast RT-qPCR Kit, ROX reaction mix was used to assess the RNA samples according to the manufacturer-recommended procedure, in which 5 μL of the 4X reaction buffer is mixed with 7.2 μL of custom TaqMan probe mixture (containing 2.691 μM of each primer and 0.5208 μM probe), 5.8 μL nuclease-free water, and 6 μL RNA sample.

3.9.2 Lung Tissue RNA Isolation and qPCR for RSV

1. Obtain the frozen lung samples from −80 °C.
2. Take ~0.3 g of tissues from each of the 4 lung vials (RCr, LCr, LM, and LCd) and put them into 12 mL of TRIzol already pre-dispensed into 50 mL tubes.
3. Each of the 50 mL tubes' contents is then OMNI TH-homogenized on low to medium speed for ~1 min.
4. Take 1.1 mL of each homogenate from each 50 mL tube and place into 1.5 mL vials for RNA isolation. Store at −80 °C or continue the process of RNA isolation **step 6**.
5. Transfer the remaining ~10 mL of each homogenate to 15 mL conical tubes, labeled, and stored at −80 °C for future use if needed.
6. If the samples are frozen, thaw them at room temperature. Vortex each tube from **step 4** for 5 s, then let them stand for 5 min.
7. Using a 1 mL pipet, add 200 μL chloroform to each vial, vortex well, and incubate for 3 min at r.t, then microfuge for 15 min at 12,000× *g* at 4 °C.

8. In one motion, carefully remove and save exactly 550 μL of each aqueous (top) layer into new 1.5 mL tubes (entirely avoid the white gDNA-protein interface layer as much as possible during transfer).
9. Add 500 μL cold (−20 °C) isopropanol to each water layer, vortex briefly, and let sit for 10 min at r.t., then microfuge for 10 min at 12,000× *g* at 4 °C (pellets should now be visible).
10. Gently dump off isopropanol, add 1 mL 75% nuclease-free ethanol to each vial, and vortex until the pellet swirls. Spin 5 min at 16,100× *g* at 4 °C.
11. Gently dump off ethanol from each vial and lay the tubes (opened) for about 35 min to evaporate the ethanol away.
12. Add 170 μL of nuclease-free water to each dried pellet, vortex each tube for 5–10 s to try to resolubilize the RNA pellet, spin tubes down briefly, heat at 60 °C for about 2.5 min, vortex each for 10 s, spin down briefly, then store at 4 °C until next procedure, or continue with DNAse treatment.
    (a) 70 μL from each 170 μL RNA is DNase treated in 200 μL thin-walled tubes:
    (b) Add 70 μL RNA, 10 μL 10X TURBO Buffer, and 20 μL TURBO DNase enzyme to each tube. Vortex gently for 2–3 s, spin down briefly and then place them in the thermocycler for 30 min.
    (c) After the 30 min incubation, add 10 μL Inactivation Reagent to each reaction (now each tube has 110 μL each), vortex every 10 s for 2 min, then spin at 10,000× *g* for 3 min.
    (d) Take 80 μL of the resulting top layer (DNase-treated RNA is in this top layer supernatant) entirely avoiding the Inactivation reagent at the bottom.
13. Place the 80 μL of each into new 1.5 mL vials and further dilute them 1:10 with 710 μL nuclease-free water and 10 μL RNAseOUT.
14. Assess the resulting 800 μL RNA concentrations using a spectrophotometer (by mixing 34.57 μL of each RNA + 75.43 μL of nuclease-free water, then measuring at A260nm and A280nm). The RNA is pre-diluted with RNAase-free water to 3.268 ng/μL such that after addition to each RT-qPCR, 0.784 ng/μL results in each final RT-qPCR (*see* **Note 9**).
15. Blank (zero) the nanodrop spectrophotometer (used in **step 14** above) with a custom buffer comprised of water being used instead of RNA throughout the entire TRIzol extraction process.

16. The final diluted RNA samples were evaluated using qScript™ One-Step Fast qRT-PCR Kit, ROX reaction mix according to the manufacturer-recommended procedure, in which 5 μL of the 4X reaction buffer is mixed with 7.2 μL of custom TaqMan probe mixture (containing 2.691 μM of each primer and 0.5208 μM probe), 5.8 μL nuclease-free water and 6 μL RNA sample.

3.10 Hematoxylin-Eosin Staining and Histological Scoring of Lung Sections

1. Stained tissue sections were examined under a light microscope.
2. Score Lung lesions according to an integer-based score of 0–4 for each parameter (bronchiolitis, syncytial cells, epithelial necrosis, epithelial hyperplasia, peribronchial lymphocytic infiltration, perivascular lymphocytic infiltration, neutrophils), with 4 as the highest score.
3. Integer-based score details as: *bronchiolitis scale*: 0 = no remarkable lesions, 1 = minimal detectable lesion (epithelial degeneration in one or a few bronchioles per 20X field), 2 = epithelial degeneration involving less than 10% of the airway lumen; minimal neutrophils and cell debris; adventitial lymphocytes in multiple bronchioles, 3 = epithelial degeneration involving more than 10–50% of the airway lumen with cell debris and neutrophils; adventitial lymphocytes; multiple bronchioles, 4 = circumferential bronchiolitis with dense adventitial lymphocytes; multiple bronchioles. *For syncytial cells scale*: 0 = none, 1 = one distinct syncytial cell, 2 = up to three in three 20× fields, 3 = more than three in three fields, 4 = numerous. *Epithelial necrosis and epithelial hyperplasia scale*: 0 = none, 1 = minimally detectable in one or a few bronchioles per 20× per field, 2 = 10% of the bronchioles in multiple airways per field, 3 = 10–50% of the bronchioles in multiple airways per field, 4 = circumferential in multiple airways. *Neutrophils scale (in bronchi, bronchioles or alveoli)*: 0 = none, 1 = minimally detectable, 2 = 10 or less neutrophils in one or a few airways/alveoli, 3 = 10 or more neutrophils in several airways/alveoli, 4 = 10 or more involving many or most airways/alveoli. The same scale was used for peribronchiolar lymphocytic infiltrates and perivascular lymphocytic infiltrates: 0 = none, 1 = earliest detectable lymphocytic infiltration in the adventitia, 2 = segmental to circumferential infiltration, 3 = circumferential infiltrates that expand more than three cells wide, 4 = circumferential infiltrates that form nodules.
4. Sum all the individual scores to form the final accumulative tissue lesion score 0–28.

3.11 Immunohistochemistry (IHC) of Lung Sections

1. Put the mounted formalin fixed paraffin-imbedded tissue section in the oven at 60 °C for 15 min to soften /melt the paraffin.
2. Immediately place the slide rack in xylene for 5 min and repeat with a second xylene bath.
3. Rehydrate the tissue sections by passing them through gradual ethanol concentrations:
   (a) Two 100% ethanol 3 min each.
   (b) Two 95% ethanol for 3 min.
   (c) One passage in 3% peroxide in methanol for 15 min to block endogenous peroxidase.
   (d) One 70% ethanol for 3 min.
   (e) Then, place the slides in ultrapure water for 5 min.
4. Proceed with the antigen retrieval process (*see* **Note 10**).
   (a) Add 500 mL of distal water to the TintoRetriever pressure cooker and turn it on.
   (b) Immerse the slides in 200 mL of antigen retrieval solution in a Coplin jar and put the Coplin jar in the TintoRetriever pressure cooker.
   (c) Close the TintoRetriever pressure cooker lid and run it for 15 min at high pressure.
   (d) At the end of the set time, leave it as it is for 15 min, then release the pressure and open the lid.
   (e) Cool down the slides by placing the Coplin jar in ice for 20 min.
5. Place the slides in the TintoDetector slides holder and start the blocking/staining process (*see* **Note 11**).
6. Add 3% BSA solution as the first blocker for 15 min at room temperature.
7. Remove the BSA and directly add 20% NSS as the second block for 15 min.
8. Remove the NSS, add the primary antibody, and incubate overnight at 4 °C.
9. The next day, rinse the slides with TBS-T 3 times.
10. Add secondary antibody and incubate for 1 hr. at room temperature.
11. Rinse the slides 3 times with TBS-T.
12. Add the 3% peroxide in TBS-T for 25 min.
13. Rinse the slides 2 times with TBS-T.
14. Add SS-HRP and incubate for 30 min and follow with 3 TBS-T rinses.

15. Prepare the Nova Red Development solution.
16. Add Nova Red staining solution and wait for the signal to develop within 1–2 min and rinse the slides with water.
17. Counterstain the slides by adding Shandon's Harris acidified hematoxylin for 2 min.
18. Rinse the slides with water, then add Scott's Tap water and incubate for 1 min, then rinse with water.
19. Dehydrate the slides through graded alcohol for 10 s each, starting with 70%, 90%, and 100%, then 2 min in 100%.
20. Follow with immersing the slides in xylene for 5 min, then mount coverslips.
21. Examine the slides under a light microscope.
22. Count the positive signal in both bronchioles and alveoli for each tissue section.
23. Assign score of 0–4 according to an integer-based scale of 0 = no positive alveoli/bronchioles, 1 = 1–10 positive alveoli/bronchioles, 2 = 11–39 positive alveoli/bronchioles, 3 = 40–99 positive alveoli/bronchioles, 4 = >100 positive alveoli/bronchioles.

3.12 RNAscope

1. Examine stained RNAscope tissue sections under the light microscope.
2. Count the number of bronchioles and alveoli containing the positive signal. If needed, assign these numbers a score according to the simple integer-based scale to be consistent with the IHC score. The integer-based scale used is 0 = no positive alveoli/bronchioles, 1 = 1–10 positive alveoli/bronchioles, 2 = 11–39 positive alveoli/bronchioles, 3 = 40–99 positive alveoli/bronchioles, 4 = >100 positive alveoli/bronchioles. However, the raw numbers can be used directly.
3. The surface area of the positive signal can be measured using the HALO image analysis program or any other image analysis platform.

4 Notes

1. The model works best with lambs aged 1–3 days old. Older lambs can be used, but they are more resistant to RSV infection due to the development of natural immunity. In our laboratory, RSV infection is induced in lambs up to 4 months of age.
2. All processes are done in the BSL2 biological safety cabinet, except when centrifuging or incubating, with strict biosafety measures.

3. During the preparation of RSV viral inoculum, the harvest time is critical in determining the highest possible virus titer. Harvests need to be started at 50% dead cells or 80% CPE. However, not all the flasks may reach that point simultaneously. Thus, the decision depends on the experience and average of all flasks.
4. Both nebulizers were tested in our Lamb model, and a similar lung infection and lesion were produced (Videos 1 and 2). However, the Aerogen nebulizer reduces the time per animal to 10–15 min and is more efficient in nebulizing all the solution in comparison to the Jet nebulizer, in which part of the solution is lost in the nebulizer itself [14, 20].
5. The lamb model was standardized for 6 days starting at the time of viral nebulization. Lambs will develop lower respiratory tract infection starting at day 2 with peak viral replication in the lung reached at day 3 post nebulization and continue till day 6 [19]. If the experiment is to evaluate the efficacy of antiviral compounds, these compounds can be tested at different time points according to the project's goal. Standard time points for starting antiviral therapeutics are day 1 post-RSV inoculation to evaluate the primary efficacy and dose response; treatment started on day 3 post-infection to evaluate the efficacy at the peak of viral replication and well-established lung lesion. At 3 days post RSV nebulization, lambs develop Lower Respiratory Tract infections comparable to RSV infection in infants at the time of admission to the hospital, which is around day 4 after the development of respiratory signs.
6. Gross lesion evaluation is done by a specialized veterinarian researcher or veterinary pathologist. The evaluation was done previously by estimating the percentage of parenchymal involvement for each lung lobe with an RSV-associated lesion, which is usually pinpointed to a few millimeters in diameter dark red areas on the lung parenchyma [14, 20].
7. There will be no DNAase treatment for BALF RNA.
8. No spectrophotometer readings are taken on BALF RNA isolates (too low to read). The interpolation of dilution factors is used along the way to interpret relative quantity and/or copy values using a BALF RSV standard curve generated from the BALF samples themselves, in addition to an external standard curve calibrated previously.
9. The 0.784 ng/μL concentration was determined by earlier experiments using Gallup, J and Ackermann, M, 2008 [21–23].
10. For IHC, sometimes antigen retrieval needs double the amount for Tween 20.
11. IHC will be the same steps whether using the TintoDetector system, IHC autostainer, or manual.

References

1. Nair H, Nokes DJ, Gessner BD et al (2010) Global burden of acute lower respiratory infections due to respiratory syncytial virus in young children: a systematic review and meta-analysis. Lancet 375:1545–1555
2. Hall CB, Weinberg GA, Iwane MK et al (2009) The burden of respiratory syncytial virus infection in young children. N Engl J Med 360: 588–598
3. Juhn YJ, Wi C-I, Takahashi PY et al (2023) Incidence of respiratory syncytial virus infection in older adults before and during the COVID-19 pandemic. JAMA Netw Open 6: e2250634
4. Blondeau JM (2024) So we now have RSV vaccines. What's our next steps? Expert Rev Respir Med 18:17–22
5. Lipp MA, Empey KM (2024) Recent advances in the prevention of respiratory syncytial virus in pediatrics. Curr Opin Pediatr 36:182
6. Coates A, Sadreameli C (2024) Updates on respiratory syncytial virus prophylaxis: the past, present and future. Curr Opin Pediatr. https://doi.org/10.1097/MOP.0000000000001339
7. Bem RA, Domachowske JB, Rosenberg HF (2011) Animal models of human respiratory syncytial virus disease. Am J Physiol-Lung Cell Mol Physiol 301:L148–L156
8. Sacco RE, Durbin RK, Durbin JE (2015) Animal models of respiratory syncytial virus infection and disease. Curr Opin Virol 13:117–122
9. Altamirano-Lagos MJ, Díaz FE, Mansilla MA et al (2019) Current animal models for understanding the pathology caused by the respiratory syncytial virus. Front Microbiol 10:873
10. Sitthicharoenchai P, Alnajjar S, Ackermann MR (2020) A model of respiratory syncytial virus (RSV) infection of infants in newborn lambs. Cell Tissue Res 380:313–324
11. Meyerholz DK, Grubor B, Fach SJ et al (2004) Reduced clearance of respiratory syncytial virus infection in a preterm lamb model. Microbes Infect 6:1312–1319
12. Garg R, Latimer L, Wang Y et al (2016) Maternal immunization with respiratory syncytial virus fusion protein formulated with a novel combination adjuvant provides protection from RSV in newborn lambs. Vaccine 34: 261–269
13. Derscheid RJ, Ackermann MR (2012) Perinatal lamb model of respiratory syncytial virus (RSV) infection. Viruses 4:2359–2378
14. McSweeney MD, Alnajjar S, Schaefer AM et al (2024) Inhaled "Muco-Trapping" monoclonal antibody effectively treats established respiratory syncytial virus (RSV) infections. Adv Sci 11:2306729
15. Roymans D, Alnajjar SS, Battles MB et al (2017) Therapeutic efficacy of a respiratory syncytial virus fusion inhibitor. Nat Commun 8:167
16. DeVincenzo JP, Wilkinson T, Vaishnaw A et al (2010) Viral load drives disease in humans experimentally infected with respiratory syncytial virus. Am J Respir Crit Care Med 182: 1305–1314
17. Kim Y-I, DeVincenzo JP, Jones BG et al (2014) Respiratory syncytial virus human experimental infection model: provenance, production, and sequence of low-passaged Memphis-37 challenge virus. PLoS One 9:e113100
18. Grosz DD, van Geelen A, Gallup JM et al (2014) Sucrose stabilization of Respiratory Syncytial Virus (RSV) during nebulization and experimental infection. BMC Res Notes 7:158
19. Larios Mora A, Detalle L, Van Geelen A et al (2015) Kinetics of respiratory syncytial virus (RSV) Memphis strain 37 (M37) infection in the respiratory tract of newborn lambs as an RSV infection model for human infants. PLoS One 10:e0143580
20. Derscheid RJ, van Geelen A, McGill JL et al (2013) Human respiratory syncytial virus Memphis 37 grown in HEp-2 cells causes more severe disease in lambs than virus grown in vero cells. Viruses 5:2881–2897
21. Gallup JM, Ackermann MR (2008) The 'PREXCEL-Q Method' for qPCR. Int J Biomed Sci 4:273–293
22. Sow FB, Gallup JM, Sacco RE et al (2009) Laser capture microdissection revisited as a tool for transcriptomic analysis: application of an excel-based qPCR preparation software (PREXCEL-Q). Int J Biomed Sci 5:105–124
23. Gallup JM, Sow FB, Van Geelen A et al (2010) SPUD qPCR assay confirms PREXCEL-Q software's ability to avoid qPCR inhibition. Curr Issues Mol Biol 12:129–134

Chapter 21

Bovine Model of Respiratory Syncytial Virus Infection

Geraldine Taylor, Ronan MacLoughlin, Sara Hägglund, and Jean François Valarcher

Abstract

As a naturally occurring bovine pathogen, bovine respiratory syncytial virus (BRSV) engages with the innate and adaptive immune responses of cattle in an evolutionarily meaningful fashion. Therefore, studying viruses in their native hosts, rather than in non-native hosts, is more likely to authentically reflect the pathogenesis of natural infection. Experimental infection of calves with BRSV provides an important tool for investigating mechanisms of pathogenesis and immunity to RSV infection, and can also be used for evaluating new strategies for the prevention and treatment of RSV infection in both cattle and man. Here, we present materials and methods used to propagate BRSV, infect calves with BRSV, and analyze samples from BRSV-infected calves.

Key words Bovine respiratory syncytial virus (BRSV), Calf, Animal model, Viral kinetics, Infection, Clinical disease, Pathology

1 Introduction

Animal models of human respiratory syncytial virus (HRSV) infection play an important role in understanding mechanisms of immunity and pathogenesis, the preclinical testing of HRSV vaccine candidates, and therapeutic intervention strategies. Although many HRSV vaccine candidates or therapeutics have shown efficacy in preclinical studies, few have progressed to clinical trials or have had only limited success. This is, at least in part, due to the lack of animal models that fully recapitulate the pathogenesis of HRSV infection in humans. There are a number of animal models of HRSV infection available [1], which have different strengths and weaknesses. One of these is HRSV infection of calves [2, 3]. Although experimental infection of BRSV-seronegative calves with HRSV results in viral replication in the respiratory tract, levels of infectious virus are low, calves did not develop macroscopic lung lesions, and microscopic lung lesions were

Bo Liang and Rachel Fearns (eds.), *Human Respiratory Syncytial Virus: Methods and Protocols*, Methods in Molecular Biology, vol. 2948, https://doi.org/10.1007/978-1-0716-4666-3_21,

comprised of mild peri-bronchiolar and alveolar cellular infiltrates [2, 3]. Despite infection being associated with an increase in neutrophils in the lower respiratory tract of HRSV-infected calves, clinical signs of disease were mild and restricted to the upper respiratory tract [3]. Taken together, these observations suggest that, like other animal models of HRSV, calves are semi-permissive for HRSV replication.

Bovine (B)RSV is a major cause of respiratory disease in young calves and is closely related to HRSV, sharing >80% amino acid identity in 7 of the 11 encoded proteins [4]. The main differences between BRSV and HRSV are in the G, SH, NS1, and M2-2 proteins, which share 30%, 38–44%, 69%, and 43% amino acid identity, respectively [5]. Furthermore, the epidemiology and pathogenesis of infections caused by HRSV in humans and BRSV in calves are similar. Like HRSV infection of infants, the spectrum of naturally occurring BRSV infection ranges from subclinical to severe bronchiolitis and pneumonia, with the peak incidence of severe disease in individuals less than 6 months of age. BRSV infection in calves reproduces many of the clinical signs associated with HRSV in infants, including fever, rhinorrhoea, coughing, harsh breath sounds, tachypnoea, and dyspnea. Both HRSV and BRSV primarily infect ciliated airway and alveolar epithelial cells and induce a similar pattern of pathology characterized by proliferative and exudative bronchiolitis and, in the most severe cases, interstitial pneumonia [6–8]. Reproducible clinical signs of disease can be produced in calves following experimental infection with 10^4 pfu of low cell-culture-passaged BRSV or with virus that has been passaged in neonatal calves. The calf allows daily sampling of the upper respiratory tract (URT), and virus replication and the inflammatory response in the lower respiratory tract (LRT) can be analyzed in samples obtained by bronchoalveolar lavage (BAL) or in lung tissue at post-mortem examination. Following experimental BRSV infection of calves, virus titres peak in the URT between 4 and 7 days post-infection (p.i.) and the virus is normally cleared by days 8–9 p.i. Clinical signs of disease usually appear between days 5 and 6 p.i. and are maximal at days 7–8. Experimental BRSV infection of calves can induce a significant decrease in arterial oxygen tension. The partial pressure of oxygen in arterial blood (PaO_2) is significantly associated with the extent of lung lesions and is highly predictive of lesion severity [9]. BRSV infection induces increased levels of proinflammatory cytokines and chemokines, and a marked increase in neutrophils in BAL similar to that seen in children with HRSV bronchiolitis. BRSV infection of calves has been used as a preclinical model to evaluate the efficacy of HRSV vaccines [10]. Furthermore, bovine neutralizing, fusion-inhibiting monoclonal antibodies (mAbs) recognize antigenic sites II and IV on the F protein of both HRSV and BRSV and can protect mice against HRSV infection [11, 12] and calves against BRSV infection [13].

The use of animals in experiments should comply with local governmental regulations. In the United Kingdom, this is regulated under the Animals (Scientific Procedures) Act 1986 (ASPA) [14]. Under ASPA, licenses are required for individuals carrying out procedures on animals (personal licenses), for the program of work (project licenses) and the place at which the work is carried out (establishment licenses). Animals should be housed according to the local governmental regulations. This includes minimum space requirements depending on the weight of the animals at the end of the experiment and standards for environmental conditions. The holding area should have bedding, water, and food dispensers, and further environmental enrichment should be supplied.

In this chapter, we describe our protocols for infection of calves with BRSV, analyzing calves after infection, and sample processing.

2 Materials

It is important to keep all materials for cell culture, virus propagation, plaque assays, and immune assays sterile by performing all procedures in a microbiological safety cabinet. In addition, cells, virus, fetal bovine serum (FBS), and calves should be tested to ensure freedom from bovine virus diarrhea virus (BVDV), which can interfere with innate host responses. BVDV RNA in blood, plasma, serum, milk and tissue samples (individual or pooled) can be detected by RT-PCR using a commercial kit (e.g. Indical Bioscience Cat. No.: VT280375). To amplify and detect live BVDV in cells or FBS, perform 5–10 cell passages and use immunostaining. Even if the selected FBS is irradiated, it needs to be checked for the absence of replicative BVDV.

All PBS used is prepared without calcium and magnesium.

2.1 Subculturing of MDBK Cells

1. Madin-Darby bovine kidney (MDBK) cells.
2. 175 cm^2 vented, nonpyrogenic flask.
3. MDBK growth media: Minimum essential medium with L-glutamine (MEM-Glut) (Gibco) containing 100,000 U penicillin & 100 μg streptomycin, 1% non-essential amino acids (NEAA) and 10% heat-inactivated (56 °C for 30 min) fetal bovine serum (HI-FBS). Store at 4 °C.
4. 37 °C water bath.
5. Phosphate-buffered saline (PBS), without calcium and magnesium.
6. Trypsin–EDTA: 0.25% trypsin, 0.9 mM EDTA.
7. Inverted microscope.
8. 50-mL conical tube.

9. Pipette boy and 10 mL pipettes.
10. Cell counting chamber.
11. 0.4% Trypan blue in isotonic saline.
12. 37 °C Incubator with 5% CO_2.
13. Waste container.

2.2 Propagation of BRSV in MDBK Cells

1. BRSV isolate (*see* **Note 1**).
2. 85% confluent MDBK cells (*see* **Note 2**) in 175 cm^2 non-vented, nonpyrogenic flasks.
3. MDBK maintenance media: Minimum essential medium with L-glutamine (MEM-Glut) (Gibco) containing 100,000 U penicillin & 100 μg streptomycin, 1% non-essential amino acids (NEAA) and 2% HI-FBS.
4. 37° C water bath.
5. Inverted microscope.
6. 50-mL conical tube.
7. Pipette boy and 1 mL, 5 mL, and 10 mL pipettes.
8. Cell scrapers.
9. 37 °C Incubator with 5% CO_2.

For preparation of the BRSV standard for qRT-PCR:

10. 1.5 mL Eppendorf tubes.
11. P200 and P1000 pipettes and tips.
12. PBS.
13. Freezer −80 °C.

2.3 Production of In Vivo-Passaged BRSV

1. One- to 2-week-old RSV-specific serum antibody-free calves (*see* Subheading 3.7).
2. Low-passaged BRSV isolate.
3. PBS (without calcium and magnesium).
4. Heat-inactivated FBS.
5. Penicillin (10,000 μg/mL) and streptomycin (10,000 μg/mL).

2.4 Characterization of In Vivo-Passaged BRSV

1. Sheep blood agar plates.
2. P100 pipette and tips.
3. Disposable bacteriology loop.
4. Pneumo4V multiplex PCR kit (DNA Diagnostic A/S).
5. Pneumo4B multiplex PCR kit (DNA Diagnostic A/S).
6. Incubator at 37 °C.

7. Plate washer/vacuum system capable of aspirating supernatant from Deepwell plates or vacuum system with pipette tips (35 mm from tip to rim stop).
8. qPCR instrument, Stratagene Mx3005 with filters for ATTO (440 nm–492 nm), FAM (492 nm–516 nm), HEX (535 nm–555 nm), ROX (585 nm–610 nm), CY5 (635 nm–665 nm).
9. Vortexer for 96 well plates.
10. Ice bucket or 0–4 °C cooling block.
11. Standard lab pipettes and sterile filter tips.

2.5 Plaque Assay for Virus Titration

1. MDBK cells: 85% confluent in 12-well tissue culture plates.
2. MDBK media with 2% heat-inactivated (HI) FBS.
3. MDBK overlay: MEM-Glut containing 1% NEAA, 200 U/mL penicillin, 200 μg/mL streptomycin, 4% HIFBS and 2% low melting point agarose (Sigma A9414) in double-distilled water.
4. PBS (without calcium and magnesium).
5. 80% cold Acetone in PBS (keep at −20 °C).
6. mAb 19 specific for the RSV F protein [12] and mAb 8 specific for the RSV M2 protein [12] (available from The Pirbright Institute) (*see* **Note 3**).
7. HRP- anti-mouse IgG (e.g., ICN, KPL, Dako).
8. 50% Sea Block blocking buffer (Life Technologies Cat No 37527) in PBS.
9. AEC substrate 50x (Merck Chemicals Cat No 152224).
10. AEC buffer 10x (Sigma Cat. No. 152226).
11. Sterile Eppendorf tubes.
12. PBS + 0.05% Tween-20 (PBS/Tw).
13. Pipette boy and pipettes.
14. P100 and P1000 pipettes and tips.

2.6 RT-qPCR for Quantification of BRSV mRNA in Samples

1. Extraction kit Qiagen RNeasy® Mini Kit (Quiagen Cat. No./ID: 74104).
2. QIAshredder homogenizer.
3. Real time PCR machine.
4. 14.3 M β-mercaptoethanol (β-ME).
5. Microcentrifuge (with rotor for 2 mL tubes).
6. Micropipettes 1000 μL, 200 μL, 20 μL and 1–10 μL.
7. Rnase free filter tips 1000 μL, 200 μL, 20 μL and 1–10 μL.
8. RT-PCR Mix (iTaq Universal Probes One-Step Kit #1725141).

9. Primers against the BRSV F gene sequences [15] (5′–3′):

   AAGGGTCAAACATCTGCTTAACTAG (25 bp, forward).
   TCTGCCTGWGGGAAAAAAG (19 bp, reverse).

10. TaqMan® probe against the BRSV F gene [15] (5′–3′), labeled with 6-carboxyfluorescein (FAM) at the 5′end and with carboxytetramethylrhodamine (TAMRA) at the 3′ end:

    AGAGCCTGCATTRTCACAATACCACCCA (34 bp, complementary).

11. PCR plate and adhesive seal for PCR.
12. Disposable gloves.
13. Computer and Excel program.

2.7 Sourcing of Calves (See Note 4)

1. Calves sourced from a commercial farm at one to 2 days of age.
2. Iodine solution.
3. 2% Marbocyl, or another broad-spectrum antibiotic.
4. Large dog rug.
5. Heat lamp.
6. Waterproof suit, wellington boots, disposable gloves and hats.
7. 1% Virkon in footbath.

2.8 BRSV Infection of Calves

2.8.1 Intranasal Challenge

1. Mucosal atomization device 300 (MAD300).
2. Interlocking 10-mL syringe.
3. 10^4 pfu of BRSV diluted in 10 mL virus diluent or PBS/calf.
4. Ice.

2.8.2 Intratracheal Challenge

1. Antiseptic wipes (70% ethanol).
2. Absorbent paper towel.
3. 20-mL syringe.
4. 16-G needle.
5. 10^4 pfu of BRSV diluted in 10 mL virus diluent or PBS/calf.
6. Ice.

2.8.3 Aerosol Challenge

1. A bespoke modified anesthetic mask referred to as the 'EOINS' customized with an arrangement of 1-way valves (Intersurgical, UK) and standard bacterial/viral filters (Respirgard 303, Vyaire, Ireland) (*see* Fig. 1) (*see* **Note 5**) (mask size depending on size of the animal (Burtons Medical Equipment Ltd., UK) [16].
2. Compressor-driven nebulizer (e.g. deVilbiss PulmoAide Compact Compressor or Philips Respironics deluxe).
3. 10^4 pfu BRSV in 2–4 mL virus diluent or PBS/calf.

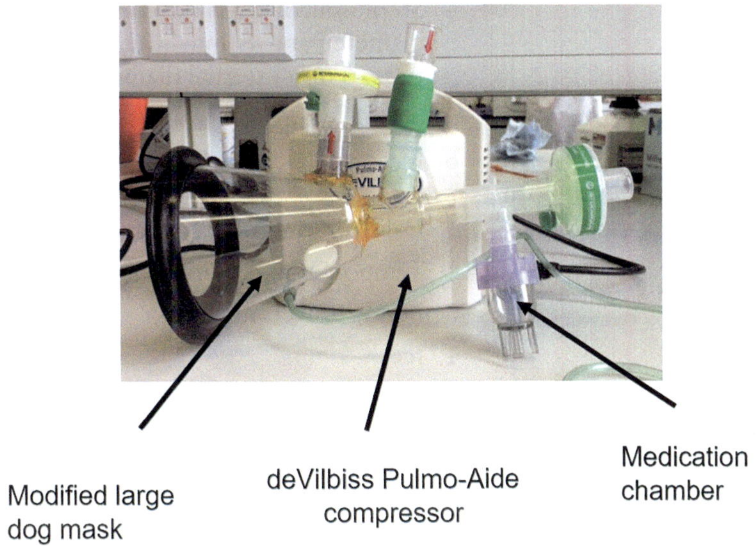

Fig. 1 Mask, nebulizer, and compressor for aerosol delivery of bRSV to calves

4. Antiseptic wipes (70% ethanol).
5. Calf halter.
6. Ice.

2.9 Clinical Monitoring

1. Rectal thermometer.
2. Clinical score sheet.
3. Pen.
4. Stopwatch.
5. Stethoscope.

2.10 In Vivo Sampling

2.10.1 Nasopharyngeal Swabs for Virus Isolation

1. Sterile 27-cm-long guarded nasopharyngeal swab with a brush bud and a twisted wire shaft (e.g., MW126 MWE Medical Wire, Corsham, Wiltshire, SN13 9RT, UK).
2. 2 mL aliquots of transport medium consisting of Hanks balanced salt solution (without calcium and magnesium), 1% Bovine serum albumin (BSA), 25 mM HEPES (pH 7.0), 200 U/mL Penicillin, 200 μg/mL Streptomycin, 50 U/mL Nystatin in screw-capped bijoux labeled with calf number.
3. 1 mL aliquots of RNA-later, labeled with calf number.
4. Ice.
5. Calf halter.

2.10.2 Bronchoalveolar Lavage

1. Calf halter.
2. 50 mL sterile saline or PBS.
3. Lidocaine 20 mg/mL spray and gel (Astra Zeneca).

4. Sterile silicone tubing: 40–50 cm-long tube, 12 mm external diameter and 8 mm internal diameter, and a 1–1.5 m-long tube, 6 mm external diameter and 4 mm internal diameter.
5. 50 mL syringe.
6. Ice.

2.10.3 Collection of Nasal Secretions for Antibody Analysis

1. Calf halter.
2. Tampon, e.g., Tampax Pearl Regular.
3. Plastic universal.
4. Ice.

2.10.4 Blood Sampling

1. Calf Halter.
2. Vacutainer system.
3. Vacutainer blood tubes (serum and heparin tubes).
4. 16–20-G needle, 1–2 in. depending on the size of the animal, compatible with vacutainer system.

2.11 Euthanasia

1. Calf halter.
2. 50-mL syringe.
3. 16-G needle 1–2 in. depending on the size of the animal.
4. Dolethal—200 mg/mL solution (Ventoquinol) (0.7 mL/kg body weight).
5. Sharp large knife.
6. Chain/rope.
7. Hoist to hang carcass and allow exsanguination.

2.12 Necropsy Sampling

1. Dissection board.
2. Camera.
3. Absorbent paper towels.
4. Post-mortem handles and blades (e.g. PM40 Swann-Morton, Sheffield, UK).
5. Clean sterile scalpel blades for each tissue.
6. Clean sterile forceps for each tissue.
7. Sterile petri dishes.
8. RNAlater™ 1 mL aliquots in cryovials.
9. 10% neutral buffered formalin in pots (60 mL per pot).
10. Plastic universals.
11. Marker pen.
12. 70% ethanol and tissue to clean the workstation between animals12″ locking straight Spencer Wells forceps (e.g., YNR® Rochester Pean Hemostat Clumps Locking Forceps Surgical Veterinary Reptile).

13. 2 mL aliquots of transport medium in plastic bijoux (*see* Subheading 2.10.1).
14. 500 mL to 1500 mL sterile PBS.
15. 500 mL sterile pots or bottles for bronchoalveolar lavage (BAL).

2.13 Processing of Samples

2.13.1 Processing of Blood for Serum

1. Centrifuge with suitable adapters for the tubes used (e.g., vacutainer tubes).
2. Pipette and pipette tips.
3. Labeled Eppendorf tube or cryovials.
4. Water bath (optional).

2.13.2 Isolation of Peripheral Blood Mononuclear Cells (PBMCs)

1. Heparinized blood.
2. PBS (without calcium and magnesium).
3. Ficoll-Paque 1.083 g/mL.
4. Ammonium chloride lysis buffer or commercially available red blood cell lysis buffer.
5. Centrifuge with adaptors suitable for the tube size used.
6. Falcon tubes (15- or 50-mL Falcon tubes depending on quantity).
7. Lymphocyte culture medium: RPMI with glutamine, 10% heated-inactivated (56 °C for 30 min) fetal bovine serum (FBS), 100 U/mL penicillin, and 100 μg/mL streptomycin.
8. Microscope.
9. Cell counting chamber and 0.4% Trypan blue in isotonic saline.

For cryopreserving cells:

10. Labeled cryovials.
11. Pipettes and pipette tips.
12. Microscope.
13. Counting chamber and 0.4% Trypan blue in isotonic saline.
14. FBS with 10% DMSO.
15. Freezer container containing isopropanol (e.g., Mr. Frosty™).
16. Ice (to keep cells cool after resuspending in FBS with 10% DMSO, before transfer to −80° C).

2.13.3 Processing of Lung Tissue Samples for Virus Isolation

1. Tissue homogenization medium: Hanks balanced salt solution (without calcium and magnesium), 25 mM HEPES (pH 7.8), 0.218 M sucrose, 4.8 mM L-glutamine, 30 mM $MgCl_2$, 1.3 g/L sodium bicarbonate, 200 U/mL Penicillin, 200 μg/mL Streptomycin, 50 U/mL Nystatin.
2. Pipette boy & 20 mL pipettes.

3. Weighing scale.
4. Petri dishes.
5. Sterile forceps & scissors.
6. Sterile pestle and mortars.
7. Sterile fine sand.
8. Universals.
9. P200 pipette and tips.
10. Labeled Eppendorf tubes or cryovials.
11. 10% Decon.
12. Hot bead sterilizer.
13. Ice.

2.13.4 Processing of Nasopharyngeal Samples and Tracheal Epithelium for Virus Isolation

1. Vortex.
2. P200 pipettes and tips.
3. Labeled Eppendorf tubes or cryovials.

2.13.5 Processing of BAL for Bacteriology, Virus Isolation and Antibody Detection

1. Sheep blood agar plates.
2. P100 and P200 pipettes and tips.
3. Bacteriology loop.
4. 50 mL Flacon centrifuge tubes.
5. Centrifuge.
6. Tissue homogenization medium (*see* Subheading 2.13.3).
7. Pipette boy and 10 mL pipettes.
8. Labeled Eppendorf tubes and/or cryovials.
9. 37 °C Incubator with 5% CO_2.
10. Ice.

2.13.6 Processing of BAL for Analysis of Pulmonary Exudate

1. P100 and P1000 pipettes and tips.
2. Cell counting chamber and 0.4% Trypan blue in isotonic saline.
3. Labeled glass microscope slides.
4. Cytocentrifuge.
5. Cytocentrifuge filter cards.
6. Cytocentrifuge funnel.
7. Cytoclip.
8. Diff quick stain.
9. Staining pots.
10. Microscope immersion oil.
11. Microscope.

2.13.7 Processing of Nasal Secretions for Antibody Detection

1. 20 mL syringe.
2. Labeled Eppendorf tubes or cryovials.

2.13.8 Processing of Lung Tissue and Lymph Nodes for Lymphoid Cell Isolation

1. Lymphocyte culture medium (*see* Subheading 2.13.2).
2. gentleMACS C-tubes containing lymphocyte culture medium.
3. gentleMACS (Miltenyi Biotec).
4. 70-μm cell strainers.
5. 50-mL Falcon tubes.
6. Centrifuge.
7. Pipette boy and pipettes.
8. Microscope.
9. Lymphocyte culture medium (*see* Subheading 2.13.2).
10. Counting chamber and 0.4% Trypan blue in isotonic saline.

    For cryopreserving cells:

11. Heat-inactivated FBS with 10% DMSO.
12. Freezer container containing isopropanol (e.g., Mr. Frosty™).
13. Ice (to keep cells cool after resuspending in FBS with 10% DMSO, before transfer to −80 °C).

3 Methods

3.1 Subculture of MDBK Cells

1. Pre-warm the PBS, trypsin-EDTA, and MDBK growth medium (*see* Subheading 2.1) in a 37 °C water bath.
2. Examine flasks macroscopically for signs of fungal or bacterial contamination. Check that the medium is not acidic and cells are not stripping off and check the cells microscopically to make sure that they are confluent and look healthy.
3. Pour the medium aseptically from the flask into a waste container, wash the cells 2–3 times with PBS to remove traces of medium (about 20–30 mL each wash for a T175). After each wash, discard the PBS into a waste container.
4. Wash the cell sheet with about 10 mL trypsin/, rock the flask to make sure the whole of the monolayer has been covered then discard into a waste container.
5. Add a further 10 mL trypsin-EDTA, rock the flask and then discard most of the T/V, leaving about 2–3 mL in the flask.
6. Incubate the flask on a level surface at 37 °C until the cells are rounded up and start to come off, then tap the side of the flask with the hand until all the cells have come off.

7. Add 10 mL MDBK growth medium to the flask to neutralize the trypsin. Pipette the cells up and down to make sure they have all detached, and to ensure a single cell suspension.
8. Remove a small aliquot of the cell suspension for counting viable cells by trypan blue exclusion.
9. Add 2–5 × 10^6 cells to each new T175 flask in 25–30 mL MDBK growth medium.
10. Incubate cells at 37 °C with 5% CO_2.

3.2 Preparation of BRSV in Cell Culture

1. Culture MDBK cells and create an appropriate number of 175-cm^2 flasks for BRSV-infected stock and mock-infected stock preparations.
2. Allow the MDBK cells to grow until roughly 80–90% confluency.
3. Pour the medium from the flasks into a beaker.
4. Warm MDBK maintenance medium (*see* Subheading 2.2) in 37 °C water bath.
5. Add 10 mL medium MDBK medium containing 2% HI-FBS to a control flask.
6. To the remaining flasks add 10 mL of virus diluted in the MDBK medium containing 2% HI-FBS to give an MOI of 0.01–0.1.
7. Tighten the tops of the flasks and incubate for 3 h at about 37 ° C.
8. Remove the virus inoculum using either an aspirator or a 10 mL pipette.
9. Add 10 or 15 mL of MDBK medium containing 2% HI-FBS.
10. Incubate the flasks at about 37 °C with the tops closed tight. Examine at intervals for cytopathic effect (cpe).
11. When cpe is about 70%, harvest the virus by scraping any remaining cells into the medium using a cell scraper.
12. Pipette the medium up and down several times using a 10 mL pipette to produce a homogenous cell/virus suspension, and pool the contents of all the virus flasks.
13. Aliquot the virus, in the desired volumes, into labeled vials suitable for storage, either 2 mL Eppendorf tubes for storage at −80 °C or cryovials for storage in liquid N_2. Generally, a virus designated for animal challenge is stored in liquid N_2, while a virus for in vitro assays may be stored at −80 °C.

For the preparation of the BRSV standard for RT-qPCR:

14. Pipette 1 mL of the BRSV cell suspension as prepared above and make a ten-fold dilution series in PBS up to 10^{-6} in 1.5 mL Eppendorf tubes.
15. Freeze the aliquot at −80 °C for until required for RNA extraction.

3.3 Preparation of In Vivo-Passaged BRSV

1. Infect BRSV antibody-free calves at about 7–14 days old either with 10^4 to 10^5 pfu of BRSV in 20 mL MDBK medium, 10 mL intranasally and 10 mL intratracheally, or with 10^4 pfu of BRSV in 5 mL by aerosol (*see* Subheadings 3.7, 3.9.1, 3.9.2).
2. Treat calves with a broad-spectrum antibiotic for the duration of the study.
3. Collect nasopharyngeal swabs for virus isolation (*see* Subheading 3.11.1).
4. Euthanise calves 5–6 days after infection and remove the lungs with the trachea from the thoracic cage (*see* Subheading 3.12).
5. Collect bronchoalveolar lavage (BAL) by filling the tracheobronchial tree with approx. 500 mL PBS, massage the lungs and tip out the BAL from the trachea into a large sterile pot.
6. Keep on ice.
7. Pipette 100 μL BAL onto a sheep blood agar plate and incubate at 37 °C overnight to test for bacterial growth.
8. Add HI-FBS to the BAL to give a final concentration of 5%.
9. Add penicillin and streptomycin to BAL to give a final concentration of 200 U/mL penicillin and 200 μg/mL streptomycin.
10. Aliquot BAL in cryovials and store either at −80 °C or in liquid N_2.

3.4 Infectious Foci Assay for Virus Titration

1. Seed 12-well plates with 2 mL MDBK cells at 6×10^4/mL in growth medium (*see* Subheading 2.1) and incubate for 3 days at 37 °C in 5% CO_2/air or until 85% confluent.
2. Serially dilute the virus samples: 100 μL undiluted sample in 900 μL MBDK maintenance medium (*see* Subheading 2.2) to the first tube to give a 10^{-1} dilution. Vortex this tube well, dispose of the pipette tip into a container partially filled with 1% Virkon, and using a new pipette tip take 100 μL from the 10^{-1} tube and place it into a second tube with 900 μL of MBDK media; this will be the 10^{-2} dilution. Repeat to a dilution of 10^{-5} for virus stocks, and 10^{-3} for samples from calves.
3. Keep samples & dilutions on ice until ready to inoculate cells.

4. Remove the media from the cells, add 100 μL of the pre-diluted inoculum to duplicate wells starting from the highest dilution.
5. Incubate for 90 min at 37 °C in 5% CO_2/air, agitating the plate every 15 min to prevent the cells drying out.
6. Remove the inoculum with a pipette and add 1.5 mL of MDBK overlay/well (*see* Subheading 2.5).
7. When overlay has set, incubate the plates for 3–4 days at 37 °C in 5% CO_2/air.
8. Remove the overlay into 1% Virkon.
9. Wash cells gently with PBS.
10. Fix with 300 μL per well of 80% cold acetone for 15 min at room temperature (RT).
11. Discard acetone & allow cells to dry at RT. Plates can be stored at 4 °C, if necessary, prior to staining.
12. Add 300 μL/well of 50% Sea Block, 1 h at RT.
13. Remove Sea Block & wash gently with PBS/Tw.
14. Add 300 μL of a mixture of mAb 19 at 1 μg/mL & mAb 8 at 1 μg/mL in 50% Sea Block, for 1 h at RT to each well of a 12-well plate (*see* **Note 3**).
15. Remove mAbs, wash 3x with PBS/Tw, gently.
16. Add 300 μL of 1/2000 dilution of HRP-GAM IgG in 50% Sea Block to each well, for 1 h at RT.
17. Prepare the AEC substrate according to the manufacturer's instructions: Prepare 1x AEC Substrate Buffer by adding 9 parts DDW to 1 part 10x AEC Substrate Buffer & mix well. Add 1 part 50x AEC Concentrate to 50 parts 1x AEC Substrate Buffer & mix well.
18. Remove the secondary antibody and wash gently 3x with PBS/Tw.
19. Add 300 μL/well of AEC substrate. Incubate at RT for 15 min to 30 min, until red/brown foci of cells are visible.
20. Wash gently with tap water, allow plates to dry, and count stained foci of cells using an inverted light microscope.
21. To calculate virus titer, multiply the number of foci of infected cells × dilution × 10 to obtain infectious foci units per mL (ifu/mL).

3.5 Characterization of In Vivo-Passaged BRSV

1. Follow the Pneumo4V kit protocol for detection of Bovine parainfluenza virus, Bovine coronavirus, Bovine Respiratory Syncytial virus, Bovine viral diarrhea virus, and Bovine Herpesvirus 1.

2. Follow the Pneumo4B protocol for detection of *Mannheimia haemolytica, Pasteurella multocida, Histophilus somni,* and *Mycoplasma bovis.*

3.6 RT-qPCR for Quantification of BRSV RNA in Samples

1. Prepare a stock of BRSV (10^6/PFU/mL) as described in Subheading 3.2.
2. Make a ten-fold dilution series of the BRSV stock in PBS up to 10^{-6}, 5 replicates of each dilution.
3. Extract RNA from 250 μL of each dilution using the Qiagen RNeasy mini kit, according to the manufacturer's instruction in RNeasy mini handbook (600 μL RLT buffer/10^7 cells).
4. Dissolve the RNA in 30 μl DEPC water.
5. Extract RNA from 250 μL of nasal swab transport medium or approximately 400 mg lung tissue homogenateby using the Qiagen RNeasy mini kit according to the manufacturers' instructions. Dissolve the RNA in 30 μL DEPC water. Avoid cross-contamination by using one tube opener per sample, and pipette fluids instead of pouring when discarding liquids.
6. Store the RNA at −80 °C and thaw RNA on wet ice.
7. Dilute primers and a TaqMan® probe against the BRSV F gene, in DEPC water to 100 μM and store at −20 °C.
8. Prepare working solutions by diluting the stock solutions with DEPC water to 10 μM for primers and 5 uM for probes.
9. Prepare a mix with (per sample).

   3.9 μL DEPC water

   10 μL iTaq reaction mix (iTaq Universal Probes One-step kit, Bio-Rad 172-5141)

   0.6 μL Forward primer

   0.6 μL Reverse primer

   0.4 μL TaqMan Probe

   0.5 μL iScript advanced reverse transcriptase (iTaq Universal Probes One-step kit, Bio-Rad 172-5141)
10. Add 4 μL of RNA to 16 μL of the mix per well in a 96-well PCR plate and close with adhesive seals for PCR.
11. Analyze in a PCR instrument with the thermodynamic profile:

    50 °C for 10 min

    95 °C for 3 min

    40 cycles of amplification: 95 °C for 15 s and 60 °C for 30 s.
12. Dilutions of the BRSV standard samples are used to generate a standard curve. The reaction efficiency (E) is calculated using the formula $E = 10^{(-1/a)} - 1$, where a is the slope value of the standard curve.

13. Assign a value correlating to $TCID_{50}$ in accordance with the dilution of the BRSV standard samples. The starting quantities of the unknown samples can be expressed as units correlating to $TCID_{50}$, according to the standard curve.

3.7 Rearing of BRSV Antibody-Free Calves

1. Calves are caught at birth, and removed from their dam so that they are not allowed to obtain any colostrum.
2. Calves should be housed in a pen separated from all other cattle on the farm until collected for transport to the experimental facility.
3. Calves to be fed with a suitable calf milk replacer.
4. Change of oversuit, boots, and gloves required for staff entering the pen containing colostrum-deprived calves.
5. Rub down the calf with towels, drying as much as possible.
6. Spray the naval with Iodine and the ear where the ear tags have punctured.
7. Calves to be given 2% Marbocyl (8 mg/kg) or other appropriate broad-spectrum antibiotic intramuscularly or subcutaneously, according to the manufacturer's instructions.
8. Calves are transported to the experimental facility according to local guidance on protecting animals during transportation.
9. Calves should be housed in CL2 facilities to minimize contact with other ruminants.
10. Calves to be kept warm for a few days in the experimental facility, e.g., heat lamp or can wear a large dog rug. Rugs can be disinfected, washed, and re-used between studies.
11. Calves to be bled on arrival at experimental facility (red top tube for serum & green top tube for BVDV PCR testing).
12. Calves may be given 500 mL colostrum containing low levels of BRSV-specific antibodies 48 h after birth, if available.
13. Calves to be bled 24–48 h after colostrum (red top tube for serum to determine levels of BRSV-specific antibodies) to determine levels of BRSV-specific serum antibodies.
14. Broad-spectrum antibiotics should be administered as the calves will have low immunity, being colostrum deprived (*see* **Note 4**).
15. Calves can be infected as per the protocol when they are approx. 1–2 weeks of age.

3.8 Rearing of Calves with Maternal Antibodies to BRSV

1. Calves are allowed colostrum for 1–2 days following birth and then can be transported to the experimental facility at 2–3 days of age.
2. Calves should be housed in CL2 facilities to minimize contact with other ruminants and reared as described in Subheading 3.7.

3.9 BRSV Infection of Calves

3.9.1 Intranasal and Intratracheal Challenge

1. Restrain the calf.
2. Keep the virus on ice.
3. Deliver 10^4 to 10^5 pfu of virus in 20 mL diluent per calf.
4. Inoculate 5 mL of virus to each nostril using a 10 mL interlocking syringe and MAD300 device on inhalation.
5. Clean skin on neck with alcohol wipe and dry with absorbent paper towel.
6. Hold the trachea to immobilize it and insert 16-G needle, attached to 20 mL syringe containing 10 mL virus, transcutaneously into the trachea.
7. Withdraw some air to make sure needle is in the trachea.
8. Inoculate 10 mL of virus suspension on inhalation and withdraw needle and syringe (*see* **Note 6**).

3.9.2 Aerosol Challenge

1. Restrain the calf.
2. Keep the virus on ice.
3. Deliver 10^4 pfu of virus in 2–4 mL diluent using the EOINS anesthetic mask and nebulizer (*see* **Note 7**).
4. Wipe the anesthetic mask with an alcohol wipe and dry with an absorbent paper towel between each calf.

3.10 Clinical Monitoring

1. Following BRSV challenge, monitor animals for clinical signs of disease such as demeanor, appetite, body temperature, nasal discharge, dyspnea, respiratory rate, and coughing (*see* Table 1), and record.
2. Daily individual clinical scores are calculated by summing these numerical values multiplied by a coefficient for each parameter (Table 1). Coefficient weights reflect parameter association with disease severity in BRSV-infected calves less than 3 months of age, based on observations during natural BRSV outbreaks [17].
3. Individual accumulated clinical scores (ACS) can be expressed as the area under daily clinical scores.
4. The humane end-point is usually determined by a combination of factors, and may vary between different institutions and the goals of the study. However, a calf should generally be euthanized immediately if it has a fever 1.5 °C above baseline at the beginning of the fourth consecutive day and does not respond to treatment with appropriate drugs (*see* **Note 8**); if it shows a severely depressed state; if it is not eating on the beginning of the third consecutive day; if it shows respiratory distress, arched back, extended neck with accompanying signs of fluid present on the lungs; if it has a persistent cough at the beginning of the third consecutive day.

Table 1
Parameters and coefficients used to calculate clinical scores in bRSV-infected calves (modified from [17])

Clinical parameter	Parameter coefficient	State description	Numerical value
General state	4	Normal	0
		Moves slowly, head down	1
		Lying down/staggers	2
		Recumbent	3
Appetite	4	Normal	0
		Reduced	1
		Absent	2
Abdominal dyspnoea	3	Normal	0
		Slight (short, rapid)	1
		Moderate (labored)	2
		Severe (very labored, stretched neck, open mouth breathing)	3
Rectal temperature	2	<39.6 °C	0
		Each 0.1° C above 39.6 scores an extra point	
Respiratory rate	2	<40/min	0
		40–45/min	1
		45–55/min	2
		55–65/min	3
		≥75/min	4
Intensity of lung sounds	1	Normal	0
		Slightly enhanced	1
		Moderately enhanced	2
		Severely enhanced	3
Added respiratory sounds (wheezing or crackles)	1	Normal	0
		Slight	1
		Moderate	2
		Severe	3
Coughing	1	Absent	0
		Only provoked	1
		Spontaneous, infrequent	2
		Persistent	3
Nasal discharge	1	Normal	0
		Slight uni−/bilateral serous	1
		Moderate bilateral serous to purulent	2
		Copious bilateral purulent	3

3.11 In Vivo Sampling

3.11.1 Nasopharyngeal/Nasal Swabs for Virus Isolation

1. Restrain the animal's head.
2. Measure the distance from the nostril to the medial canthus of the eye.
3. Insert the 27 cm guarded culture swab into the ventral meatus of the nose and advance it the premeasured distance from the nostril to the medial canthus of the eye.
4. Retract the guarded culture swab approximately 1–2 inches.
5. Push the brush-tipped swab out of the outer swab sheath for a distance of roughly 1–2 inches.
6. Push the swab up and down against the pharyngeal mucosa 4–5 times.
7. Retract the brush-tipped swab into the outer swab sheath.
8. Remove the entire guarded swab from the animal's nose.
9. Push the brush-tipped swab out of the outer sheath and agitate the brush in 2 mL transport medium (*see* Subheading 2.10.1) for about 15 secs (*see* **Note 9**).
10. Dispense the nasopharyngeal swab sample into two aliquots and store at −80 °C until required for analysis of BRSV titres.

3.11.2 Bronchoalveolar Lavage

1. Spray local anesthesia (lidocaine spray 20 mg/mL) in one nostril of the calf.
2. Five to ten minutes later, restrain the calf by holding the head with the nose up.
3. Cover the extremity of the first sterile silicane tube (external diameter of 12 mm and internal diameter 8 mm; 35–45 cm long according to the animal size) with anesthetic ointment (lidocaine gel 20 mg/mL) and introduce the tube in the nostril of the calf.
4. Progress blindly until the larynx and the proximal trachea is reached (feel the air when the calf breaths in the tubing). A cough reflex often confirms successful introduction in the trachea.
5. Introduce the second sterile silicane tube (external diameter of 6 mm and internal diameter of 4 mm; 110–150 cm long according to the animal size) into the first tube and stop when the tubing cannot progress any further.
6. Gently (slowly) inject 50 mL of sterile isotonic saline with a sterile syringe in the smallest tube and aspirate immediately. Repeat the operation with a second sterile syringe containing 50 mL of sterile isotonic saline. Finally, aspirate gently the BAL liquid that might remain in the lung and the BAL tube with an empty, third 60 mL sterile syringe. Between 50 and 80 mL may be recovered in total.
7. Remove both tubes from the animal and release it from restraint (*see* **Note 10**).

3.11.3 Collection of Nasal Secretions for Antibody Analysis

1. Restrain calf.
2. Insert tampon into one nostril and leave for 5 min.
3. Remove tampon from the nostril and place it in a plastic universal tube.
4. Keep tubes on ice.

3.11.4 Blood Sampling

1. Restrain the calf.
2. The blood sample can be obtained from the jugular vein using an appropriately sized needle on a vacutainer system with the desired collection tubes.
3. Position the needle over the jugular groove cranial to where your thumb is raising the vein.
4. Hold the needle at an angle of about 45 degrees to the skin surface.
5. Push the needle through the skin and into the jugular vein following the direction of the jugular with the needle.

3.12 Necropsy Sampling

1. Euthanisation is performed by an overdose of general anesthesia (e.g., 5 mg/kg ketamine and 15 mg/kg pentobarbital sodium) administered IV, followed by exsanguination by cutting the jugular veins and bleeding out by hanging the carcass vertically.
2. The trachea is clamped as near to the larynx as possible using large Spencer Wells forceps (*see* **Note 11**) and the trachea and lungs are removed from the thoracic cavity, taking care to avoid cutting the trachea or lungs.
3. Clean the outside of the lungs with sterile PBS and absorbent paper towels.
4. Photograph the whole of the dorsal and the whole of the ventral aspects of the lungs together with the calf identity tag (*see* Fig. 2).
5. Clamp 3 cranial lobes with Spencer Wells forceps (*see* **Note 12**).
6. Cut the trachea below the clamp and remove 2–3 cm of trachea for virus isolation and a tracheal ring for fixation in 10% buffered formalin for histology.
7. Cut the 2–3 cm piece of trachea vertically to expose the epithelium and using a scalpel blade scrape the epithelium. Agitate the blade to which the epithelium is attached in a bijoux containing 2 mL of transport medium (*see* Subheading 2.10.1), labeled with the calf number, and put the sample on ice.
8. While holding the lungs by the trachea in a vertical position, pour ~1000 mL (depending on size of the lungs) of sterile PBS into the trachea to inflate the lungs, massage the lungs, and

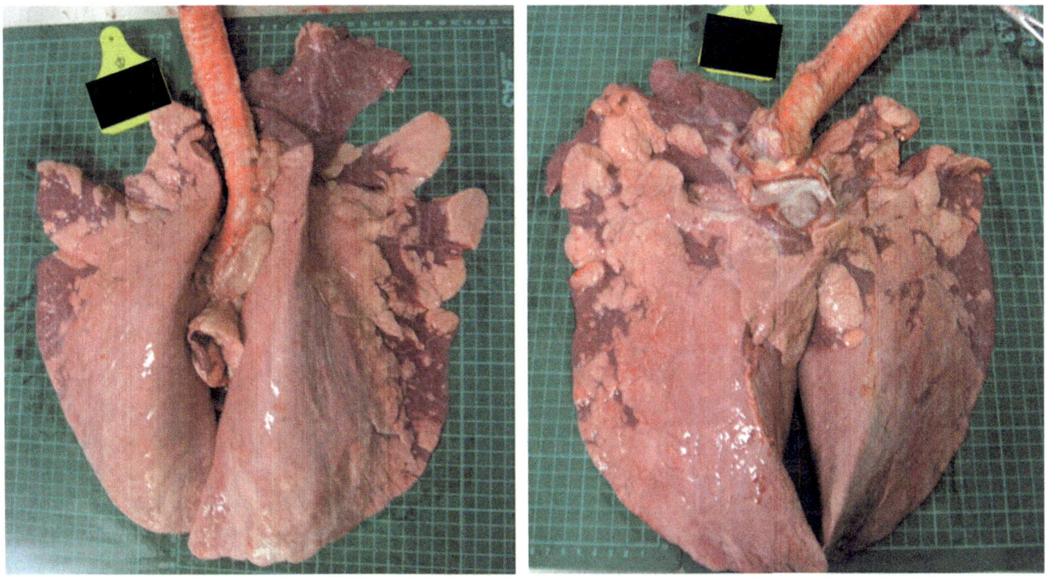

Fig. 2 Dorsal and ventral aspects of a calf lung showing macroscopic lung lesions, 7 days after aerosol infection with BRSV. Two-month-old calf with BRSV-specific maternal antibodies (neutralizing antibody titer = $\log_{10} 2.57$), at the time of aerosol infection with 10^4 pfu of the Snook strain of BRSV in lung lavage fluid. The calf had developed clinical signs of respiratory disease characterized by coughing, raised respiratory rate (72 breaths/min), raised temperature (39 °C), and dyspnea

harvest the fluid containing BAL cells (~40–50% of the volume infused) into sterile pots. Put the labeled pots containing the BAL on ice (*see* **Note 13**).

9. Using sterile post-mortem blades and forceps, collect a sample of lung tissue from each of 3 different lung lobes that were not subjected to BAL. Samples should be obtained preferentially from consolidated areas of each lobe. Each tissue sample should be placed in a different labeled universal and stored on ice.
10. Put a thin slice of each lung tissue sample from each of the 3 different lung lobes, in 10% buffered formalin for histology.
11. Small samples of lung tissue (<1 cm^2) may also be immersed in tubes containing 1 mL RNA-later for RT-qPCR.

3.13 Sample Processing

3.13.1 Processing of Blood for Serum

1. Allow blood to clot. (Optional: Leave sample in fridge overnight).
2. Clarify samples by centrifugation at 2000× *g* for 5 min.
3. Harvest serum (be careful not to disturb the blood clot).
4. Heat-inactivate sera for 30 min at 56 °C.
5. Use serum for ELISA, and or virus neutralization assays, or store at −20 °C until required.

3.13.2 Isolation of Peripheral Blood Mononuclear Cells (PBMC)

1. Dilute heparinized blood with an equal volume of PBS in a 50 mL Falcon tube, e.g. 15 mL blood and 15 mL PBS.
2. Underlay each 30 mL diluted blood with 15 mL Ficoll-Paque 1083 by putting the pipette into the bottom of the tube and, with the tube at an angle of about 45°, expelling the Ficoll-Paque gently under the blood, forming a gradient, with the blood on top of the Ficoll-Paque.
3. Centrifuge tubes at 1100× *g* at 20 °C for 30–45 min depending on the age of the calf (blood from young animals needs to be spun for longer and faster, e.g., 1800× *g* for 60 min).
4. Aspirate the mononuclear cells from the interface, taking care to take as little of the plasma or Ficoll-Paque as possible, into cold PBS in a plastic universal **or** remove the upper layer that contains plasma and most of the platelets and discard. Then using a clean pipette transfer the cells at the interface to a fresh universal (1 universal for each gradient) and make up to 20 mL with cold PBS. If making more than one gradient from one calf, the cells at the interface from 2 or 3 gradients may be pooled into one Falcon 50 mL tube and made up to 45 mL with PBS for washing.
5. Incubate cells with red blood cell lysis buffer if required.
6. Centrifuge at 500× *g* for 10 min. at 4 °C. Remove supernatant and replace with 20 mL fresh cold PBS. Spin at 300× *g* at 4 °C. Repeat this step once more.
7. Resuspend cells in 5–10 mL lymphocyte culture medium (*see* Subheading 2.13.2), count viable cells by trypan blue exclusion, and use cells for downstream applications such as proliferation assays, flow cytometry, ELISPOT, or cryopreserve in FBS with 10% DMSO.

3.13.3 Processing of Lung Tissue Samples for Virus Isolation

1. Keep lung samples on ice.
2. Using sterile scissors and forceps, place a sample of lung into a pre-weighed petri dish, weigh and subtract the weight of the petri dish to calculate the weight of the sample. Aim to take approximately 2–4 g of lung per petri dish.
3. Add appropriate amount of tissue homogenization medium (*see* Subheading 2.13.3) to a labeled universal to give a 20% w/v suspension (i.e. for 3 g of tissue add 15 mL medium to a universal).
4. Transfer tissue to a mortar and chop finely with sterile scissors.
5. Add some sterile fine sand and a small amount of medium from the pre-determined volume in the universal.
6. Homogenize with the pestle until the sample looks like yogurt.

7. Add the remainder of the medium from the universal and mix using the pestle.
8. Pour the lung homogenate into the universal and allow large lumps and sand to settle.
9. Remove 2 × 2 mL of supernatant and aliquot to labeled vials for freezing at −80 °C until required for titration of infectious virus as described in Subheading 3.4 or for RT-qPCR as described in Subheading 3.6.
10. Use fresh sterile pestles and mortars for each sample.
11. Put pestles & mortars into an autoclave tin.
12. Clean scissors and forceps in 10% Decon and wipe with tissue.
13. Sterilize scissors and forceps in hot bead sterilizer for ~15 s.

3.13.4 Processing of BAL for Virus Isolation and Antibody Detection

1. Keep the BAL at 4 °C until it can be processed.
2. Streak 100 μL BAL onto blood agar plates which have been dried for about 30 min in the MBSC. Incubate overnight at 37 °C and look for bacterial growth.
3. Transfer about 90–100 mL of BAL to 2 × 50 mL Falcon tubes. Spin 500 x*g* for 15 min, retain some BAL supernatant, aliquot to 2× labeled vials/calf and freeze at −20 °C until required for analysis of BRSV-specific antibodies by ELISA and/or virus neutralization assays.
4. Discard the rest of the supernatant into 1% Virkon.
5. Resuspend the pellets in a total of 5 mL tissue homogenization medium (*see* Subheading 2.13.3) or in RNA-later.
6. Aliquot 2 × 2 mL to labeled vials and store at −80 °C until required for titration of infectious virus as described in Subheading 3.4 or RT-qPCR as described in Subheading 3.6.

3.13.5 Processing of BAL for Analysis of Pulmonary Exudate

1. Keep the BAL at 4 °C until it can be processed.
2. Take a sample and count viable leukocytes by trypan blue exclusion.
3. Label back of slide with calf number so that the number can be seen when the slides are in the cytocentrifuge.
4. Use 2 slides per calf.
5. Assemble the cytology funnel with the filter paper and microscope slide into the cytoclip. Place this into the cytocentrifuge.
6. Add 4×10^4 cells to each cytospin chamber in a 400 μL volume. (Dilute sample if necessary to 1×10^5/mL in PBS and add 400 μL /chamber). Centrifuge in cytocentrifuge at 70× *g* for 10 min.
7. After the cytocentrifuge stops spinning, remove each sample one at a time and carefully remove the cytology funnel by

peeling it away from the microscope slide. Next, peel the filter card away from the microscope slide. Repeat this step for each sample.

8. Allow slides to dry at RT.
9. Fix and stain with Diff Quick (6 dips of 1 s in each reagent).
10. Allow to drain.
11. Determine differential cell counts (macrophages, lymphocytes polymorphonuclear neutrophils, eosinophils) of approx. 300 cells/slide under oil immersion microscopy (*see* **Note 14**).

3.13.6 Processing of Nasal Secretions for Antibody Detection

1. Using forceps, transfer the tampon to a 20 mL syringe and depress syringe plunger to squeeze out nasal secretions into the universal.
2. Dispense the nasal secretions into one to 2 aliquots and store at −20 °C until required for antibody analysis.

3.13.7 Processing of Lung Tissue and Lymph Nodes for Lymphoid Cell Isolation

1. Chop lung tissue or lymph nodes into small pieces with sterile scissors.
2. Put C-tube containing chopped tissue and lymphocyte culture medium (*see* Subheading 2.13.2) on gentle MACS Dissociator and use the appropriate program.
3. Filter dissociated tissue using 70 μm cell strainer into a 50-mL Falcon tube.
4. Dispose strainer and put lid on 50 mL Falcon tube.
5. Pellet cells by centrifugation at 300× *g* for 10 min.
6. Wash once in PBS or cell culture medium, and pellet cells by centrifugation at 300× *g* for 10 min.
7. Resuspend cells in lymphocyte culture medium (*see* Subheading 2.13.2).
8. Take an aliquot of the cells, stain with Trypan blue, transfer stained cells to counting chamber and count viable cells by trypan blue exclusion.
9. Resuspend pellet in complete cell culture medium and use cells for downstream application such as proliferation assays, flow cytometry, ELISPOT, or cryopreserve in HI-FBS with 10% DMSO.

3.14 Determination of the Extent of Macroscopic Lung Lesions

1. Using imageJ (https://imagej.net/ij/), open the photograph of the dorsal aspect lung.
2. Draw round the whole lung, not counting the trachea, and determine total lung area.
3. Draw round each area of macroscopic consolidation and determine the area.

4. Add the individual areas of macroscopic lesions together and express as a % of the total lung area.
5. Repeat for the ventral aspect of the lung.
6. Express the % macroscopic lesions as the mean of the dorsal and ventral aspects (*see* **Note 15**).

4 Notes

1. BRSV to be used to infect calves should have been passaged only 4–6 times in cell culture.
2. Other cells, such as bovine nasal turbinate cells, primary calf kidney or calf testes cells may be used for culturing BRSV.
3. Other commercially available mAbs or polyclonal antibody specific for RSV proteins that are conserved between HRSV and BRSV can be used.
4. In order to minimize infection with BRSV and other respiratory pathogens, calves should be removed from the commercial farm preferably at one to 3 days of age to the experimental facility where biosecurity measures should be implemented. In the United Kingdom, movement of calves at this age has to be carried out under the regulations of the Home Office Scientific Procedures Act (1986) of the United Kingdom. For BRSV antibody-free calves, animals should be deprived of colostrum, or fed with approx. 400 mL colostrum containing either no or only low levels of BRSV-specific antibody on arrival at the experimental facility. Colostrum-deprived calves should be treated for other infections as required e.g. Halocur for cryptosporidia infection.
5. For large calves, a larger equine nebulizer mask may potentially be modified and used.
6. Calves may cough after intratracheal inoculation of virus suspension.
7. Make sure the jet nebulizer medication chamber is kept upright. Aerosol delivery of 3–4 mL virus may take up to 10 min per calf. A small amount of virus suspension may remain at the end of the delivery period and must be disposed of appropriately.
8. Calves with a raised rectal temperature can be treated with a non-steroidal anti-inflammatory drug, such as Flunixin meglumine.
9. For evaluation of levels of BRSV replication in nasopharyngeal swabs by RT-qPCR, the nasopharyngeal brush can be agitated in RNA-later and stored at −80 °C until required.

10. BAL can be also performed with an endoscope instead of tubes, using the biopsy channel to perform the lavage. The advantage of this technique it that it enables visualization of the part of the respiratory tract where the BAL is performed. However, the cost is higher and the technique of disinfection of the endoscope needs to be validated before use.
11. Clamping the trachea as near to the larynx as possible before removal of the lungs from the thoracic cavity helps to prevent contamination of the respiratory tract with blood and ruminal contents.
12. The same 3 apical lobes on each calf should be clamped. Alternatively, just the left bronchus could be clamped. Clamping parts of the lung allows lung tissue samples to be obtained for virology and histology that have not been subject to bronchoalveolar lavage (BAL).
13. Collecting the BAL usually involves two people, one to hold the lungs and the other to pour the PBS into the lungs and then hold the pot beneath the lungs when they are inverted to pour the PBS out of the lungs into the pot.
14. The proportion of polymorphonuclear neutrophils in BAL is a good indicator of the extent of the inflammatory response in the lower respiratory tract.
15. The extent of macroscopic lung lesions is a good indicator of the severity of BRSV infection.

References

1. Taylor G (2017) Animal models of respiratory syncytial virus infection. Vaccine 35:469–480
2. Thomas LH, Stott EJ, Collins AP et al (1984) Infection of gnotobiotic calves with a bovine and human isolate of respiratory syncytial virus. Modification of the response by dexamethasone. Arch Virol 79:67–77
3. Cortjens B, de Jong R, Bonsing JG et al (2019) Human respiratory syncytial virus infection in the pre-clinical calf model. Comp Immunol Microbiol Infect Dis 65:213–218
4. Rima B, Collins P, Easton A et al (2017) ICTV virus taxonomy profile: pneumoviridae. J Gen Virol 98:2912–2913
5. Valarcher JF, Taylor G (2007) Bovine respiratory syncytial virus infection. Vet Res 38:153–180
6. Viuff B, Uttenthal A, Tegtmeier C, Alexandersen S (1996) Sites of replication of bovine respiratory syncytial virus in naturally infected calves as determined by in situ hybridisation. Vet Pathol 33:383–390
7. Viuff B, Tjornehoj K, Larsen LE et al (2002) Replication and clearance of respiratory syncytial virus. Apoptosis is an important pathway of virus clearance after experimental infection with bovine respiratory syncytial virus. Am J Pathol 161:2195–2207
8. Johnson JE, Gonzales RA, Olson SJ et al (2007) The histopathology of fatal untreated human respiratory syncytial virus infection. Mod Pathol 20:108–119
9. Ellis J, Waldner C, Gow S, Jackson M (2013) Relationship of the extent of pulmonary lesions to the partial pressure of oxygen and the lactate concentration in arterial blood in calves experimentally infected with bovine respiratory syncytial virus. Can J Vet Res 77:205–210
10. Taylor G, Thom M, Capone S et al (2015) Efficacy of a virus-vectored vaccine against human and bovine respiratory syncytial virus infections. Sci Transl Med 7:300ra127
11. Taylor G, Stott EJ, Furze J et al (1992) Protective epitopes on the fusion protein of

respiratory syncytial virus recognized by murine and bovine monoclonal antibodies. J Gen Virol 73:2217–2223
12. Taylor G, Stott EJ, Bew M et al (1984) Monoclonal antibodies protect against respiratory syncytial virus infection in mice. Immunology 52:137–142
13. Thomas LH, Cook RS, Wyld SG et al (1998) Passive protection of gnotobiotic calves using monoclonal antibodies directed at different epitopes on the fusion protein of bovine respiratory syncytial virus. J Infect Dis 177:874–880
14. Homeoffice, Code of practice for the housing and care of animals bred, supplied or used for scientific purposes. ISBN 9781474112390, 19111403 12/14 44389. (2014)
15. Hakhverdyan M, Hagglund S, Larsen LE et al (2005) Evaluation of a single-tube fluorogenic RT-PCR assay for detection of bovine respiratory syncytial virus in clinical samples. J Virol Methods 123:195–202
16. Martini V, Hinchcliffe M, Blackshaw E et al (2020) Distribution of droplets and immune responses after aerosol and intra-nasal delivery of influenza virus to the respiratory tract of pigs. Front Immunol 11:594470
17. Blodorn K, Hagglund S, Gavier-Widen D et al (2015) A bovine respiratory syncytial virus model with high clinical expression in calves with specific passive immunity. BMC Vet Res 11:76

Chapter 22

Serial Adaptation of Viruses for Reduced Susceptibility Profiling

Michael R. Sirrine and Richard K. Plemper

Abstract

Serial adaptation of viruses to inhibitors enhances the understanding of the mechanism of inhibition, helps to identify physical target sites of compounds, and serves as a test for the genetic barrier to reducing virus susceptibility. Virus adaptations can be carried out in vivo or in tissue culture using sub-inhibitory concentrations of compounds that are increased with each passage of the virus or kept static. Here we describe the method of serially adapting a virus in tissue culture.

Key words Virus adaptation, Antiviral development, In vivo, Virus inhibition, Serial adaptation

1 Introduction

Virus adaptation is commonly used for antiviral development to identify the genetic barriers to reducing viral susceptibility and to increase understanding of the mechanism of action of novel antivirals. Serial passaging of the virus has been performed to describe the specific antiviral activities of many compounds [1–5]. These studies serve many purposes through adaptation we understand the barrier to reduced susceptibility and the fitness penalties associated with resistant viruses; we can also learn more about how a compound directly interacts with the viral target. In this chapter, we discuss several methods of virus adaptation and the associated techniques. Dose response assays alongside $TCID_{50}$ and plaque assays support and inform decisions made during adaptation. The most common method for adaptation is serial adaptation. This method helps to identify upper limits of reduced susceptibility and can give a preliminary idea of the genetic barrier to reduced susceptibility.

In vivo adaptation is possible and informative; however, it comes with many limitations, for example, recombinant Sendai virus (SeV) was passaged in mice with sub-inhibitory

Bo Liang and Rachel Fearns (eds.), *Human Respiratory Syncytial Virus: Methods and Protocols*, Methods in Molecular Biology, vol. 2948, https://doi.org/10.1007/978-1-0716-4666-3_22,

concentrations of inhibitor GHP-88309 [1]. After the second passage, two lineages were not recovered from the mice, and the remaining three were still susceptible to GHP-88309. This can be attributed to a high barrier of adaptation and to the complex situation in vivo that is not present in vitro.

2 Materials

2.1 Dose Response

1. Purified virus stocks.
2. 96-well clear tissue culture plates.
3. 96-well Deep well plates (Sterile).
4. DMEM supplemented with 7.5% FBS.
5. DMEM un-supplemented (Serum-Free).

2.2 Determining Starting Concentration

1. Purified reporter virus stock (Titer at or above 1×10^6).
2. DMEM supplemented with 7.5% FBS.
3. DMEM un-supplemented (Serum-Free).
4. 12-well clear tissue culture plate.
5. 15 mL conical tubes.
6. Di-methyl sulfoxide (DMSO).

2.3 Dose Escalation Adaptation

1. Purified virus stock (Titer at or above 1×10^6).
2. DMEM supplemented with 7.5% FBS.
3. DMEM un-supplemented (Serum-Free).
4. 6-well clear tissue culture plates.
5. 2 mL Cryogenic vials with screw caps.
6. 1.5 mL Eppendorf tubes.
7. DMSO.

2.4 Static-Dose Adaptation

1. Purified virus stock (Titer at or above 1×10^6).
2. DMEM supplemented with 7.5% FBS.
3. DMEM un-supplemented (Serum-Free).
4. 6-well clear tissue culture plates.
5. 2 mL Cryogenic vials with screw caps.
6. 1.5 mL Eppendorf tubes.
7. DMSO.

2.5 $TCID_{50}$

1. Purified virus stocks.
2. 96-well clear tissue culture plates.
3. 96-well deep well plates (Sterile).

4. DMEM supplemented with 7.5% FBS.
5. DMEM un-supplemented (Serum-Free).

2.6 Plaque Assay

1. Fixing solution: 10% Formaldehyde in dH_2O.
2. Crystal violet stain: 1% Crystal violet in 20% EtOH and H_2O.
3. 2× DMEM.
4. Immobilizing overlay: 2.4% Avicel in dH_2O.
   (a) Add Avicel powder to water that is rapidly mixing with a stir bar. Autoclave to sterilize and store sealed at room temperature.
5. 6 or 12 well tissue culture plates.

2.7 Virus Rescue

1. DMEM supplemented with 7.5% FBS.
2. DMEM un-supplemented (Serum-Free).
3. 6-well tissue culture plates.
4. 150 mm tissue culture dishes.
5. GeneJuice transfection reagent.
6. OptiMEM.
7. 1.5 mL Eppendorf tubes.

3 Methods

3.1 Dose Response Assay

This assay is performed with newly identified compounds and serves as a good starting point for adaptation experiments (Fig. 1). From dose responses, we can calculate the concentration of the compound at which 50% (EC_{50}) and 90% (EC_{90}) of virus growth are inhibited.

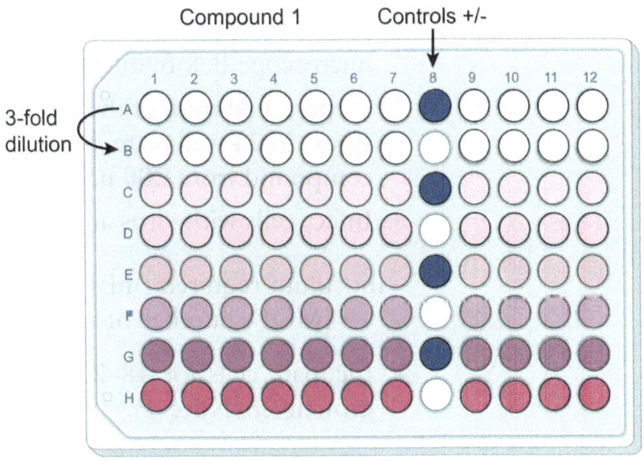

Fig. 1 Layout of dilution plate for dose response. Positive wells are with DMSO only, negative wells have cycloheximide, and the other wells have compound dilutions

1. Prepare a cell suspension 10 mL cell suspension at 1×10^5 cells/mL. From the cell suspension seed one 96-well white/clear tissue culture plate with 50 μL of cell suspension per well. Incubate at 37 °C 5% CO_2 overnight to allow cells to attach.
2. After incubation, prepare dilutions of the compound in DMSO, starting compound solution is 100 mM. Dilute to 150 μM in serum-free DMEM. In clear 96 well plate, add 80 μL DMEM to every well. To wells in the first row, add 40 μL of 150 μM compound. Perform threefold serial dilutions down the plate to dilute the compound. In column 8, add cycloheximide to alternating wells to serve as a negative control. Transfer 50 μL from each well of the dilution plate to the cell plate (*see* **Note 1**).
3. Prepare dilution of virus in 1 mL to MOI of 0.01, assuming 2e5 cells per plate. After treating cells, infect them with 10 μL of virus dilution. Incubate for 24–48 h and measure reporter expression.
4. After measuring reporter expression, determine the percent inhibition by comparing the signal of treated wells to the average of the signal in the positive control wells.

$$\text{Percent inhibition} = ((\text{Signal of treated well–background signal})/(\text{Positive signal–background signal})) \times 100$$

5. Using the percent inhibition at each compound concentration, calculate EC_{50} of the compound using GraphPad Prism non-linear regression modeling.

3.2 Determine the Highest Tolerated Compound Concentration

An alternative starting point for adaptations is with the highest tolerated concentration of compound. The purpose of this assay is to determine the maximum concentration of compound that can be added to infected cells without completely sterilizing the inoculum.

1. Prepare a 15 mL cell suspension at 1×10^5 cells/mL, add 1 mL per well of 12-well tissue culture plate. Allow cells to attach and grow overnight, at 37 °C 5% CO_2. Confirm confluency under a microscope before infecting.
2. Prepare a dilution of the compound in DMEM, using 100 mM stock. In 24 well plate, perform eight threefold dilutions of the compound from 200 μM down (*see* **Note 2**).
3. Infect cells with virus at an MOI of 0.01 (*see* **Note 3**).

$$\text{Inoculum required (mL)} = (\text{MOI} \times \text{number of cells})/\text{Titer of viral stock (pfu/mL)}$$

4. Incubate cells for 48–72 h at 37 °C 5% CO_2, monitor virus growth (*see* **Note 4**).

3.3 Dose Escalation Adaptation

1. Prepare a 25 mL cell suspension at 1×10^5 cells/mL, add 2 mL of cell suspension to each well of two 6-well plates. Allow cells to attach and grow overnight, at 37 °C 5% CO_2. Confirm confluency under microscope before infecting.
2. Before infecting cells, prepare compound and virus dilutions. Dilute the compound in DMSO to 0.5–1× EC_{90}, or the highest tolerated concentration, in 5 mL 7.5% FBS DMEM, also prepare 5 mL of 7.5% FBS DMEM with only DMSO at the same percent as treated wells for control wells. Dilute virus in 10 mL serum-free DMEM to MOI of 0.01.

$$\text{Inoculum required (mL)} = (\text{MOI} \times \text{number of cells})/\text{Titer of viral stock (pfu/mL)}$$

3. Before infecting, remove media from cells and add 1 mL of virus dilution to wells in columns one and three of each plate. Replace the media in column two with fresh 7.5% FBS DMEM.
4. To the infected wells on plate 1 add 1 mL of diluted DMSO, these will be control wells. To the infected wells on plate 2, add 1 mL of diluted compound, these are test wells (Fig. 2). Incubate plates at 37 °C 5% CO_2 until virus is ready for harvest.
5. Monitor the plates beginning ~48 h post-infection to check for signs of virus replication. When ~80% of cells in each well are infected harvest virus, for cell-associated viruses, remove the cells with a cell scraper and transfer to 2 mL Eppendorf tube, isolate virus by two freeze-thaw cycles and pellet by spinning at 3000× *g* 15 min at 4 °C, transfer clarified supernatant to fresh

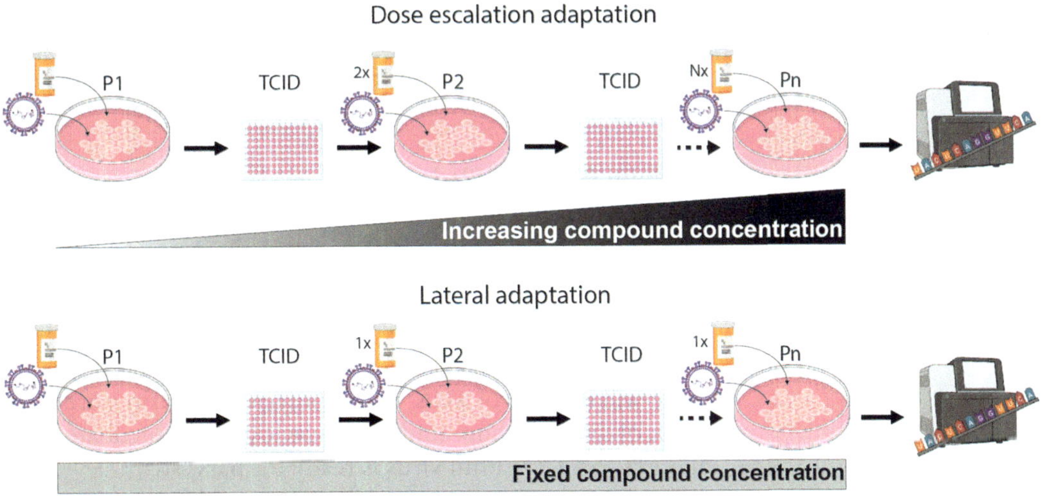

Fig. 2 Schematic overview of dose escalation adaptation and lateral adaptation. Dose escalation adaptation has increasing compound across each passage and lateral adaptation is done at a consistent concentration of compound

tube. For supernatant viruses, remove supernatant and centrifuge to remove cell debris, then put into fresh 2 mL Eppendorf tubes (*see* **Notes 5** and **6**).

6. After harvesting virus aliquot 200 μL of each lineage at each passage to store in −80 as a backup. Also extract RNA from 50 μL of each lineage to track allele frequency of mutations across passages.
7. Determine viral titer via $TCID_{50}$ or plaque assay (*see* **Note** 7).
8. After titration, prepare three 6-well plates with permissive cells for the second batch of infections. One plate will be a direct re-infection with control viruses, each lineage represented one time, plate 2 will be a repeat of adaptation passage one with the same concentration of compound, and plate 3 will be infected with adapted viruses and treated with 2× compound (*see* **Note 8**).
9. Every second or third passage, check for changes in susceptibility by performing dose-response assays with compound-adapted and DMSO adapted viruses. Changes in EC_{50} and EC_{90} of five to tenfold are indicative of successful adaptation (*see* **Note 9**).

3.4 Static-Dose Adaptation

1. Prepare a 25-mL cell suspension at 1×10^5 cells/mL, add 2 mL of cell suspension to each well of two 6-well plates. Allow cells to attach and grow overnight, at 37 °C 5% CO_2. Confirm confluency under a microscope before infecting.
2. When ready to infect cells, prepare compound and virus dilutions. Dilute the compound in DMSO to 1× EC_{90} determined by dose response in 5 mL 7.5% FBS DMEM, also prepare 5 mL of 7.5% FBS DMEM with only DMSO for control wells. Dilute the virus in 10 mL serum-free DMEM to MOI of 0.01 (*see* **Note 10**).

$$\text{Inoculum required (mL)} = (\text{MOI} \times \text{number of cells})/\text{Titer of viral stock (pfu/mL)}$$

3. Before infecting remove media from cells and add 1 mL of virus dilution to wells in columns one and three of each plate. Replace the media in column two with fresh 7.5% FBS DMEM.
4. To the infected wells on plate 1, add 1 mL of diluted DMSO, these will be control wells. To the infected wells on plate 2 add 1 mL of diluted compound, these are test wells. Incubate plates at 37 °C 5% CO_2 until virus is ready for harvest.
5. Monitor the plates beginning ~48 h post-infection to check for signs of virus replication. When ~80% of cells in each well are infected harvest virus, for cell-associated viruses, remove the cells with a cell scraper and transfer to 2 mL Eppendorf tube,

isolate virus by two freeze-thaw cycles and pellet by spinning at 3000× *g* 15 min at 4 °C, transfer clarified supernatant to fresh tube. For supernatant viruses, remove supernatant and centrifuge to remove cell debris, then put into fresh 2 mL Eppendorf tubes (*see* **Notes 5** and **6**).

6. After each passage, extract RNA from 50 μL of each lineage to use for whole genome sequencing.
7. After harvesting the virus determine viral titer via $TCID_{50}$ or plaque assay. With these titers, prepare two 6-well plates with permissive cells for second batch of infections. One plate will be a direct re-infection with control viruses, each lineage represented one time, and plate 2 will be a repeat of the adaptation from passage one (*see* **Notes 7** and **11**).
8. Repeat checking for changes in susceptibility by performing dose-response assay. When reduced susceptibility is observed, extract RNA from adapted and vehicle lineages and sequence the whole viral genome for mutations that are unique to the adapted lineages (*see* **Note 12**).

3.5 Virus Titration by $TCID_{50}$

Titration in this way is determined by the number of wells that are counted positive for either expression of fluorescence or the presence of cytopathic effects (CPE).

1. Prepare 22 mL cell suspension at 1×10^5 cells/mL. From the cell suspension, seed four 96-well clear tissue culture plates with 50 μL of cell suspension per well. Incubate at 37 °C 5% CO_2 overnight to allow cells to attach (*see* **Note 13**).
2. In sterile biosafety cabinet prepare sterile deep well plate for adaptation. Perform six tenfold dilutions of each virus lineage in 600 μL from 10^{-1} to 10^{-6} (Fig. 3).
3. Transfer 50 μL of diluted virus to 96-well plate to infect and incubate for 48–72 h. Calculate titer by counting wells that are infected by fluorescence or CPE (*see* **Notes 14** and **15**).
4. After the virus has reached peak infection count, the number of wells at each dilution that are infected with virus and used to calculate the titer of each lineage in pfu/mL.

3.6 Virus Titration by Plaque Assay [6]

1. Prepare cell suspension at 1.5×10^5 cells/mL and plate 2 mL per well for 6-well plates or 1 mL per well for 12-well plates. Incubate overnight (*see* **Note 16**).
2. Perform tenfold serial dilution of samples in DMEM. From 10^{-1} to 10^{-6}. Include one negative control to ensure cell viability.
3. Remove media from wells and infect cells for 45 min–1 h with 400 μL per well on 6-well or 200 μL per well on a 12-well (*see* **Note 17**).

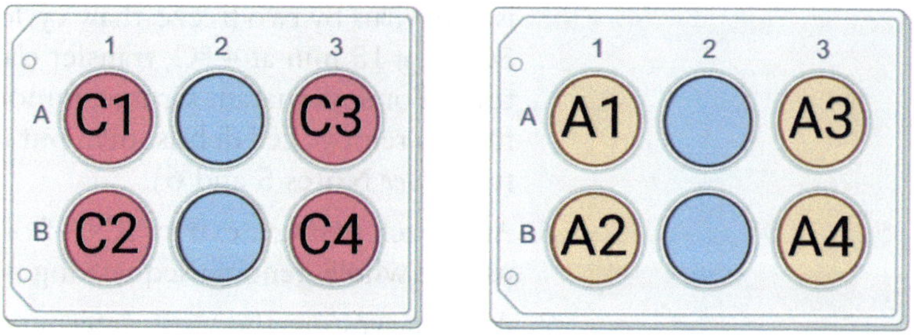

Fig. 3 Graphical representation of plates from first round of infection. Pink wells on the left plate are control wells, infected with virus and treated with DMSO. Light orange wells on the right plate are the test wells treated with compound

4. After infection overlay with 3 mL 1:1 2× DMEM and 2.4% Avicel 6-well or 1.5 mL 1:1 2× DMEM 2.4% Avicel 12-well. Gently rock to mix.
5. After the addition of overlay, incubate plates for 2–14 days until distinct, clearly countable plaques are visible (*see* **Note 18**).
6. After plaques are visible, remove Avicel overlay and fix cells with 10% formaldehyde solution for 30 min to overnight, use ~1 mL per well in 6-well and 0.5 mL/well in 12-well (*see* **Note 19**).
7. After fixing, discard formaldehyde and rinse plates with water to remove the overlay or formaldehyde (*see* **Note 20**).
8. Stain with crystal violet solution, use just enough to cover cells and rock for 15–20 min to ensure even coverage.
9. Gently wash off the stain with water.
10. Count plaques in each well to determine titer. PFU/mL = (Number of plaques)/(dilution × volume virus added to well).

3.7 Rebuild and Rescue Recombinant Resistant Viruses

1. Design primers for the mutagenesis of recombinant virus plasmid (*see* **Note 21**).
2. Perform linear amplification to generate full mutant transcripts and use DpnI to remove any parent plasmid.
3. Reamplify the mutant virus plasmids or BACs in bacteria and confirm the presence of desired mutations by Sanger sequencing (*see* **Note 22**).
4. With purified virus genomes prepare for transfection. Mix helper plasmids and genome together in 1.5 mL Eppendorf tube at appropriate ratios. Separately mix GeneJuice at 3 μl/μg plasmid DNA in 100 μl OptiMEM/construct (*see* **Note 23**).
5. Mix GeneJuice and DNA solutions and incubate 15 min at room temperature.
6. Add GeneJuice/DNA mixture to cells dropwise and incubate at 37 °C.

7. Incubate cells 36–72 h, harvest virus when cells are ~75–80% infected (*see* **Note 24**).
8. After harvesting P1 stock of virus determine titers by $TCID_{50}$ or plaque assay and infect cells for P2 stock with MOI of 0.1 of P1 virus (*see* **Note 25**).
9. Harvest and purify P2 stock of recombinant virus and verify presence of desired mutations by whole genome sequencing (*see* **Note 26**).

3.8 Test Rebuilt Viruses for Increases in Pathogenicity or Infectivity

1. Infect permissive animal models with purified recombinant viruses (*see* **Note 27**).
2. Monitor the animals for clinical signs of disease especially any changes in disease onset or progression, i.e., more rapid symptom onset, length and severity of symptoms, increased lethality, or clinical signs that are not seen with parent virus.
3. In parallel treat animals with a known efficacious dose of compound and track titer of shed virus from wildtype and mutants to determine the effect of compound on virus replication in vivo.
4. To see if mutant viruses are more infectious than wildtype infect one group of naïve animals with virus.
5. After 24 h co-house infected animals with animals that have not been exposed to virus.
6. Co-house for 12 h and then separate the co-housed and infected animals.
7. Track the amount of shed virus from nasal lavage for both infected and co-housed animals. At study end harvest organs and determine organ titers of virus in infected and co-housed animals. Compare the titers in exposed animals between wildtype and resistant viruses.

4 Notes

1. Infect cells with the virus in columns of three. One plate can be used to test three different viruses, using the parent virus for positive and negative controls.
2. On the dilution plate, your concentration should be 2× the final desired concentration of the compound. You will add 500 μL of dilution to 500 μL of diluted virus. The highest final concentration will be 100 μM compound.
3. When preparing the virus thaw in 37 °C water bath until only a small chunk of ice remains, over-warming the virus will reduce the infectious titer. It is also best to prepare the virus dilution last so that you reduce the amount of time between thawing and infecting cells.

4. The starting concentration for adaptation should be a concentration where you have some virus after 48–72 h that is five to tenfold less than virus observed without the compound. This ensures that you are putting selective pressure on your adaptation and that the virus will be able to grow.
5. When harvesting the virus be careful that each lineage remains separate, any cross-contamination between wells will mask results and can obscure mutations that may arise. It is also important during each step to keep track of each lineage and ensure they are labeled with the passage number and the concentration of compound they were exposed to, as well as the date of harvest. The RNA that is extracted from each lineage at each passage is used for whole-genome sequencing to track allele frequency of mutations across multiple passages.
6. The length of time it takes until virus is ready to harvest can vary, often viruses that are growing in sub-inhibitory concentrations of compound will take longer to reach peak infection. If this is the case replace the media every 72 h with fresh 7.5% FBS DMEM with compound to ensure cells can continue to grow and that virus is still under selective pressure. If cell confluency gets to more than 100% you can remove the media, add 200 μL of trypsin to detach the cells, and dilute them to about 60% of the starting concentration to keep the cells replicating and allow for the proliferation of virus.
7. Blind passaging can be performed without titration of harvested virus, this often results in decreased titer from subsequent passages and can result in a complete loss of infectious particles.
8. From each passage, you want to keep the adapted lineages that are growing at the highest concentration of compound. This can vary between your adapted lineages, so be sure to keep track of what concentration of compound each lineage is growing at to ensure proper selective pressure is applied during each infection.
9. For allosteric compounds, it is often possible for your virus to become completely insensitive to the inhibitor, for other compound classes and depending on the area of the virus that is inhibited by your compound of interest sometimes moderate resistance is all that can be achieved, represented by a two to fivefold increase in EC_{50} and EC_{90}.
10. For lateral adaptation, choose a concentration that reduces progeny virus by about 90%. You will be passaging the same lineages at a constant level of selective pressure to see what mutations arise.

11. For lateral adaptation, you will move the same virus lineages forward, and the primary readout indicating resistance will be that titers begin to reach vehicle-treated levels.
12. Adaptation mutations are those that are observed only in the compound adapted viruses, any mutation with an allele frequency greater than 50% should be considered for further testing to determine if it is a primary adaptation or a secondary adaptation.
13. When seeding plates for titration, ensure that the bottom of every well is fully covered by media and that the cells are evenly distributed in the well. To do this, gently tap the sides of the plate and check under a microscope to verify cell distribution. If the cells are too clumped or exposed to air virus will be unable to infect evenly, and your titer will be underrepresented.
14. When preparing dilutions, use one row to dilute two viruses starting from wells one and twelve, dilute toward the center (Fig. 4). You can run two titrations on one 96-well plate. It is important when doing dilutions that you replace the tips between each dilution to not misrepresent the total amount of virus in the stocks.
15. It is important that there is some phenotypic change in your cells with infection for TCID, without a way to determine under a microscope if your cells are infected a plaque assay should be performed.
16. Depending on how many replicates you want to have or how many plates you feel comfortable handling, you can choose a 6-well or 12-well plate.

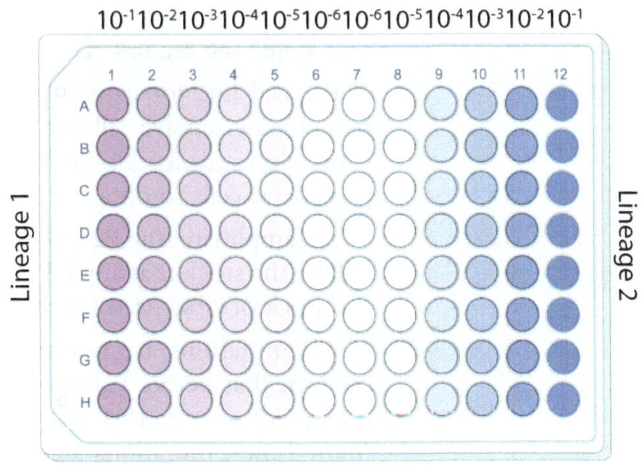

Fig. 4 Graphical representation of titration plate. Each column is infected with the same dilution of virus. In this way one plate is used to calculate the titer of two independent virus lineages

17. Rock the cells and make sure the entire monolayer is covered with media so the cells don't dry out during the infection step.
18. If using liquid overlay DO NOT move the plates, this can result in smeary plaques that are not easily countable. You can instead use a semisolid overlay of Agarose or CMC that can be moved gently and checked periodically for plaque development.
19. If using agarose or CMC overlay, add formaldehyde solution directly to cells from 1 h to overnight.
20. If using agarose or CMC overlay, remove with either running water or spatula.
21. Fold changes greater than 5 are usually indicative of resistance. For allosteric inhibitors, it is often possible to observe complete resistance.
22. When designing mutagenesis primers, the best is to mutate two nucleotides of a codon to give the mutation the most likely chance of remaining after virus rescue. Mutagenesis primers should anneal to 15–20 nucleotides up and downstream of the mutation site.
23. Stbl2 cells can amplify large plasmids without recombination events. If your DNA is particularly unstable amplification at 32 °C instead of 37 can help. After amplification and verification of DNA by restriction digest sanger sequencing is the best to verify that your point mutations are present in the plasmid without any additional non-specific changes.
24. The ratios of helper plasmids to genome can vary depending on virus lineage. You can determine what works for your virus by performing minigenome assays with the helpers and a reporter construct flanked by the viral untranslated regions (UTRs). The optimal ratio of helper plasmids and genome is the one that yields the highest reporter signal.
25. If the cell line for expression of virus constructs is different from the permissive cell line for virus replication. Transfect the expression cells for 24–36 h and then overlay those cells onto permissive cells to obtain a P1 stock of rescued virus.
26. When amplifying the P2 stock, make a larger stock of virus, this will be the stock that any P3 stock is based on and will often be the stock of virus used for testing in animal studies.
27. If the mutations are not stable by themselves, it is best to confirm that before you blindly test a recombinant virus and incorrectly assume that a mutation does not confer resistance. Many times, mutations are not well tolerated in certain viral proteins or at key locations without significant selective pressure.

References

1. Cox RM, Sourimant J, Toots M, Yoon JJ, Ikegame S, Govindarajan M et al (2020) Orally efficacious broad-spectrum allosteric inhibitor of paramyxovirus polymerase. Nat Microbiol 5(10):1232–1246
2. Lieber CM, Kang HJ, Aggarwal M, Lieberman NA, Sobolik EB, Yoon JJ et al (2024) Influenza A virus reduced susceptibility to 4′-fluorouridine coincides with viral attenuation in vitro and in vivo. PLoS Pathog 20(2):e1011993
3. Sourimant J, Lieber CM, Yoon JJ, Toots M, Govindarajan M, Udumula V et al (2022) Orally efficacious lead of the AVG inhibitor series targeting a dynamic interface in the respiratory syncytial virus polymerase. Sci Adv 8(25): eabo2236
4. Yoon JJ, Krumm SA, Ndungu JM, Hoffman V, Bankamp B, Rota PA et al (2009) Target analysis of the experimental measles therapeutic AS-136A. Antimicrob Agents Chemother 53(9):3860–3870
5. Tiong-Yip CL, Aschenbrenner L, Johnson KD, McLaughlin RE, Fan J, Challa S et al (2014) Characterization of a respiratory syncytial virus L protein inhibitor. Antimicrob Agents Chemother 58(7):3867–3873
6. Baer A, Kehn-Hall K (2014) Viral concentration determination through plaque assays: using traditional and novel overlay systems. J Vis Exp 93: e52065

Chapter 23

Quantification of RSV Antibodies in Human Plasma by Enzyme Immunoassays

Ranjini Sankaranarayanan, Binh Ha, Samadhan J. Jadhao, and Larry J. Anderson

Abstract

Enzyme immunoassays (EIAs) to detect and quantify RSV IgG antibodies and RSV IgA antibodies in human plasma or sera are described. The first IgG EIA described uses RSV lysate antigens produced in the HEp-2 cell line to detect and quantify RSV IgG antibodies. The second IgG EIA uses the RSV F ectodomain secreted into the serum-free media of the 293-F expressing cell line for detecting antibodies against the RSV F protein. The third and fourth IgG EIAs use the media from 293-F cells expressing the ectodomain of RSV subgroup A G protein or subgroup B G protein to detect and quantify IgG antibodies in patient sera. The IgA EIA uses either the lysate antigens or the affinity-purified RSV F/G protein ectodomains expressed in 293-F cells, in serum-free media, for antibody detection. These EIAs have been characterized and evaluated for their ability to detect the respective RSV antibodies.

Key words Respiratory syncytial virus, Enzyme immunoassay, RSV F protein, RSV G protein

1 Introduction

Antibodies against RSV proteins play an important role in preventing disease through various mechanisms, including virus neutralization, antibody-dependent cellular cytotoxicity (ADCC), and complement-mediated neutralization [1, 2]. The RSV attachment (G) and fusion (F) proteins are the major targets of RSV-specific neutralizing antibodies [3] and, along with antibodies targeting other RSV proteins, may be helpful for serological diagnosis of infection. IgG- and IgA-specific enzyme immunoassays (EIAs) detect both neutralizing and non-neutralizing antibodies, and the use of RSV lysate antigen provides a way to detect antibodies generated against multiple RSV proteins in patient sera. Additionally, the use of F or G gene-expressed antigens provides a way to determine protein-specific antibody responses. Since IgA antibodies are inefficiently transferred across the placenta, they provide a

Bo Liang and Rachel Fearns (eds.), *Human Respiratory Syncytial Virus: Methods and Protocols*, Methods in Molecular Biology, vol. 2948, https://doi.org/10.1007/978-1-0716-4666-3_23,

way to detect infection-induced antibodies in young children who still have maternal IgG antibodies.

2 Materials

1. Class II biosafety cabinet.
2. 175 cm^2 vented cell culture flasks.
3. HEp-2 cells.
4. Dulbecco's Modified Eagle Medium (DMEM).
5. Fetal Bovine Serum (FBS).
6. Penicillin and Streptomycin antibiotics.
7. Incubator set at 37 °C and 5% CO_2.
8. RSV A2 and RSV B1 viruses.
9. Microscope.
10. −80 °C freezer.
11. Lysis buffer (20 mM Tris–HCl (pH 7.5), 150 mM NaCl, 1 mM Na_2-EDTA, 1 mM EGTA, 1% Triton, 2.5 mM sodium pyrophosphate, 1 mM beta-glycerophosphate, 1 mM Na_3VO_4, 1 μg/mL leupeptin; Cell Signaling Technology; Catalog no: 9803S).
12. Sonicator.
13. 1x PBS.
14. 96-well microtiter plate.
15. Sealing tape/Adhesive film.
16. 4 °C fridge/cold room.
17. Automated plate washer (ELx405, BioTek Instrument Inc., USA).
18. Gelatin from cold-water fish skin.
19. Casein.
20. Non-fat dried milk powder.
21. Tween-20.
22. Human RSV antisera (BEI resources, NIAID/NIH; Cat no: NR-4021).
23. HRP-conjugated goat anti-human IgG antibody.
24. HRP-conjugated goat-anti-human IgA antibody.
25. His-tagged GAPDH.
26. Citric acid buffer: 36.83 mM citric acid monohydrate, 126.3 mM sodium phosphate dibasic, pH 6.0–6.2.
27. Hydrogen peroxide.

28. o-Phenylenediamine dihydrochloride (OPD).
29. Sulfuric acid.
30. Microplate reader (Synergy H1, BioTek Instrument Inc., USA).

3 Methods

3.1 Quantification of Anti-RSV IgG Antibodies in Human Plasma or Serum Using RSV Lysate EIA

The RSV lysate antigen-based EIA helps detect antibodies against RSV proteins including F, G, and N proteins. This EIA can be performed using either the RSV subgroup A or B lysate preparations or by combining lysate preparations from both subgroups. The EIA's sensitivity for detecting past infection may be improved by using both RSV A and RSV B subgroup lysates.

3.1.1 Production of RSV Virus and HEp-2 Lysate Antigens

The procedure involving handling and production of infectious RSV virus in cell culture is performed in a Class II biosafety laminar flow cabinet, adhering to the biosafety guidelines.

1. Seed 3×10^6 live count HEp-2 cells in 175 cm^2 vented cell culture flask in 30 mL DMEM/5% FBS/1% L-Glutamine and 1% Penicillin and Streptomycin antibiotics (complete media). Incubate the flask at 37 °C/5%CO_2. At 24 h post-seeding, wash cells with 10 mL of serum-free DMEM medium.
2. Thaw the RSV A2 or RSV B1 virus strain stock at RT for 5 min. Infect HEp-2 cells in 175 cm^2 flask with RSV A2 or RSV B1 virus strain at 0.1 m.o.i. using a total volume of 3 mL serum-free DMEM at 37 °C/5%CO_2 for 2 h. The infected flasks are rocked every 20 min to spread the virus inoculum on the cell monolayer.
3. Remove the virus inoculum and wash cells 3 times with 10 mL of serum-free DMEM medium.
4. Add 30 mL of complete media and incubate flasks at 37°C/5% CO_2.
5. On day 3 (72 h) post-infection, at 3–4+ viral cytopathic effect (CPE), remove the medium and wash the flask thrice with 10 mL of serum-free DMEM medium. The RSV virus CPE is characterized by the formation of syncytia, the formation of giant cells, and the rounding and sloughing of HEp-2 cells.
6. Freeze the 175 cm^2 flasks at −80 °C and thaw them on ice. Cells are thawed, scraped from the flasks, pooled together, and a 1:10 dilution of lysis buffer is added. The cells are then sonicated (*see* **Note 1**) and centrifuged at 850× *g* for 20 min at 4 °C. The lysate is divided into aliquots and stored at −80 °C.

HEp-2 Control Antigen for Lysate EIA

Uninfected HEp-2 cells grown in 175 cm^2 vented cell culture flask is processed in an identical fashion as the RSV-infected cells and used as the control antigen.

3.1.2 Procedure for Quantification of Antibodies to RSV in Patient Plasma/Serum

Antigen Titration

1. Thaw an aliquot of the RSV A2 and B1 lysate antigens and the HEp-2 lysate controls and perform serial two-fold dilutions in 1x PBS starting at 1:12.5. HEp-2 lysate control is serially diluted two-fold starting at 1:6.25 (*see* **Note 2**). The wells are coated with 100 μL of the diluted control/antigens, sealed with an adhesive film, and incubated at 4 °C overnight.
2. The plate is incubated in blocking buffer (*see* Subheading 3.1.2.2) for 2 h. Human RSV antisera (standard sera; *see* **Note 3**) is added for 1.5 h, then following a wash step, HRP-conjugated goat-anti-human IgG secondary antibody is added for 1 h followed by OPD, and absorbance read as described below.
3. The highest dilution of RSV lysate antigen that maintains near maximum P (mean absorbance against RSV antigens)—N (mean absorbance against HEp-2 control antigen) value is used for the subsequent assays.

Quantification of RSV-Specific Antibodies in Patient Plasma/Serum

A general overview of the steps involved in EIAs is depicted in Fig. 1. Antigen concentration to be coated onto the microtiter plate for EIA is determined based on antigen titration (as explained above). For example, a volume of the RSV A2 and the B1 lysate is added to 1x PBS to give 1:50 of each and 1:25 antigen dilution (Fig. 2). The HEp-2 lysate control is diluted at 1: 25. Mix by vortexing.

1. Dispense 100 μL of diluted control and lysate antigen solution into appropriate wells of a 96 well EIA plate. Seal the plate with an adhesive film and incubate overnight at 4 °C (*see* **Note 4**). Wash the plates with 300 μL 1x PBS, twice using an automated plate washer (*see* **Note 5**).
2. Prepare blocking buffer: 0.33% gelatin +0.33% casein +0.33% non-fat dried milk powder in PBS. This solution needs to be heated (90–95 °C) for 30–45 min to help dissolve the solids. The blocking buffer is freshly prepared for each EIA (*see* **Note 6**).
3. Dispense 200 μL of blocking buffer per well and incubate for 2 h at RT.
4. Wash the plates with 300 μL of 1x PBS, twice, using the plate washer.

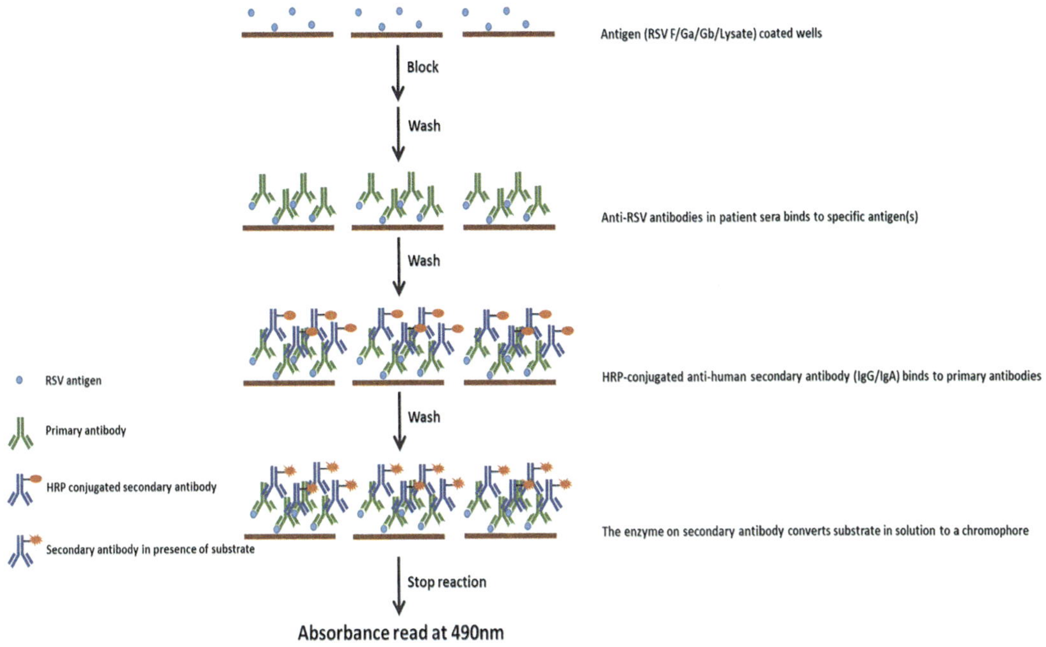

Fig. 1 Schematic of steps involved in EIA. Antigens are diluted at a set concentration and coated on a 96-well plate. After blocking, the diluted primary antibody (patient sera/standard sera) is added to the wells, followed by the anti-human secondary antibody (against either IgG or IgA). Substrate solution (OPD) is then added and, following an incubation of 30 min in the dark, absorbance is read at 490 nm

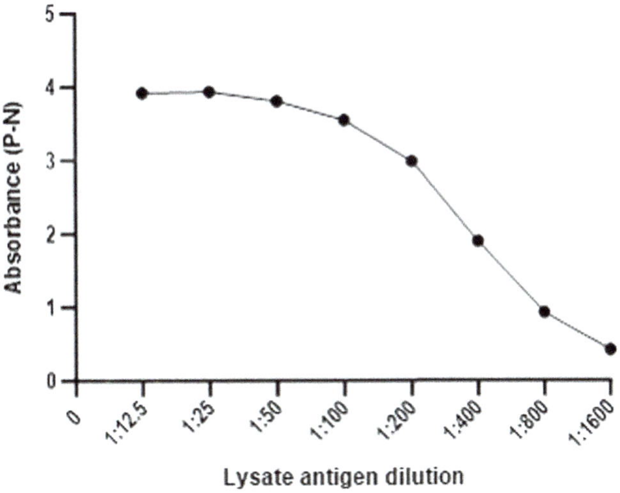

Fig. 2 Antigen titration curve for lysate EIA. Antigens are diluted as indicated and coated on a 96-well plate. After blocking, BEI sera are added to the plate, followed by goat-anti-human IgG secondary antibody. OPD is added, and absorbance is read at 490 nm. The curve is antigen dilutions versus P-N absorbance. The 1:50 dilution of this antigen was used for all subsequent experiments.

5. Add 100 μL of patient sera, diluted 1:200 in blocking buffer supplemented with 0.15% Tween-20 [BBT] to the plates. The standard plasma specimen is also serially diluted in BBT, and 100 μL of each dilution is added to the appropriate wells. Incubate the plates for 1.5 hrs at RT.
6. Wash the plates with 300 μL 1x PBS + 0.05% Tween-20 [PBST], 5 times using the plate washer.
7. Add 100 μL HRP-conjugated goat-anti-human IgG secondary antibody (*see* **Note** 7) at 1:5000 dilution in BBT and incubate at RT for 1 h.
8. Wash the plates with 300 μl PBST, 5 times using the plate washer.
9. Prepare substrate solution (25 mL citric acid buffer (*see* **Note 8**) + 25 mg OPD + 25 μL H_2O_2) before use (*see* **Note 9**). Add 100 μL of the substrate and incubate the plates for 30 min in the dark at RT.
10. Stop reaction by adding 50 μL of 4 N H_2SO_4 per well and gently rocking the plates to mix.
11. Read optical density/absorbance at 490 nm within 10–15 min.

3.2 Quantification of Anti-RSV G or F Protein-Specific IgG Antibodies in Human Plasma or Serum

The assay is based on the expression of the ectodomain of RSV G or RSV F proteins in 293-F cells. 293-F 6/TR cells are grown in Freestyle media to a density of 3–4 × 10^6 cells/ml, then transfected with F or G protein expressing or empty (control) plasmids. The cells are selected using appropriate antibiotics and protein expression is induced through the tetracycline promoter. The secreted protein is then collected in the supernatant following centrifugation and stored at −80 °C until further use for the EIA, as described in the sections below.

3.2.1 Transfection of FreeStyle 293-F Cells with pcDNA/5TO-G-His and pcDNA/5TO-F-His

1. Refer to the chapter titled "Secretory Expression and Purification of Respiratory Syncytial Virus G and F Ectodomain in Human Cells" for the transfection and expression of F and G ectodomain in 293-F 6/TR cells.
2. After the cells have been induced for 72 h to secrete the RSV F or G ectodomain protein, centrifuge the medium at 850× *g* for 20 min at 4 °C. Filter the cell supernatant through a 0.45 μm filter to remove cell debris. Aliquot the medium containing secreted RSV G or RSV F protein and store at −80 °C for further use.

3.2.2 Procedure for Quantification of Antibodies to RSV G or F Ectodomain in Patient Plasma/Serum

1. Antigen concentration for coating is determined through antigen titration as explained in Subheading 3.1.2.1.
2. For the F, Ga or Gb EIAs, 293-F control antigen, RSV Ga/Gb antigen or RSV F antigen are diluted at 1:50 in 1x PBS and mixed by vortexing.

3. Dispense 100 µL of the control or antigen solution to appropriate wells, seal the plate using an adhesive film, and store the plate overnight at 4 °C.
4. The blocking solution is prepared as described in Subheading 3.1.2. Dispense 200 µL of blocking solution per well and incubate at RT for 2 h.
5. Wash the plates twice with 300 µL of 1x PBS using the plate washer. Add 100 µL of plasma diluted at 1:200 in blocking buffer supplemented with 0.15% Tween-20 [BBT]. The serum standard is also serially diluted in BBT. Incubate the plates at RT for 1.5 h.
6. Wash the plates with 300 µL 1x PBS + 0.05% Tween-20 [PBST], 5 times using the plate washer. HRP-conjugated goat-anti-human IgG secondary antibody is then diluted at 1: 5000 in BBT. Dispense 100 µL per well and incubate at RT for 1 h.
7. Wash the plates with 300 µL PBST, 5 times using the plate washer. Add 100 µL OPD/H_2O_2 substrate to each well (25 mL Citric acid buffer +25 mg OPD + 25 µL H_2O_2) and incubate the plates for 30 min in the dark at RT.
8. Stop reaction by adding 50 µL of 4 N H_2SO_4 and gently rocking the plates to mix.
9. Read optical density/absorbance at 490 nm.

3.3 Detection of Anti-RSV IgA Antibodies in Human Plasma or Serum Using RSV Lysate, Purified RSV G or F Protein EIA

IgA EIA using patient serum or plasma can help identify the RSV-infected young child, when IgG titer data is insufficient to confirm an active/past infection. We developed IgA assays using the lysate antigen and RSV F or G protein ectodomains.

3.3.1 Production of RSV Virus and HEp-2 Lysate Antigens

Refer Subheading 3.1.1 of this chapter for production of RSV lysate antigen and HEp-2 control.

3.3.2 Production of Purified F and G Protein Ectodomain

Refer to the chapter titled "Secretory Expression and Purification of Respiratory Syncytial Virus G and F Ectodomain in Human Cells" for the production and purification of the His-tagged RSV F, Ga, or Gb antigen. His-tagged GAPDH was obtained commercially from Sino Biological to serve as a control antigen.

3.3.3 Procedure for Quantification of IgA Antibodies to RSV in Patient Plasma/Serum

The antigen concentration for coating was initially optimized for the IgG lysate EIA as explained in Subheading 3.1.2.1. The same antigen concentrations were used for the IgA lysate EIA.

RSV A2 lysate antigen and RSV B1 lysate antigen stocks are diluted at 1:50 in 1x PBS (or other dilution to achieve optimized concentration). The HEp-2 lysate control is diluted 1:25. Purified recombinant His-tagged F or G proteins and GAPDH are diluted to a concentration of 1 μg/mL in 1x PBS. Mix by vortexing.

1. Dispense 100 μL of diluted control/antigen solution into appropriate wells of a 96 well EIA plate. Seal the plate with an adhesive film and incubate overnight at 4 °C.
2. Wash the plates with 300 μL 1x PBS, twice using the plate washer.
3. Prepare blocking buffer as described in Subheading 3.1.2. Dispense 200 μL of blocking buffer per well to the plates and incubate at RT for 2 h.
4. Wash the plates with 300 μL of 1x PBS twice using the plate washer.
5. Add 100 μL of patient sera, diluted 1:100 in blocking buffer supplemented with 0.15% Tween-20 [BBT] to the plates. Incubate the plates for 1.5 h at RT.
6. Wash the plates with 300 μL PBS + 0.05% Tween-20 [PBST], 5 times using the plate washer.
7. Add 100 μL HRP-conjugated goat-anti-human IgA secondary antibody at a dilution of 1:2000 in BBT and incubate at RT for 1 h.
8. Wash the plates with 300 μl PBST, 5 times using the plate washer.
9. Prepare OPD/H_2O_2 (25 mL Citric acid buffer +25 mg OPD + 25 μL H_2O_2) substrate solution before use. Add 100 μL of the prepared substrate and incubate the plates for 30 min in the dark at RT.
10. Stop reaction by adding 50 μL of 4 N H_2SO_4 and gently rock the plates to mix it.
11. Read optical density/absorbance at 490 nm within 10–15 min.

4 Notes

1. If a sonicator is not available, the virus lysate can be obtained by performing 2–3 freeze thaw cycles of the cells in the presence of lysis buffer.
2. Standardization of the antigen and antibody/sera dilutions for each assay is initially done using a checkerboard analysis involving different dilutions of the antigen and antibody. Starting dilution(s) for this analysis is usually based on results obtained from previous experiments. Alternatively, antigen titrations can

be achieved by starting at the lowest dilution of the antigen and antibodies through serial dilution. The final dilution of the antigen/antibodies is determined by the combination of antigen and antibodies that maintains near maximum P (mean absorbance against antigens)—N (mean absorbance against control) value.

3. In our studies, we used the standard RSV antisera to determine the IgG titer against different RSV antigens in patient sera. Alternatively, commercial recombinant-total human IgG protein can be used to determine IgG concentrations in samples. In this method, the recombinant protein is coated at different concentrations on the microtiter plate, the plate is blocked using blocking buffer, and after a wash step, HRP-conjugated anti-human secondary antibody is added and protein is detected using OPD. Quantification of the IgG antibodies in patient sera is done by extrapolation from the standard curve.
4. Coating the plate with the antigens is usually done overnight (~16 h) at 4 °C, but coating can also be done at least 4 h before the start of the EIA, either at RT or 4 °C.
5. If an automated plate washer is not available, the washing step can be performed manually.
6. Blocking buffer is prepared by stirring the components in 1x PBS at 90–95 °C. Heating helps to effectively dissolve the solids. Make sure to let the blocking buffer cool down before adding it to the microtiter plate. The blocking buffer is usually prepared on the day of the assay but can also be prepared and stored in the fridge up to 16 h before the start of the assay. Storing at 4 °C helps reduce the possibility of microbial growth in the blocking buffer.
7. We used HRP-conjugated secondary antibody for the detection of anti-RSV antibodies in our assays. Other enzyme-substrate systems (including fluorescent substrates) can also be used for detection using appropriate wavelength/instruments.
8. Citric acid buffer can be prepared as mentioned in the materials section. Filtering the solution after preparation will ensure the buffer remains stable for a prolonged period of time.
9. Substrate solution is usually prepared ≤5 mins before its use. Store the prepared substrate solution in a dark place and prevent exposure to light. If exposed to light for a prolonged time, color will automatically develop and increase background absorbance readings.

Acknowledgments

This study was funded in part through the Vaccine and Treatment Evaluation Units (VTEU) awarded to Emory University (HHSN272201300018I). The content is solely the responsibility of the authors and does not necessarily reflect the official views of the National Institutes of Health.

Conflict of Interest Statement In the last 3 years, Dr. Larry Anderson has done paid consultancies on RSV vaccines for GSK, Janssen, and AstraZeneca and on RSV antiviral drugs for Enanta and on influenza virus vaccines for Pfizer and Dr. Anderson's laboratory is currently receiving funding through Emory University from Pfizer for laboratory studies for RSV surveillance studies in adults and maternal antibody transfer to the infant and Sciogen for animal studies of RSV vaccines and previously received funding from Vernagen, LLC for animal studies of RSV and other vaccines. Dr. Anderson is co-inventor on several CDC patents on the RSV G protein and its CX3C chemokine motif relative to immune therapy and vaccine development and on a device to capture coughed droplets from the lung and co-inventor on Emory patent filing for use of RSV platform VLPs with the F and G proteins for vaccines and G protein constructs for RSV vaccines.

References

1. Hiatt A, Bohorova N, Bohorov O, Goodman C, Kim D, Pauly MH, Velasco J, Whaley KJ, Piedra PA, Gilbert BE, Zeitlin L (2014) Glycan variants of a respiratory syncytial virus antibody with enhanced effector function and in vivo efficacy. Proc Natl Acad Sci USA 111(16):5992–5997. https://doi.org/10.1073/pnas.1402458111
2. Miao C, Radu GU, Caidi H, Tripp RA, Anderson LJ, Haynes LM (2009) Treatment with respiratory syncytial virus G glycoprotein monoclonal antibody or F(ab′)2 components mediates reduced pulmonary inflammation in mice. J Gen Virol 90(Pt 5):1119–1123. https://doi.org/10.1099/vir.0.009308-0
3. McLellan JS, Ray WC, Peeples ME (2013) Structure and function of respiratory syncytial virus surface glycoproteins. Curr Top Microbiol Immunol 372:83–104. https://doi.org/10.1007/978-3-642-38919-1_4

Index

Bo Liang and Rachel Fearns (eds.), *Human Respiratory Syncytial Virus: Methods and Protocols*, Methods in Molecular Biology, vol. 2948, https://doi.org/10.1007/978-1-0716-4666-3,

9781071646656